Rechargeable Organic Batteries

Rechargeable Organic Batteries

Materials, Mechanisms, and Prospects

Yongzhu Fu
Xiang Li
Shuai Tang
Wei Guo

WILEY-VCH

Authors

Prof. Yongzhu Fu
Zhengzhou University
100 Science Avenue
Zhengzhou
China 450001

Prof. Xiang Li
Zhengzhou University
College of Chemistry
100 Science Avenue
Zhengzhou
China 450001

Prof. Shuai Tang
Zhengzhou University
College of Chemistry
100 Science Avenue
Zhengzhou
China 450001

Prof. Wei Guo
Zhengzhou University
College of Chemistry
100 Science Avenue
Zhengzhou
China 450001

Cover: © Yagi Studio/Getty Images

Library of Congress Card No.: applied for

British Library Cataloguing-in-Publication Data
A catalogue record for this book is available from the British Library.

Bibliographic information published by the Deutsche Nationalbibliothek
The Deutsche Nationalbibliothek lists this publication in the Deutsche Nationalbibliografie; detailed bibliographic data are available on the Internet at <http://dnb.d-nb.de>.

Print ISBN: 978-3-527-35080-3
ePDF ISBN: 978-3-52783911-7
ePub ISBN: 978-3-52783912-4
oBook ISBN: 978-3-52783913-1

Typesetting Straive, Chennai, India

Printed and bound by CPI Group (UK) Ltd, Croydon, CR0 4YY

C9781118874387_041124

Contents

Preface

Energy plays a vital role in our society, and energy storage is important for our mobile lives and the usage of renewable energies such as solar and wind. As the state-of-the-art electrochemical energy storage technology, lithium-ion batteries (LIBs) are widely used in cell phones, laptops, and electric vehicles. LIBs were commercialized in 1991, and then their share in the energy storage market grew rapidly. However, with the rapid development of modern society, LIBs face a shortage of energy density and constraints of transition metals such as cobalt and nickel, which will become a challenge for a sustainable society. Alternative materials and systems are highly needed. Over the past years, organic batteries have attracted much attention because organic compounds are used as positive electrode materials, which are renewable and show a large difference from the traditional positive electrode materials, e.g. layered oxides, spinel oxides, and olivines. The energy storage mechanisms are well known for the ion-intercalation electrodes of LIBs. In contrast, organic materials undergo breakage/reformation of bonds, radical reactions, and anion compensation, which are quite different from those of transition metal oxides. Due to the variety of structures, organic electrode materials include carbonyl compounds, organosulfides, and N-containing organic compounds.

This book aims to provide a comprehensive "textbook" that covers most organic materials that have been studied as electrodes for rechargeable batteries. It includes various organic compounds, which can be positive and/or negative electrode materials. Beneficial to the variety of organic functional groups, the working conditions of organic batteries can be modified such as working voltage, open/sealed systems, and nonaqueous/aqueous systems. Some knowledge learned from the study of LIBs is also suitable for organic batteries such as the electrolyte design and interphase modifications between the electrode and electrolyte. Therefore, this book is suitable for readers with an electrochemistry and organic chemistry background including graduate students, postdoctoral fellows, and young faculty members. We hope this book brings new comprehension and possibilities in the field of electrochemical energy storage.

Finally, we gratefully acknowledge the substantial contribution from Dr. Qiliang Chen, Dr. Dan-Yang Wang, Dr. Qianqian Fan, Mr. Pengfei Sang, Mr. Fulong Zhu, Mr. Xiuqing Zhang, Ms. Siyuan Ren, Ms. Pei Yu, Ms. Hanshuang Song, and Ms. Qianhan Chen to this book.

Yongzhu Fu, Xiang Li, Shuai Tang, and Wei Guo
Zhengzhou, PR China

1

Necessity and Advantages of Developing Rechargeable Organic Batteries

1.1 Current Electrochemical Energy Storage Technologies

Li-ion battery (LIB) is well known as one of the electrochemical energy storage (EES) technologies, which can be seen in our daily lives, such as portable equipment and electric vehicles. LIBs have made great progress in the last 30 years, which can be traced back to 1991, when the first reversible LIB was commercialized by Sony Corp. [1]. The battery is based on $LiCoO_2$, graphite, and ester-solvents with $LiPF_6$ [2, 3]. Afterward, a series of ternary $LiNi_xCo_yMn_zO_2$ (NCM, $x+y+z=1$) and $LiFePO_4$ spring out, considering the aspects of energy density and security [4–11]. Prior to LIBs, actually, lithium metal batteries (LMBs) were commercialized by Moli Energy Corp., based on lithium metal and metal sulfide as negative and positive electrodes, respectively [12]. However, the battery was in a tailspin after several safety incidents due to the lithium dendrites, which are easily generated after cycling [13, 14]. Note that for the discovery and development of LIBs, John B. Goodenough, M. Stanley Whittingham, and Akira Yoshino were awarded the 2019 Nobel Prize in Chemistry.

Similar to LIB, sodium-ion battery (SIB) is also one of the state-of-the-art EES technologies. Actually, SIBs have a longer history compared with LIBs, with the layered oxides discovered toward the end of the 1960s [15, 16]. Considering the limited Li resource, SIB is a suitable alternative EES due to the relatively abundant Na resource (420 times more than Li). It is noted that the oxides Na_xMO_2 (M = 3d element) have some special structure by regulating deficient sodium [17–20], such as O2, O3, P2, and P3 types, according to the structural packing described by Delmas [17]. Therefore, the electrochemical performance can be modified in Na-based oxides, which has an evident advantage compared with the Li-based oxides used for LIBs. The energy density of SIB could climb to 200 Wh kg^{-1}, as reported by Hu's group, which is a breakthrough [21]. However, the energy density is still limited with respect to LIBs, which could deliver over 300 Wh kg^{-1} [22].

The good news is that SIBs have been commercialized by some Chinese companies such as CATL Corp. and HiNa Battery Corp. Given the energy density difference among the typical EES, the different battery systems aim for different

Rechargeable Organic Batteries: Materials, Mechanisms, and Prospects, First Edition.
Yongzhu Fu, Xiang Li, Shuai Tang, and Wei Guo.

market orientation to share the energy pressure. For example, lead-acid batteries, Ni–Cd batteries, and supercapacitors are used for devices with short mileage or low energy density, which are still required by the market. Nonetheless, both LIBs and SIBs cannot satisfy our demand in the long term, considering the resource crisis accompanied by high costs and pollution. Moreover, the traditional batteries are restricted to a sealed system and organic electrolytes (aqueous electrolytes are still facing great difficulties [23, 24]). Herein, we need a new EES without (or with mitigatory) the concerns.

A rechargeable organic battery is a good choice because the active materials are low cost, and the battery has comparable energy density when compared with LIB and SIB [25–31]. Moreover, the properties of organic materials can be controlled by different functional groups, such as the charge/discharge potential, the reaction dynamics, and the structural stability [32–37]. Furthermore, the system is unrestricted which can be used in an aqueous system, typically redox flow batteries (RFBs) [38–42]. Actually, the organic battery has been studied for over 60 years [43]. At the initial stage, the electrochemical performance of the organic material is poor with an ambiguous reaction mechanism, which impedes the development of the battery. In recent years, the corresponding published papers have a manifest rising tendency which can be seen in Figure 1.1, which partially benefits from technological advancements and several outstanding contributions made by Chen's group and Schubert's group since 2012 [44–69]. Now, there are many kinds of organics with active centers based on O, N, and S, enriching the family of organic batteries, which could compete with the traditional metal-ion batteries.

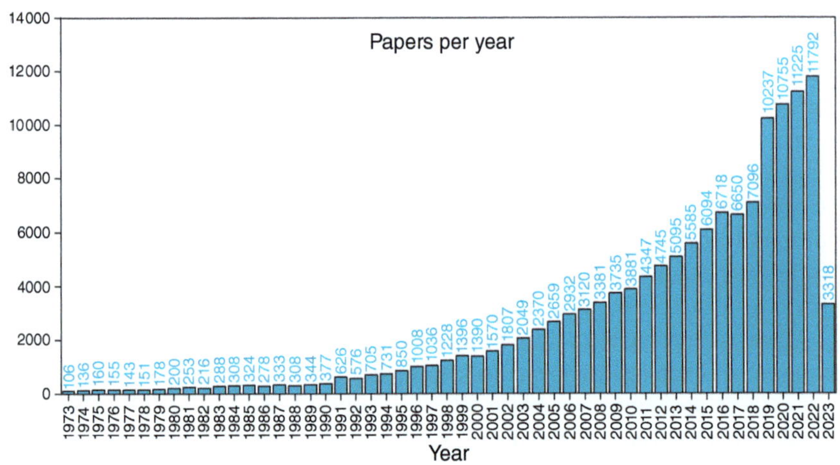

Figure 1.1 Published papers per year for the rechargeable organic batteries with keywords such as organic electrode, organic cathode, and organic battery. The time of the statistics is April 2023.

1.2 Rechargeable Organic Batteries

Organic batteries show different mechanisms from the typical LIBs, which mainly include insertion/extraction ($LiCoO_2$, NCM, etc.), alloying (Al, Si, etc.), and conversion mechanism (O, P, S-based composites, etc.) [70–79]. For the organic electrodes, the mechanism usually contains the repeated breaking and bonding of a bond (carbonyl, organosulfide, and radical materials in Chapter 2). A single bond (typically S—S bond) is broken during discharge, after which the broken bond receives an electron and bonds with a metal ion for charge balance. A double bond (typically carbonyl units) shows a similar mechanism. Note that during the reaction, radical materials are usually generated which has been applied in RFBs due to the fast kinetics [80–82]. There seems to be another mechanism not involved in bond breaking, which is based on electron transfer and anion compensation (typically N-containing active materials) [83].

The first investigation of organic materials is carbonyl compounds, which can be traced back to the 1960s [43]. However, the material shows high solubility in aprotic electrolytes, restricting the application although some other carbonyl composites are constructed [32]. Afterward, the direction was turned to conductive polymers in the 1970s because of their less solubility, such as polyacetylene and polypyrrole [83–89]. Unfortunately, these electrodes suffered from limited capacity due to incomplete reaction [88, 89]. A revival emerged when Armand and Tarascon depicted a bright future for organic batteries, attracting more attention [28]. Encouragingly, molecules with popular functional groups (quinones, carboxylates, radical centers, etc.) and other redox-active centers (imines, alkenes, alkynes, azo, etc.) have been investigated [90–97].

Another typical organic material is organosulfide with S as the redox center. Visco et al. initially studied tetraethyl thiuram disulfide (TETD) in 1988 [98]. However, the electrode delivered poor electrochemical performance, which cannot be used in a battery. Actually, the research of organosulfide battery mainly focused on polysulfides from the 1980s to 2015, such as naphtho[1,8-*cd*][1,2]dithiol and dibenzo[*c,e*][1,2]dithiin, which have not attracted full attention [99, 100]. The polysulfides have a large specific capacity (over 300 $mAh\,g^{-1}$), however, with poor stability due to the rigid framework in which the S—S bonds suffer from breaking and painful bonding, deteriorating the original structure which can be only used for lithium primary battery. Afterward, organosulfide with small molecule was investigated. However, the materials were considered hopeless because they easily dissolved into the electrolyte, leading to a shuttle effect that normally appeared in Li–S batteries. Until 2016, dimethyl trisulfide (DMTS) [101] was successfully applied in organic batteries with a reversible charge/discharge process (with 849 $mAh\,g^{-1}$) and cycling performance (50 cycles) with the assistance of a carbon paper which was also proposed for polysulfide in 2013 [102]. The creative idea has opened a broad perspective for the research of small-molecule organosulfur

[103–111]. Hereafter, molecules with more sulfur content (–Sn–) were studied to modify the specific capacity [112–116]. The performance of organosulfur can be regulated by heteroatom doping (such as Se, Te), combination with metal sulfides for good conductivity and stability, and the application of RFBs [117–126].

1.3 Goal, Scope, and Organization of this Book

It is obvious that a consequent and growing amount of literature is now easily available on organic batteries after years of silence. There is room for reversible electroactive organic systems in the future EES landscape in view of the application. However, it must be noted that there exists a certain disciplinary boundary between inorganic and organic compounds because the redox chemistry of organics is different from that of typical LIBs, making it challenging for nonspecialist readers when dealing with organic batteries. Therefore, it would be timely to provide a kind of "tutorial"-oriented book for a broader audience. Based on the latest selected and reliable data from both general and specialized scientific literature, this contribution also aims at providing the readers with a better understanding of the consecutive global demand for electrical energy sources and the evolution trends of organic batteries.

The following approach will be stepwise introduced in this book.

1.3.1 Working Principles and Fundamental Properties

First, the reader will have a concept of some prototypical organic materials, including the mechanisms and cell configurations. The organic electrodes as cathodes are the key point in this book, including the introduction of the mechanism (Chapter 2), carbonyl-based organic cathodes (Chapter 3), sulfur-containing organic cathodes (Chapter 4), radical-based organic cathodes (Chapter 5), organometallic complex-based cathodes (Chapter 6), polymer-based organic cathodes (Chapter 7), and other organic cathodes. The reader will know the basic types and reaction mechanisms of these organic electrodes. Afterward, some typical organic anodes are introduced (Chapter 8), which can be assembled as an all-organic battery counter to an organic cathode (Chapter 9).

1.3.2 A Selection of an Organic Electrode

After our introduction, the reader will know that the function of an organic electrode could be modified by applying a wealth of functional groups. Compared with the typical inorganic electrodes, organic electrodes can be considered cathodes (even working at high potential like P-type electrodes), anodes, and soluble agents for RFBs. Moreover, with the assistance of some special functional groups (such as N-based group), a battery with ultra-high-rate performance (e.g. thiuram monosulfide [TM]) and eutectic solution at low temperature as a special electrolyte (such as the interaction between 2,2′-dipyridyl disulfide (DPyDS) and lithium

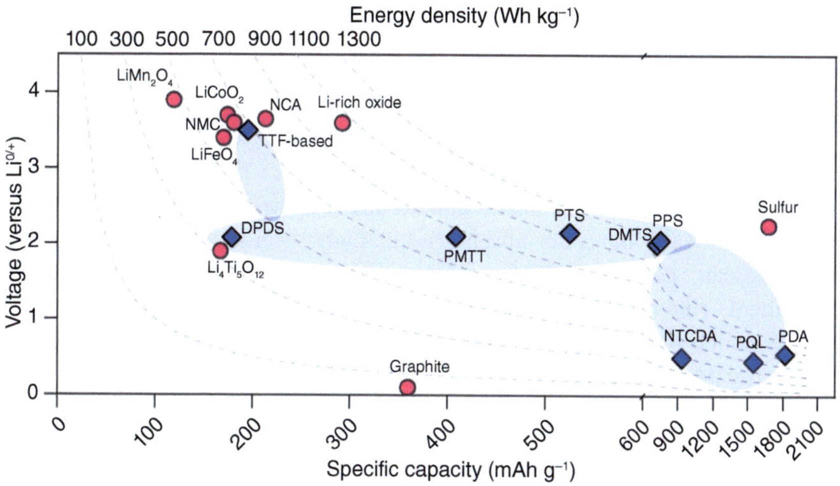

Figure 1.2 Typical organic materials with voltage, specific capacity, and energy density are compared with some typical traditional electrode materials.

bis(trifluoromethanesulfonyl)imide [LiTFSI]) can be achieved [107, 127]. The organic materials hold a wide range of voltage, specific capacity, and energy density (Figure 1.2). Thus, the reader will know how to choose an organic electrode based on the demand.

Note that TTF-based, DPDS, PMTT, PTS, DMTS, PPS, NTCDA, PQL, and PDA denote tetrakis(methylthio)-derivative cyclohexene-1,4-diylidenes (TTF-based) [128], diphenyl disulfide (DPDS) [119], dipentamethylenethiuram tetrasulfide (PMTT) [129], phenyl tetrasulfide (PTS) [118], dimethyltrisulfide (DMTS) [101], phenyl hexasulfide (PPS) [92], 1,4,5,8-naphthalenetetracarboxylic dianhydride (NTCDA) [95], poly(1,6-dihydropyrazino[2,3g]quinoxaline-2,3,8-triyl-7-(2*H*)-ylidene-7,8-dimethylidene) (PQL) [130], and polydopamine (PDA) [131], respectively.

1.3.3 EES Applications

After the introduction, the reader will understand the concept of the organic electrodes in terms of the cost, resource availability, and stability. For the cost, the typical DMTS and TM have an obvious price advantage compared with $LiCoO_2$ and NCM. Notably, TM has an ultrahigh redox activity with long cycling performances (over 8000 cycles), making it a potential candidate [127]. As for availability, organic materials are based on plentiful elements such as C, N, O, and S. It is obvious from the price comparison that the organic electrodes have rich resources. Herein, taking the long view, the organic electrodes have a low cost. As for chemical stability, the typical inorganic electrodes always suffer from a phase transition during cycling; hereafter, the structure deteriorates until the end of the cycle life. By contrast, the solubility property of organic electrodes makes the electron transfer process faster through a new mechanism different from the insertion/extraction mechanism. Note that a moderate dissolution of the active material benefits the rate performance and mitigates the evolution of the bulk structure of the electrode.

1.3.4 Practical Applications

Although research on the organic electrodes is earlier than the typical LIBs, the development of LIBs is faster due to their successful commercialization by Sony Corp. LIBs have the advantages of high specific capacity and stable cycling lifespan. However, LIBs encounter the bottleneck at the point of long development. Herein, other types of batteries emerge as alternatives, such as SIBs and Zn-ion batteries. As a prototype post-LIB, the battery based on organic materials holds obvious advantages. On the one hand, a P-type organic electrode (e.g. ethyl viologen iodide) can deliver a capacity of 230 $mAh\,g^{-1}$, with a high output potential of 3.7 V. On the other hand, the TM electrode can hold a robust cycling lifespan, with a high capacity retention of 70% even after 8000 cycles. As for security, the electrolyte based on DPyDS and LiTFSI consists of a eutectic solution just by grinding together, being of incombustibility, which potentially matches the demands of safety for electrolytes. Moreover, RFBs can be designed based on organic electrodes applying their solubility. Therefore, it should be declared that organic electrodes have a high potential for practical application.

1.3.5 Key Challenges

- The solubility of the organic electrodes is one key challenge. Because the soluble organic material has a shuttle effect, the organic electrodes keep the silence for a long time. With the application of carbon paper for organosulfide in 2016, a reversible and long cycling performance of organic battery was achieved. A carbon paper has strong adsorption ability which restricts the dissolution of the organic electrodes, mitigating the shuttle effect to some extent. Recently, a functional MOF-based separator has been designed to hinder the shuttle of the active material for LIBs based on the organic electrode (5,5′-dimethyl-2,2′-*bis*-*p*-benzoquinone), prolonging the cycle life to 2000 cycles (capacity retention of 82.9%) [132]. However, the mentioned separator raises the cost. From the view of practical use, a more facile and low-cost method should be applied.
- The cycling stability is another challenge for the application of organic electrodes. Although the soluble property accelerates the reaction kinetics, it also leads to the loss of the active material, resulting in a capacity loss and a short cycle life. The dissolution degree of organic electrodes should be carefully regulated. Moreover, some organic electrodes have the intrinsic properties of electrochemical inertness, resulting in a large polarization, which should be a concern.
- Various mechanisms are involved in the organic reactions, which are different from the traditional mechanisms in LIBs. The reaction always involves the break and recombination of the active bond, free radical reaction, charge recombination,

and so on. Herein, the detailed reactions step by step are still indistinct due to the complex electron transfer, dissolution–diffusion, element, and steric-dependent properties involved in the processes. The details also depend on the advanced equipment.

- As discussed above, some organic electrodes have their advantages, such as long lifespan, high working potential, and robust stability. However, it is still hard to find one organic electrode that has its full advantages. The organic electrode should be comparable in the fields of energy density, lifespan, and cost.

Acknowledgments

This work was supported by the National Natural Science Foundation of China (Grant Nos. U2004214 and 21975225) and the National Key Research and Development Program of China (2022YFB2402200).

References

1 Li, M., Lu, J., Chen, Z., and Amine, K. (2018). *Adv. Mater.* 30: 1800561.
2 Mizushima, K., Jones, P.C., Wiseman, P.J., and Goodenough, J.B. (1980). *Mater. Res. Bull.* 15: 783–789.
3 Yazami, R. and Touzain, P. (1983). *J. Power Sources* 9: 365–371.
4 Padhi, A.K., Nanjundaswamy, K.S., and Goodenough, J.B. (1997). *J. Electrochem. Soc.* 144: 1188.
5 Choi, J. and Manthiram, A. (2005). *J. Electrochem. Soc.* 152: A1714.
6 Jiang, J., Shi, W., Zheng, J. et al. (2014). *J. Electrochem. Soc.* 161: A336.
7 Kang, B. and Ceder, G. (2009). *Nature* 458: 190–193.
8 Kim, U.-H., Park, G.-T., Son, B.-K. et al. (2020). *Nat. Energy* 5: 860–869.
9 Li, W., Erickson, E.M., and Manthiram, A. (2020). *Nat. Energy* 5: 26–34.
10 Logan, E.R., Hebecker, H., Eldesoky, A. et al. (2020). *J. Electrochem. Soc.* 167: 130543.
11 Sun, H.H., Ryu, H.-H., Kim, U.-H. et al. (2020). *ACS Energy Lett.* 5: 1136–1146.
12 Whittingham, M.S. (1976). *J. Electrochem. Soc.* 123: 315.
13 Liu, Y., Lin, D., Liang, Z. et al. (2016). *Nat. Commun.* 7: 10992.
14 Ye, H., Xin, S., Yin, Y.-X. et al. (2017). *J. Am. Chem. Soc.* 139: 5916–5922.
15 Thery, J. and Briancon, D. (1964). *Rev. Int. Hautes Temp. Refract.* 1: 221.
16 Delmas, C. (2018). *Adv. Energy Mater.* 8: 1703137.
17 Delmas, C., Fouassier, C., and Hagenmuller, P. (1980). *Phys. B+C* 99: 81–85.
18 Parant, J.-P., Olazcuaga, R., Devalette, M. et al. (1971). *J. Solid State Chem.* 3: 1–11.
19 Delmas, C., Braconnier, J.-J., Fouassier, C., and Hagenmuller, P. (1981). *Solid State Ionics* 3–4: 165–169.
20 Braconnier, J.-J., Delmas, C., Fouassier, C., and Hagenmuller, P. (1980). *Mater. Res. Bull.* 15: 1797–1804.

21 Li, Y., Zhou, Q., Weng, S. et al. (2022). *Nat. Energy* 7: 511–519.
22 Qiao, Y., Yang, H., Chang, Z. et al. (2021). *Nat. Energy* 6: 653–662.
23 Yue, J., Zhang, J., Tong, Y. et al. (2021). *Nat. Chem.* 13: 1061–1069.
24 Suo, L., Borodin, O., Gao, T. et al. (2015). *Science* 350: 938–943.
25 Sato, K., Ichinoi, R., Mizukami, R. et al. (2018). *J. Am. Chem. Soc.* 140: 1049–1056.
26 Kim, D.J., Yoo, D.-J., Otley, M.T. et al. (2019). *Nat. Energy* 4: 51–59.
27 Han, C., Li, H., Shi, R. et al. (2019). *J. Mater. Chem. A* 7: 23378–23415.
28 Armand, M. and Tarascon, J.M. (2008). *Nature* 451: 652–657.
29 Friebe, C., Lex-Balducci, A., and Schubert, U.S. (2019). *ChemSusChem* 12: 4093–4115.
30 Wang, S., Li, F., Easley, A.D., and Lutkenhaus, J.L. (2019). *Nat. Mater.* 18: 69–75.
31 Xie, J. and Zhang, Q. (2019). *Small* 15: 1805061.
32 Wang, D.-Y., Guo, W., and Fu, Y. (2019). *Acc. Chem. Res.* 52: 2290–2300.
33 Wang, D.-Y., Liu, R., Guo, W. et al. (2021). *Coord. Chem. Rev.* 429: 213650.
34 Wang, D.-Y., Si, Y., Guo, W., and Fu, Y. (2020). *Adv. Sci.* 7: 1902646.
35 Wang, D.-Y., Si, Y., Guo, W., and Fu, Y. (2021). *Nat. Commun.* 12: 3220.
36 Wang, D.-Y., Si, Y., Li, J., and Fu, Y. (2019). *J. Mater. Chem. A* 7: 7423–7429.
37 Wang, D.-Y., Wang, W., Li, F. et al. (2022). *J. Energy Chem.* 71: 572–579.
38 Alotto, P., Guarnieri, M., and Moro, F. (2014). *Renewable Sustainable Energy Rev.* 29: 325–335.
39 Sum, E. and Skyllas-Kazacos, M. (1985). *J. Power Sources* 15: 179–190.
40 Sum, E., Rychcik, M., and Skyllas-kazacos, M. (1985). *J. Power Sources* 16: 85–95.
41 Rychcik, M. and Skyllas-Kazacos, M. (1988). *J. Power Sources* 22: 59–67.
42 Xu, Y., Wen, Y., Cheng, J. et al. (2009). *Electrochem. Commun.* 11: 1422–1424.
43 Williams, D.L., Byrne, J.J., and Driscoll, J.S. (1969). *J. Electrochem. Soc.* 116: 2.
44 Liang, Y., Tao, Z., and Chen, J. (2012). *Adv. Energy Mater.* 2: 742–769.
45 Häupler, B., Wild, A., and Schubert, U.S. (2015). *Adv. Energy Mater.* 5: 1402034.
46 Janoschka, T., Hager, M.D., and Schubert, U.S. (2012). *Adv. Mater.* 24: 6397–6409.
47 Lee, S., Kwon, G., Ku, K. et al. (2018). *Adv. Mater.* 30: 1704682.
48 Leung, P., Shah, A.A., Sanz, L. et al. (2017). *J. Power Sources* 360: 243–283.
49 Luo, J., Hu, B., Hu, M. et al. (2019). *ACS Energy Lett.* 4: 2220–2240.
50 Muench, S., Wild, A., Friebe, C. et al. (2016). *Chem. Rev.* 116: 9438–9484.
51 Peng, H., Yu, Q., Wang, S. et al. (2019). *Adv. Sci.* 6: 1900431.
52 Schon, T.B., McAllister, B.T., Li, P.-F., and Seferos, D.S. (2016). *Chem. Soc. Rev.* 45: 6345–6404.
53 Song, Z. and Zhou, H. (2013). *Energy Environ. Sci.* 6: 2280–2301.
54 Wang, H. and Zhang, X. (2018). *Chem. Eur. J* 24: 18235–18245.
55 Wei, X., Pan, W., Duan, W. et al. (2017). *ACS Energy Lett.* 2: 2187–2204.
56 Winsberg, J., Hagemann, T., Janoschka, T. et al. (2017). *Angew. Chem. Int. Ed.* 56: 686–711.
57 Wu, Y., Zeng, R., Nan, J. et al. (2017). *Adv. Energy Mater.* 7: 1700278.

58 Xie, J., Gu, P., and Zhang, Q. (2017). *ACS Energy Lett.* 2: 1985–1996.
59 Xu, Y., Zhou, M., and Lei, Y. (2018). *Mater. Today* 21: 60–78.
60 Zhao, Q., Lu, Y., and Chen, J. (2017). *Adv. Energy Mater.* 7: 1601792.
61 Zhao, Q., Zhu, Z., and Chen, J. (2017). *Adv. Mater.* 29: 1607007.
62 Zhu, L., Ding, G., Xie, L. et al. (2019). *Chem. Mater.* 31: 8582–8612.
63 Zhu, Z. and Chen, J. (2015). *J. Electrochem. Soc.* 162: A2393.
64 Armstrong, C.G. and Toghill, K.E. (2018). *Electrochem. Commun.* 91: 19–24.
65 Casado, N., Hernández, G., Sardon, H., and Mecerreyes, D. (2016). *Prog. Polym. Sci.* 52: 107–135.
66 Ding, Y., Zhang, C., Zhang, L. et al. (2018). *Chem. Soc. Rev.* 47: 69–103.
67 Gong, K., Fang, Q., Gu, S. et al. (2015). *Energy Environ. Sci.* 8: 3515–3530.
68 Gracia, R. and Mecerreyes, D. (2013). *Polym. Chem.* 4: 2206–2214.
69 Oltean, V.A., Renault, S., Valvo, M.L., and Brandell, D. (2016). *Materials* 9: 142.
70 Abel, P.R., Lin, Y.-M., de Souza, T. et al. (2013). *J. Phys. Chem. C* 117: 18885–18890.
71 Gao, H., Zhou, T., Zheng, Y. et al. (2016). *Adv. Energy Mater.* 6: 1601037.
72 Li, L., Seng, K.H., Li, D. et al. (2014). *Nano Res.* 7: 1466–1476.
73 Li, X., Guo, S., Jiang, K. et al. (2018). *ACS Appl. Mater. Interfaces* 10: 16–20.
74 Mao, J., Fan, X., Luo, C., and Wang, C. (2016). *ACS Appl. Mater. Interfaces* 8: 7147–7155.
75 Wang, Y.-X., Seng, K.H., Chou, S.-L. et al. (2014). *Chem. Commun.* 50: 10730–10733.
76 Wu, L., Hu, X., Qian, J. et al. (2014). *Energy Environ. Sci.* 7: 323–328.
77 Xu, Y., Zhu, Y., Liu, Y., and Wang, C. (2013). *Adv. Energy Mater.* 3: 128–133.
78 Zheng, Y., Zhou, T., Zhang, C. et al. (2016). *Angew. Chem. Int. Ed.* 55: 3408–3413.
79 Zhou, T., Pang, W.K., Zhang, C. et al. (2014). *ACS Nano* 8: 8323–8333.
80 Nakahara, K., Iwasa, S., Satoh, M. et al. (2002). *Chem. Phys. Lett.* 359: 351–354.
81 Suga, T., Pu, Y.-J., Oyaizu, K., and Nishide, H. (2004). *Bull. Chem. Soc. J.* 77: 2203–2204.
82 Liu, M., Visco, S.J., and De Jonghe, L.C. (1990). *J. Electrochem. Soc.* 137: 750.
83 Ma, T., Liu, L., Wang, J. et al. (2020). *Angew. Chem. Int. Ed.* 59: 11533–11539.
84 Shirakawa, H., Louis, E.J., MacDiarmid, A.G. et al. (1977). *J. Chem. Soc., Chem. Commun.* 16: 578–580.
85 Chance, R.R., Shacklette, L.W., Miller, G.G. et al. (1980). *J. Chem. Soc., Chem. Commun.* 8: 348–349.
86 Ivory, D.M., Miller, G.G., Sowa, J.M. et al. (1979). *J. Chem. Phy.* 71: 1506–1507.
87 MacInnes, D., Druy, M.A., Nigrey, P.J. et al. (1981). *J. Chem. Soc., Chem. Commun.* 7: 317–319.
88 Novák, P., Müller, K., Santhanam, K.S.V., and Haas, O. (1997). *Chem. Rev.* 97: 207–282.
89 Jia, X., Ge, Y., Shao, L. et al. (2019). *ACS Sustainable Chem. Eng.* 7: 14321–14340.
90 Lu, Y., Hou, X., Miao, L. et al. (2019). *Angew. Chem. Int. Ed.* 58: 7020–7024.

91 Iordache, A., Delhorbe, V., Bardet, M. et al. (2016). *ACS Appl. Mater. Interfaces* 8: 22762–22767.

92 Bhargav, A., Bell, M.E., Karty, J. et al. (2018). *ACS Appl. Mater. Interfaces* 10: 21084–21090.

93 Hansen, K.-A., Nerkar, J., Thomas, K. et al. (2018). *ACS Appl. Mater. Interfaces* 10: 7982–7988.

94 Matsunaga, T., Kubota, T., Sugimoto, T., and Satoh, M. (2011). *Chem. Lett.* 40: 750–752.

95 Han, X., Qing, G., Sun, J., and Sun, T. (2012). *Angew. Chem. Int. Ed.* 51: 5147–5151.

96 Renault, S., Oltean, V.A., Araujo, C.M. et al. (2016). *Chem. Mater.* 28: 1920–1926.

97 Luo, C., Borodin, O., Ji, X. et al. (2018). *Proc. Natl. Acad. Sci. U.S.A.* 115: 2004–2009.

98 Visco, S.J. and DeJonghe, L.C. (1988). *J. Electrochem. Soc.* 135: 2905.

99 Liu, M., Visco, S.J., and De Jonghe, L.C. (1989). *J. Electrochem. Soc.* 136: 2570.

100 Inamasu, T., Yoshitoku, D., Sumi-otorii, Y. et al. (2003). *J. Electrochem. Soc.* 150: A128.

101 Wu, M., Cui, Y., Bhargav, A. et al. (2016). *Angew. Chem. Int. Ed.* 55: 10027–10031.

102 Fu, Y., Su, Y.-S., and Manthiram, A. (2013). *Angew. Chem. Int. Ed.* 52: 6930–6935.

103 Sun, J., Zhang, K., Fu, Y., and Guo, W. (2022). *Nano Res.* 16: 3814–3822.

104 Guo, W., Zhang, W., Si, Y. et al. (2021). *Nat. Commun.* 12: 3031.

105 Lian, J., Guo, W., and Fu, Y. (2021). *J. Am. Chem. Soc.* 143: 11063–11071.

106 Gao, M., Lan, J., Fu, Y., and Guo, W. (2022). *ChemSusChem* 15: e202200423.

107 Song, J., Si, Y., Guo, W. et al. (2021). *Angew. Chem. Int. Ed.* 60: 9881–9885.

108 Tang, S., Chen, Q., Si, Y. et al. (2021). *Adv. Mater.* 33: 2100824.

109 Tang, S., Guo, W., and Fu, Y. (2021). *Adv. Energy Mater.* 11: 2000802.

110 Li, F., Si, Y., Liu, B. et al. (2019). *Adv. Funct. Mater.* 29: 1902223.

111 Zhao, B., Si, Y., Guo, W., and Fu, Y. (2022). *Adv. Funct. Mater.* 32: 2112225.

112 Li, F., Guo, W., Si, Y. et al. (2021). *Electrochim. Acta* 370: 137757.

113 Zhao, J., Si, Y., Han, Z. et al. (2020). *Angew. Chem. Int. Ed.* 59: 2654–2658.

114 Li, F., Si, Y., Li, Z. et al. (2020). *J. Mater. Chem. A* 8: 87–90.

115 Sang, P., Si, Y., and Fu, Y. (2019). *Chem. Commun.* 55: 4857–4860.

116 Wu, M., Bhargav, A., Cui, Y. et al. (2016). *ACS Energy Lett.* 1: 1221–1226.

117 Chen, Q., Si, Y., Guo, W., and Fu, Y. (2022). *Chem. Commun.* 58: 10993–10996.

118 Lv, X., Guo, W., Song, J., and Fu, Y. (2022). *Small* 18: e2105071.

119 Ren, S., Si, Y., Guo, W., and Fu, Y. (2022). *ACS Sustainable Chem. Eng.* 10: 7526–7535.

120 Xie, K., Sun, J., Lian, J. et al. (2022). *Appl. Phys. Lett.* 121: 133904.

121 Cui, Y., Ackerson, J.D., Ma, Y. et al. (2018). *Adv. Funct. Mater.* 28: 1801791.

122 Wang, Z., Li, X., Guo, W., and Fu, Y. (2021). *Adv. Funct. Mater.* 31: 2009875.

123 Fan, Q., Guo, W., Si, Y. et al. (2021). *Adv. Funct. Mater.* 31: 2001493.

124 Chen, Q., Wang, W., Li, X. et al. (2022). *Proc. Natl. Acad. Sci. U.S.A.* 119: e2202449119.

125 Chen, Q., Guo, W., and Fu, Y. (2022). *Adv. Sci.* 9: 2104036.

126 Chen, Q., Guo, W., Wang, D., and Fu, Y. (2021). *J. Mater. Chem. A* 9: 12652–12658.

127 Chen, Q., Li, L., Wang, W. et al. (2022). *J. Am. Chem. Soc.* 144: 18918–18926.

128 Kato, M., Senoo, K.-I., Yao, M., and Misaki, Y. (2014). *J. Mater. Chem. A* 2: 6747–6754.

129 Bhargav, A., Ma, Y., Shashikala, K. et al. (2017). *J. Mater. Chem. A* 5: 25005–25013.

130 Wu, J., Rui, X., Long, G. et al. (2015). *Angew. Chem. Int. Ed.* 54: 7354–7358.

131 Sun, T., Li, Z.-J., Wang, H.-G. et al. (2016). *Angew. Chem. Int. Ed.* 55: 10662–10666.

132 Bai, S., Kim, B., Kim, C. et al. (2021). *Nat. Nanotechnol.* 16: 77–84.

2

Redox Mechanisms and Characterization Methods of Organic Electrode Materials

2.1 Introduction

In general, organic electrode materials of rechargeable batteries are classified into n-type (e.g. carbonyls, organosulfides, azos, and imines), p-type (e.g. thioethers and phenothiazines), and bipolar-type (e.g. nitrogen radicals) according to their charge state changes of redox-active groups on the electrochemical reaction process [1, 2]. Figure 2.1 gives three types of redox mechanisms and some representative molecular structures of organic electrode materials. For n-type organic electrode materials which can change their state between a neutral state and a negatively charged state. Typically, n-type organic materials first undergo reduction from neutral states to form anions, and then compensate anions by cations, e.g., Li^+, Na^+, or K^+, in the electrolyte. In turn, the anions experience oxidation in their original states. As the redox-active component for rechargeable batteries, n-type materials often display high specific capacity, high energy density, and low redox potential [3–5].

Conversely, p-type organic materials usually change their state between a neutral state and a positively charged state. They first lose electrons to form cations and then compensate cations via anions (e.g. $TFSI^-$ and ClO_4^-) from supporting electrolytes to maintain the charge neutral. Considering their high redox potentials and fast redox kinetics, p-type organic materials are usually used as the cathode component in rechargeable batteries [6, 7]. Nonetheless, the low specific capacity of p-type organic materials remains a formidable challenge [8, 9]. Bipolar-type materials can shift their state not only between a neutral state and a negatively charged state but also between a neutral state and a positively charged state. Therefore, they have characteristics of both n-type and p-type organic electrode materials.

In the following, redox mechanisms of carbonyls, organosulfides, radicals, and N-containing active materials are separately introduced in detail. Meanwhile, some typical advanced characterization techniques, including Fourier transform infrared spectroscopy (FTIR), X-ray diffraction (XRD), Raman spectroscopy, ultraviolet–visible (UV–vis) spectroscopy, nuclear magnetic resonance (NMR), electron paramagnetic resonance (EPR) spectroscopy, and fluorescence spectroscopy, for verifying their redox mechanisms are selected and discussed, which give us a comprehensive understanding of the application of advanced characterization

Rechargeable Organic Batteries: Materials, Mechanisms, and Prospects, First Edition.
Yongzhu Fu, Xiang Li, Shuai Tang, and Wei Guo.

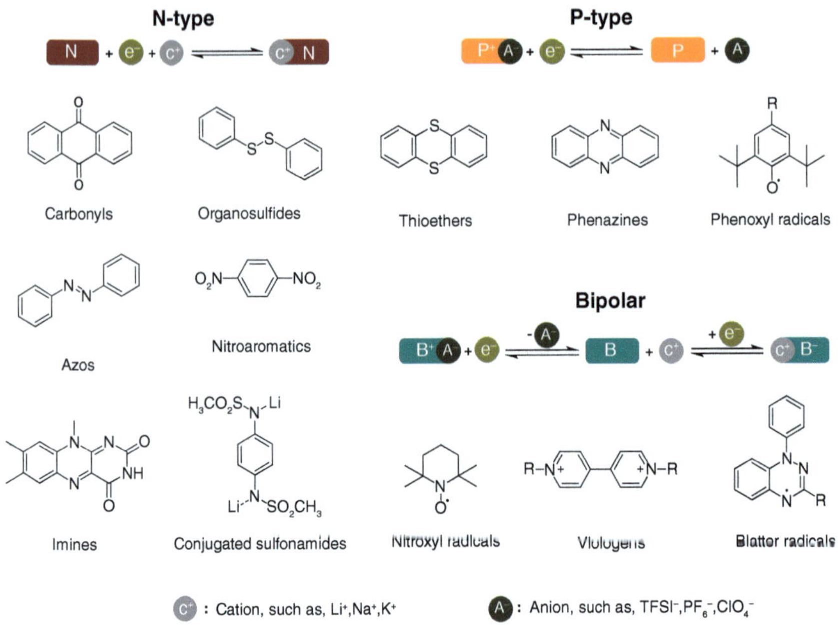

Figure 2.1 Three types of reaction mechanisms and some representative molecular structures of organic electrode materials.

methods in organic rechargeable batteries and lead to building better organic rechargeable batteries.

2.2 Carbonyl Materials

2.2.1 Redox Mechanisms

Carbonyl redox-active materials commonly undergo a n-type redox behavior, which includes one or more carbonyl units as redox centers to store/release electric charges [10]. Upon the electrochemical reduction, a C=O unit can accept one electron to form a monoanion of $C–O^-$ (Figure 2.2). Then, a cation, e.g., Li^+ or K^+, offered in the electrolyte combined with the $C–O^-$ to realize electrically neutral. This redox process is reversible in the presence of various appropriate organic functional groups. In addition, carbonyl redox-active materials are also the most widely studied in organic batteries, which have four categories, i.e., imides, quinones, anhydrides, and carboxylates till now [11–13]. Therein, quinone electrode materials consist of two adjacent carbonyl units directly connected to conjugated rings, which undergo

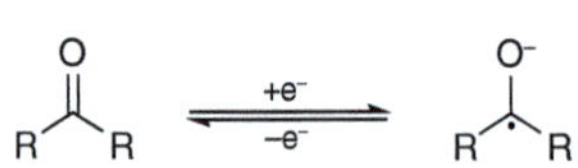

Figure 2.2 Redox behavior of a carbonyl unit, R represents one organic functional group.

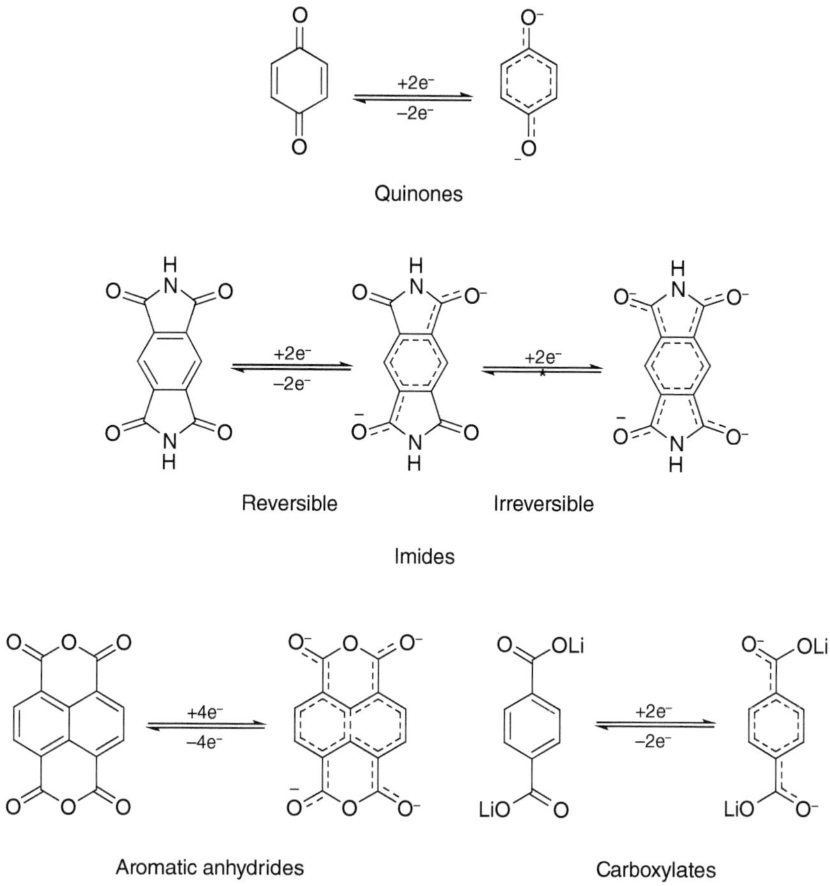

Figure 2.3 Redox behaviors of four categories of carbonyl redox-active materials.

well-known reversible two-electron redox reactions (Figure 2.3) that enable their high theoretical capacities, e.g., 1,4,5,8-naphthodiquinone affords a high theoretical value of 570 mAh g^{-1} [14–18].

Imide electrode materials have imino moieties around the carbonyl groups. Theoretically, an imide molecule with four carbonyl units can undergo four electron transfers reversibly. However, the reversible reaction behavior only realizes the first two electron transfers [19]. If continue to accept two electrons, the charge repulsion will emerge that damages the imide structure irreversibly. Therefore, the practical capacities of imide redox-active materials are contributed by their first two electron transfers (Figure 2.3) [20]. For example, a lithiation pyromellitic diimide material undergoes two one-electron redox reactions in rechargeable lithium batteries [21]. Anhydride electrode materials carry four carbonyl units distributed around an aromatic core. They are able to undergo four electron transfers reversibly for a monomer molecule (Figure 2.3) [22–24].

Carboxylate materials have an aromatic ring connected with two/multiple metal carboxylates, which are capable of undergoing reversible two-electron conversion reactions (Figure 2.3) [4]. During the reduction, the generated dianion is stabilized via a conjugated system. The loss of resonance or aromaticity in the core enables them to have low redox potential, which is beneficial for them as the anode redox-active component for organic rechargeable batteries. For example, a sample of carboxylate material, i.e., di-lithium terephthalate, delivered a low cell voltage of 0.8 V in lithium batteries [4, 5].

2.2.2 Characterization Methods

FTIR is a powerful analytical tool to explore electrode reactions of carbonyl materials, which due to the C=O group have a strong stretching vibration peak at 1900~1650 cm^{-1} of the spectrum [26–28]. For example, the reaction mechanism of 1,4-bis(*p*-benzoquinonyl) benzene (BBQB) in a rechargeable lithium-organic battery is analyzed by ex situ FTIR (Figure 2.4) [25]. The characteristic stretching vibration peak of C=O units in the pristine BBQB electrode appears at 1652 cm^{-1}. Upon discharging to ~2.6 V, the peak intensity becomes weak. It almost disappears at the end of discharge, proving that most C=O units in BBQB are reduced to –C–O^-. In recharge, the characteristic peak at 1652 cm^{-1} shows again, suggesting that one BBQB molecule can accept four electrons and four Li^+ reversibly.

The above example indicates that the ex situ FTIR spectra can effectively respond to the disappearance/appearance of the characteristic peak of C=O bonds during

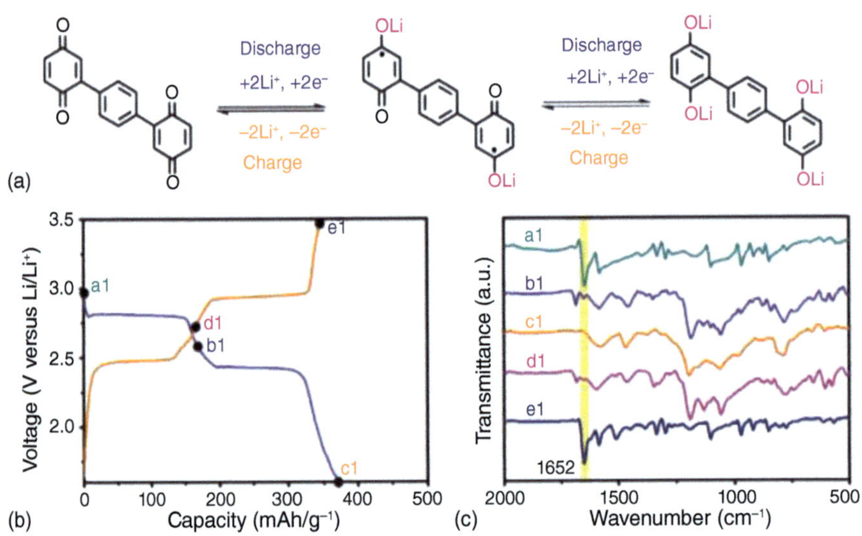

Figure 2.4 (a) Multielectron redox process of the BBQB electrode in lithium batteries. (b) Discharge/charge profiles of a lithium-BBQB cell. (c) Ex situ FTIR spectra of BBQB at various states of discharge/charge as marked in (b). Source: Reproduced with permission from Yang et al. [25]. Copyright 2020, Wiley-VCH.

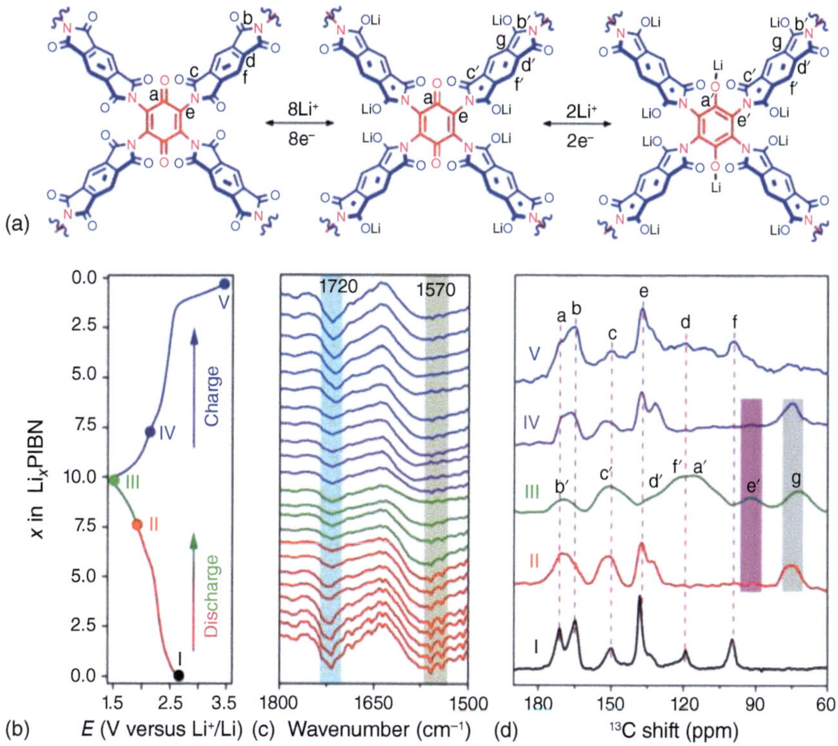

Figure 2.5 (a) Electrochemical redox reactions of PIBN in lithium batteries. (b) Discharge–charge profile of the PIBN-G electrode. (c) In situ FTIR spectra collected in real time. (d) Ex situ ^{13}C NMR spectra collected at different discharge–charge states, as marked (b). Source: Reproduced with permission from Luo et al. [29]. Copyright 2018, Wiley-VCH.

electrochemical oxidation/reduction. In spite of this, to gain insight into the reaction condition of electrodes at different states of charge-discharge by ex situ methods, assembling and disassembling multiple cells is always required, which may bring some inevitable experimental errors [30]. To avoid this vulnerability, in situ characterization methods emerge that only assemble a test cell to monitor electrode reactions in real time. Luo et al. used in situ FTIR spectra to explore the reaction process of poly(imide-benzoquinone)-graphene (PIBN-G) in lithium-organic batteries (Figure 2.5a)[29]. During cycling (Figure 2.5b), real-time monitoring of the PIBN-G electrode was carried via in situ FTIR spectra (Figure 2.5c). Before discharging, there are two characteristic signals of the PIBN-G electrode: one is ascribed to the =O units of imides and appears at 1720 cm^{-1}, and another signal appears at 1570 cm^{-1} is the C=O units of quinones. During discharging, the characteristic peak of 1720 cm^{-1} gradually disappears first, and then the peak of 1570 cm^{-1} fades. This indicates the redox process of the PIBN-G electrode involves two steps.

Ex situ solid-state ^{13}C NMR was also used to investigate the redox process of PIBN-G (Figure 2.5d) [29]. There are six signals of the pristine PIBN-G cathode. When discharging to stage II, as marked in Figure 2.5b, the signals of imide groups disappear and a new signal at 75.7 ppm (g) appears, suggesting that the C=O units of imides have been reduced to –C–O$^-$. After fully discharging, another new signal at 92.0 ppm (e) emerges with the loss of the characteristic signals of quinones. In recharge, all signals return to the natural state. Therefore, the ex situ solid-state ^{13}C NMR is also an effective tool for studying the redox reactions of carbonyl materials.

Organic electrode materials are also used as the redox-active component for redox flow batteries. The active species are stored in tanks and flow into an electrochemical cell for oxidation and reduction via a pump. To gain insight into the redox mechanisms of soluble organic electrode materials in flow batteries, researchers proposed in situ NMR and in situ EPR to probe changes in active species [31, 32]. Figure 2.6a shows an in situ NMR setup where a flow cell is outside the NMR magnet, and the redox-active materials flow through an NMR probe via the pump. The 2,6-dihydroxyanthraquinone (DHAQ) anolyte changes in flow batteries were investigated in real time via coupling in situ NMR and in situ EPR during oxidation and reduction [32]. The molecular structures of DHAQ and two corresponding reduction products are presented in Figure 2.6b. In the NMR spectra, the characteristic peak of A closest to the C=O redox center basically disappears upon reduction, whereas the peak signal B is broadened and only just visible above the baseline (Figure 2.6c, d). The formation of the intermediate product of DHAQ$^{3-\bullet}$ radical anion and the intermolecular electron transfer between DHAQ^{2-} and DHAQ$^{3-\bullet}$ are responsible for the chemical shift changes. As the intermediate product of DHAQ$^{3-\bullet}$ continues to be reduced, the signals A and B reach maximum to become the characteristic peaks of the final product DHAQ^{4-}. Upon oxidation, the characteristic peaks of DHAQ^{2-} appear reversibly. The radical anion of DHAQ$^{3-\bullet}$ was verified via in situ EPR (Figure 2.6e). Before charging to 50% SOC, the signal intensity of DHAQ$^{3-\bullet}$ increases with the charging process, and then it decreases from 50% to 100% SOC. Reversible changes emerge during discharging.

In addition to characterizing the changes in chemical bonds to explore redox mechanisms of organic electrode materials, phase transformation is also considered for some unique organic electrodes with crystal structures [34]. For instance, disodium rhodizonate ($Na_2C_6O_6$) has a typical crystal structure, where the anions and cations are alternately arranged to form a hexagonal stack of layers (Figure 2.7a) [35]. The adjacent C_6O_6 layers are connected via the coordination of Na^+ and O atoms. As the active material for Na-organic batteries, the phase transformation of the $Na_2C_6O_6$ electrode during cycling was investigated by in situ XRD (Figure 2.7b) [33]. In the initial discharge stage, the characteristic peak of α-phase decreases corresponding to the γ-phase peak increases. When discharging to 1.0 V, the α-phase almost disappears, and another phase transition peak at q of 2.74 appears. This indicates that the α-phase is converted into γ-phase during reduction. After charging to 2.4 V, multiple steps of peak evolution are shown in XRD patterns, but the α-phase peaks do not appear at this stage. When the charging voltage reaches 2.9 V, the characteristic peaks belonging to the γ-phase have a positive shift. After full charging, the characteristic peaks of pristine α-$Na_2C_6O_6$ emerge again, indicating the reversible transformation between α-phase and γ-phase occurs during the Na^+ insertion/removal (Figure 2.7c).

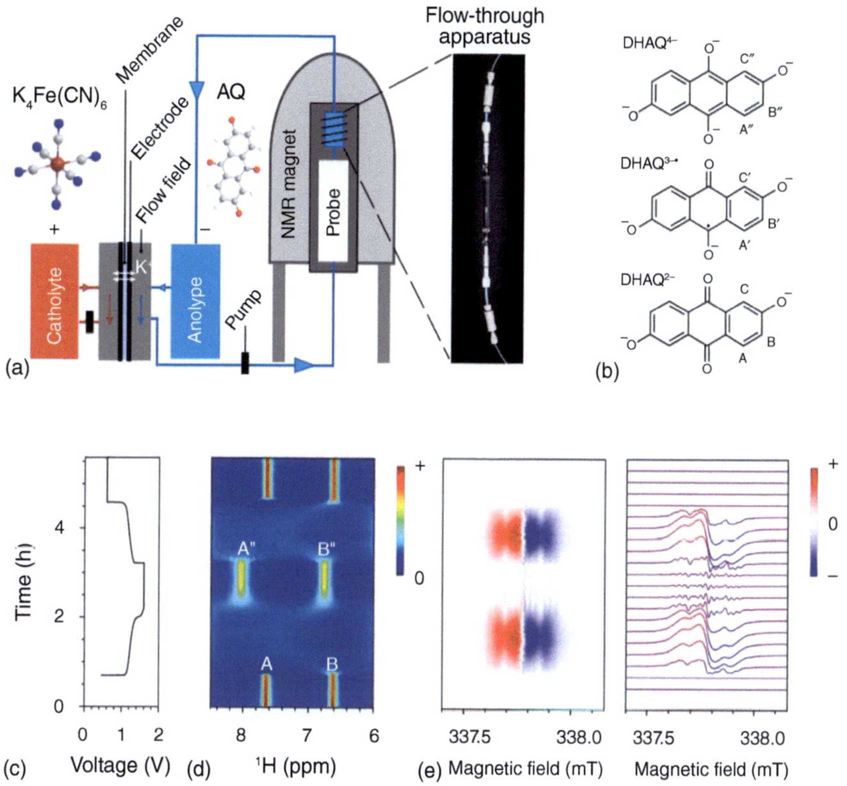

Figure 2.6 (a) Schematic diagram of in situ NMR analysis. Source: Reproduced with permission from Zhao et al. [31]. Copyright 2020/Springer Nature. (b) Molecular structures of $DHAQ^{2-}$, $DHAQ^{3-\bullet}$, and $DHAQ^{4-}$. (c) Voltage curves of a flow cell, with DHAQ as the anolyte active material and $K_4Fe(CN)_6/K_3Fe(CN)_6$ as the catholyte active material. (d) ^1H NMR spectra of the anolyte upon cycling. (e) EPR spectrum of the anolyte upon cycling. Source: Reproduced with permission from Zhao et al. [32]/John Wiley & Sons, Inc.

2.3 Organosulfide Materials

2.3.1 Redox Mechanisms

Organosulfide electrode materials for energy storage applications include n-type and p-type redox reactions. The redox centers of n-type organosulfides are based on S—S bonds. Upon redox reaction, they are capable of undergoing the cleavage/formation of S—S bonds. For p-type organosulfides that are without S—S bonds, which typically contain two single bonds between S and C atoms, and their redox reactions rely on the transformation between -S- and $-S^+-$ species [36]. Therefore, both show significant differences in electrochemical behaviors. In general, p-type organosulfides exhibit faster reaction kinetics and higher voltage than n-type organosulfides, but the capacities of p-type organosulfides are often lower than n-type organosulfides [37, 38].

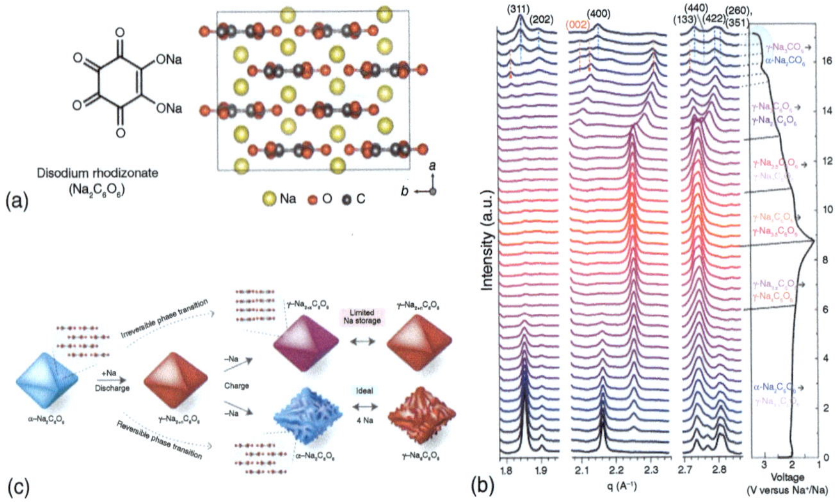

Figure 2.7 (a) Molecular structure of $Na_2C_6O_6$ and its corresponding crystal. (b) In situ XRD patterns and the corresponding voltage profile of $Na_2C_6O_6$. (c) Redox mechanisms for the $Na_2C_6O_6$ electrode. Source: Reproduced with permission from Lee et al. [33]. Copyright 2017, Nature Publishing Group.

2.3.1.1 Redox Mechanisms of n-type Organosulfides

The n-type organosulfide electrode materials contain small molecules, oligomers, and polymers. Organodisulfides (RSSR, R represents one organic functional group) are representatives of simple small n-type organosulfides, which only consist of one S—S bond. The reversible redox reaction of RSSR relies on the cleavage/formation of the S—S bond [39]. For example, one S—S bond in phenyl disulfide is broken to generate sulfur-containing anions after obtaining two electrons (Figure 2.8, Scheme 1). Asymmetric organodisulfides (RSSA, A is an allyl) are a special class of organodisulfides, which are prepared by mixing organodisulfides and ASSA in the organic solvent [40]. In addition to the S—S bond cleavage, the S—C bond also breaks to generate one allyl free radical during the electrochemical reduction. Therefore, one molecule of RSSA can accept three electrons in theory (Figure 2.8, Scheme 2). For organopolysulfide electrode materials (Figure 2.8, Scheme 3), which are able to accept more $2n$ electrons than RSSR to produce $2n$ S^{2-} that enables their high capacities [41]. In addition, organosulfur-based oligomers are also used as the redox-active component for organic rechargeable batteries [42]. Unlike RSSR, one organic function group of oligomers connects multiple S atoms, and the molecular mass of oligomers is between small molecules and polymers. For example, an oligomer contains four benzene rings and one benzene ring connects two S atoms, and its redox reaction is a two-step eight-electron process (Figure 2.9) [43].

2.3.1.2 Redox Mechanisms of p-Type Organosulfides

In addition to n-type redox reactions, organosulfide electrode materials also undergo p-type redox reactions. Thianthrene is a p-type organosulfide material,

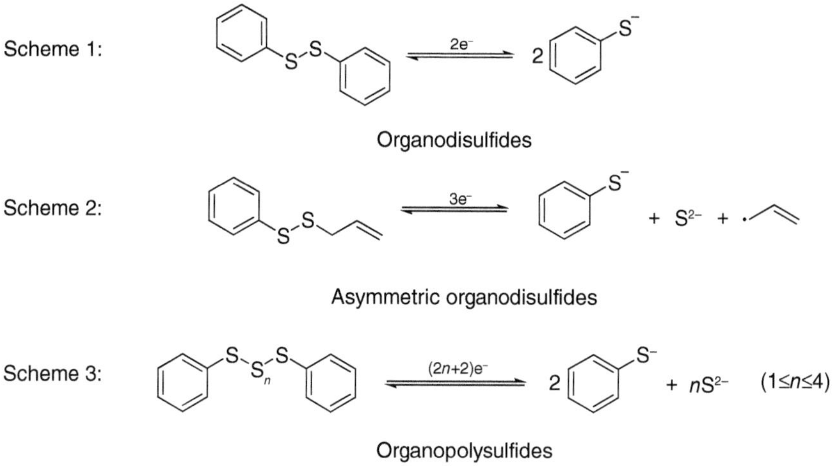

Figure 2.8 Redox mechanisms of three representative n-type organosulfides.

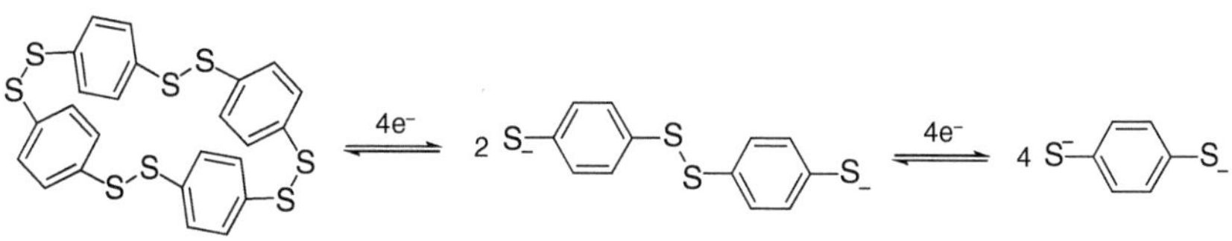

Figure 2.9 The redox reaction process of an oligomer.

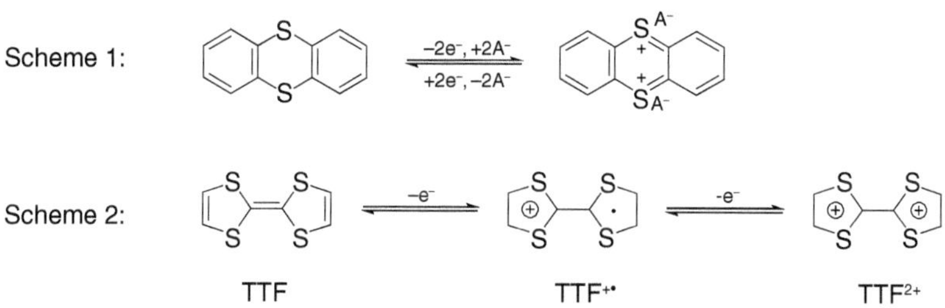

Figure 2.10 Redox mechanisms of two representative p-type organosulfides.

which undergoes a two-step oxidation process to generate the dication of $-S^{+}-$, and then reduction to the original form of –S– during a full cycle (Figure 2.10, Scheme 1) [38]. But not all –S– units of p-type organosulfide materials act as redox centers, e.g., tetrathiafulvalene (TTF) has four S atoms per molecule, but it only permits the two-electron reactions (Figure 2.10, Scheme 2) [44].

2.3.2 Characterization Methods

Various advanced characterization technologies, including MS, UV-vis, NMR, and Raman, have been used to investigate the redox mechanisms of organosulfide

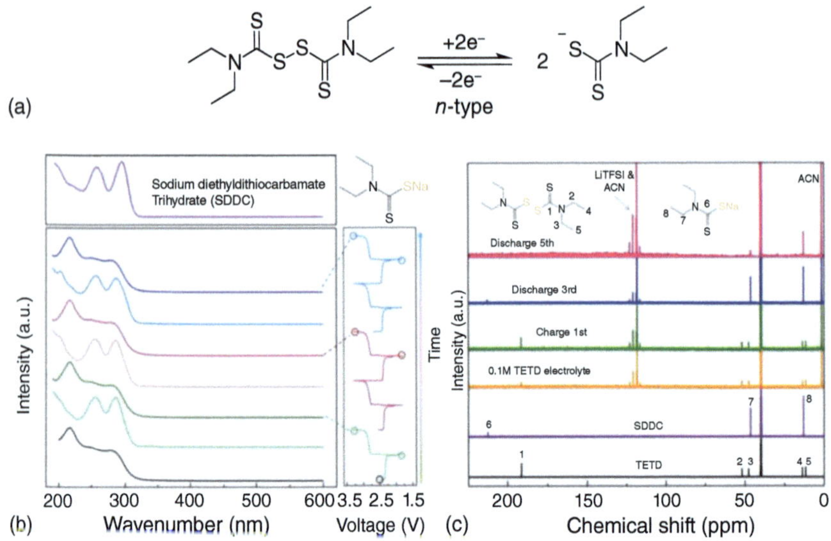

Figure 2.11 (a) Redox mechanism of the TETD electrode. (b) UV–vis spectra of TETD-based catholyte during cycling. (c) ^{13}C NMR spectra of TETD-based catholyte during cycling. Source: Reproduced with permission from Zhang et al. [45]. Copyright 2020, Wiley-VCH.

electrode materials. For example, the redox reaction of an organodisulfide (Figure 2.11a), i.e. tetraethyl thiuram disulfide (TETD), was investigated via UV–vis and NMR [45]. The TETD-based catholyte shows three absorption peaks at about 218, 251, and 282 nm, in the UV/Vis spectra, respectively (Figure 2.11b). Therein, the n-π* transition peak of S—S bonds appears at 218 nm. During reduction, the absorption peak of S—S bonds disappears, implying the S—S bond is broken to generate thiolate anions. After losing two electrons, the peak at 218 nm reappears, indicating the formation of S—S bonds reversibly. Furthermore, the redox reaction was also confirmed by ^{13}C NMR (Figure 2.11c) [45]. Compared to pristine TETD, the number of C signals in thiolate anions decreases due to the anion having a symmetric structure. And, the C position connected with S atoms in thiolate anions displays a downfield shift due to the resonance effect. In recharge, the C signals return to the pristine state. All in all, the reversible redox behavior between TETD and thiolate anions is well proved by the above two characterization technologies.

Organosulfide polymers are difficult to use UV–vis or NMR to study their redox reactions due to their limited solubility in solvents. To confirm the cleavage of S–S bonds in a disulfide polymer after discharging (Figure 2.12a), ex situ Raman analysis was used [46]. The S—S bond has a strong characteristic peak at about 480 cm^{-1} in the Raman spectra (Figure 2.12b). During discharging, the peak at 480 cm^{-1} disappears and the peak of –S–Li at 270 cm^{-1} appears, suggesting that the S—S bonds of the polymer cleavage. The signal at 480 cm^{-1} reappears after charging. Therefore, Raman spectra can identify the cleavage/formation of S—S bonds in organosulfide polymers upon cycling.

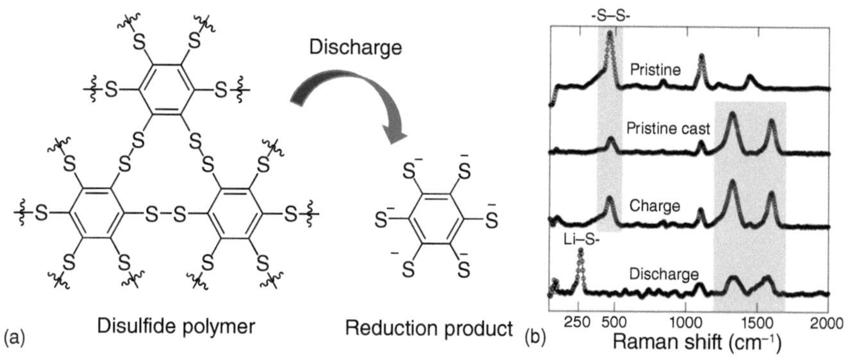

Figure 2.12 (a) Molecular structure of a disulfide polymer and its reduction product. (b) Raman spectra of the disulfide polymer before and after discharging. Source: Reproduced with permission from Preefer et al. [46]. Copyright 2017, Wiley-VCH.

2.4 Radical Materials

2.4.1 Redox Mechanisms

Organic radicals for rechargeable batteries have three types till now, i.e., nitroxyl radicals, phenoxyl radicals, and blatter radicals [47–49]. Nitroxyl radicals include four methyl units at the carbons near the nitrogen, which can be stable in the air and feature unique redox properties [50]. One of the most studied examples of nitroxyl radicals is 2,2,6,6-*tetra* methylpiperidin-*1*-yl)oxyl (TEMPO), which is based on the conversion between the initial state TEMPO and the oxoammonium cation/aminoxy anion (Figure 2.13, Scheme 1) [7]. During oxidation, the initial state TEMPO loses one electron to form an oxoammonium cation, and the anion in the electrolyte, e.g., $TFSI^-$ and PF_6^-, combine the oxoammonium cation to ensure charge balance. So, the redox process of TEMPO conforms to the properties of p-type reactions. In addition, it also meets the characteristics of n-type reactions, i.e., one TEMPO accepts an electron to reduce an aminoxy anion. Therefore, TEMPO is a bipolar-type electrode material.

Besides nitroxyl radicals, there are also two other types of organic radical materials reported for rechargeable batteries. One type is phenoxyl radicals, which can allow one reversible electron transfer between the neutral state of a phenoxyl radical and the reduction state of a phenolate anion (Figure 2.13, Scheme 2). Therefore, phenoxyl radicals are capable of n-type reactions [51]. Last is blatter radicals, which have similar redox behaviors to nitroxyl radicals, i.e., to become a cationic after losing one electron or to become an anionic after accepting one electron (Figure 2.13, Scheme 3)[52].

2.4.2 Characterization Methods

EPR spectroscopy is a powerful analytical tool to detect the presence of free radicals. The TEMPO catholyte at various charge states is taken out to test EPR (Figure 2.14).

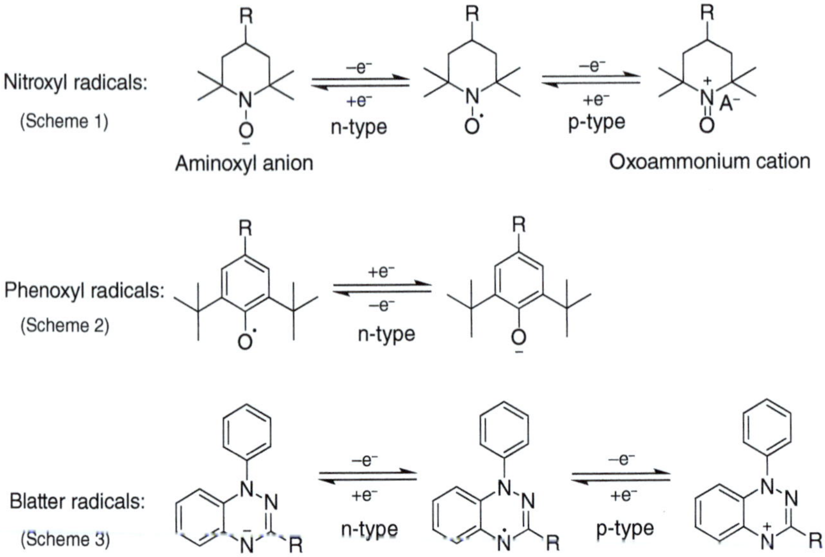

Figure 2.13 Redox behaviors of nitroxyl, phenoxyl, and blatter radicals (R is one organic functional group).

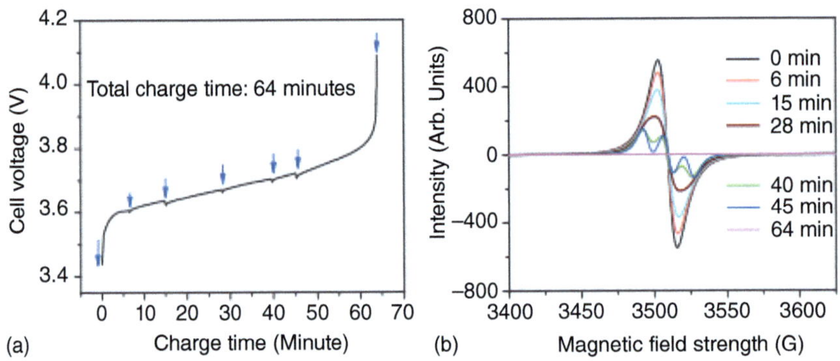

Figure 2.14 (a) The voltage curve of lithium-TEMPO cell during charging. (b) EPR spectra of the TEMPO catholyte at different states of charge. Source: Reproduced with permission from Wei et al. [53]. Copyright 2014, Wiley-VCH.

After the TMEPO radical is converted into oxoammonium cation, the characteristics of free radicals gradually decrease, and then disappear with the increase of the charging capacity [53]. Therefore, the EPR technology can judge the degree of electrode reactions in radical-based batteries.

2.5 N-Containing Active Materials

2.5.1 Redox Mechanisms of Azo Materials

Azo electrode materials with one or more azo (N=N) units display the characteristics of n-type redox reactions [54, 55]. The N=N unit plays the role of the redox center.

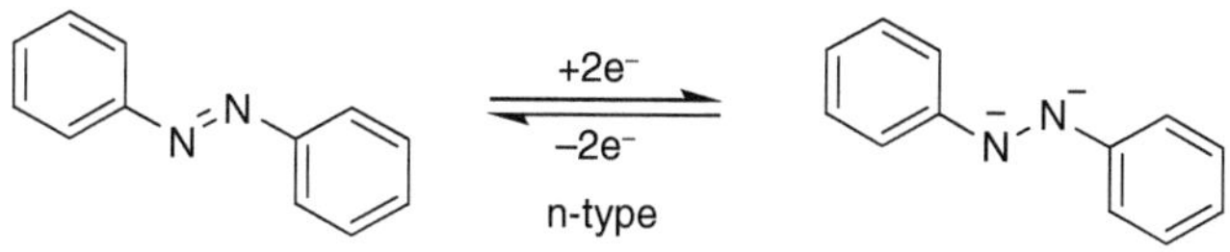

Figure 2.15 The redox mechanism of an azo material, azobenzene.

For benzene-based azo materials, one N=N group gains two electrons and is converted to a single bond during reduction (Figure 2.15). And then to reform the N=N group after losing two electrons, i.e., the single bond is reconverted to a double bond.

2.5.2 Redox Mechanisms of Imine Materials

Imine materials for rechargeable batteries include the following categories, phenazine derivatives, pteridine derivatives, and Schiff bases. The redox center of imine materials is based on the C=N bond that generally triggers a n-type redox reaction. Phenazine derivatives have two benzene rings and one pyrazine group, which are a class of electron-deficient aromatic compounds [56, 57]. For a phenazine molecule, the C=N bond breaks to generate two nitrogen-negative ions after accepting two electrons (Figure 2.16, Scheme 1) [58]. Meanwhile, the aromaticity of the phenazine molecule is also lost. The charge process is the leaving of electrons to reform the C=N bond. Interestingly, phenazine derivatives can be transformed into a p-type multi-electron redox material after the N atoms are substituted via one organic group, e.g., methyl and phenyl (Figure 2.16, Scheme 2) [59, 60]. Unlike the phenazine molecule, the N atoms can be oxidized to generate positive cations via forming the C=N bond. Hexaazatrinaphthylene (3Q) is a special phenazine derivative, which contains six C=N bonds. Therefore, per 3Q can achieve six electron transfers (Figure 2.16, Scheme 3) [61–63]. **Flavins** are a family of pteridine derivatives, which rely on the isoalloxazine ring as their redox centers (Figure 2.16, Scheme 4) [64].

2.5.3 Redox Mechanisms of Conjugated Sulfonamides

Conjugated sulfonamides are the ninth class of organic electrode materials, which are stable in the atmospheric environment and contain Li-ion [65]. Figure 2.17 shows the redox behavior of a conjugated sulfonamide, i.e., dilithium 1,4-phenylenebis((methylsulfonyl)amide), which enables a reversible two-electron redox reaction. Under oxidation, the conjugated sulfonamide changes to the quinone imine, and two C=N bonds are formed.

2.5.4 Redox Mechanisms of Nitroaromatic Materials

In 2020, the nitro group was proposed to be used as the redox center of electrode materials [66, 67]. The typical reaction behavior of a nitroaromatic compound is shown in Figure 2.18, one nitro group in aromatic compounds can reversibly accept/loss two electrons. Therefore, these materials usually allow high theoretical specific capacities, e.g., the theoretical specific capacity of 1,4-dinitrobenzene is 638 mAh g^{-1}, which is higher than most reported other organic electrode

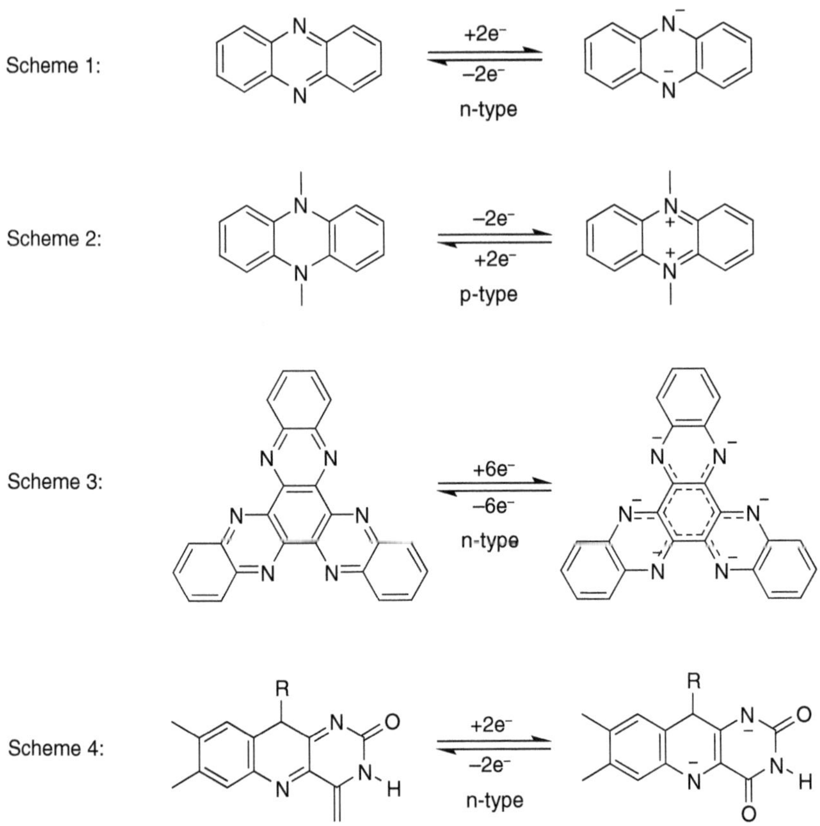

Figure 2.16 Schemes 1 and 2, redox mechanisms of phenazine materials. Scheme 3, the redox mechanism of 3Q with six electron transfers. Scheme 4, the redox mechanism of pteridine derivatives (R represents one organic functional group).

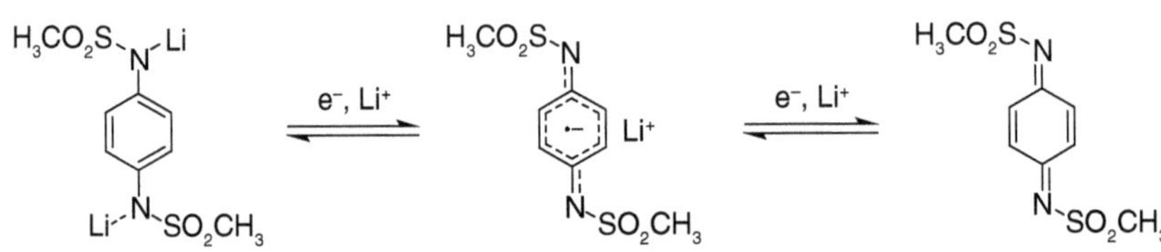

Figure 2.17 The redox behavior of the dilithium 1,4-phenylenebis((methylsulfonyl)amide).

materials [66]. According to the classification of organic electrode materials, the reaction of the nitro group conforms to the characteristics of n-type organic materials. In addition, the nitro group is also regarded as the active site in primary batteries, but this process is irreversible [68].

The redox process of triphenylamine belongs to a p-type reaction, which first undergoes an oxidation process to turn into a cation of N^+. After that, the N^+

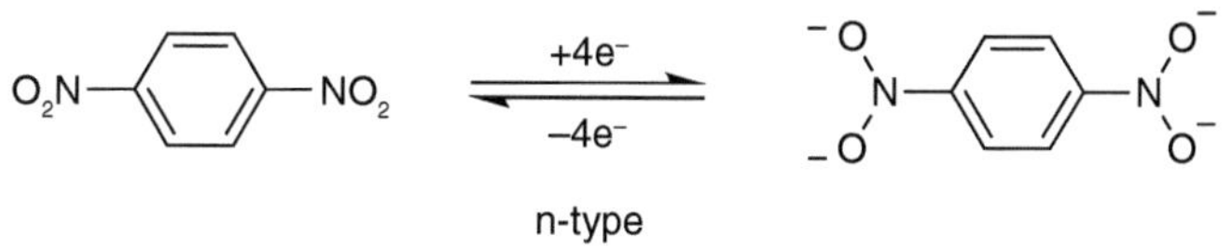

Figure 2.18 The redox reaction of a representative nitroaromatic material, 1,4-dinitrobenzene.

combines with anions, e.g., ClO_4^-, to maintain the charge neutral, and the process is reversible in the following cycling (Figure 2.19, Scheme 1). Viologens are p-type organic electrode materials, but their redox behaviors are different from other p-type organic materials that start from the neutral state to cations at high voltage oxidation reactions. Viologens start from the dicationic form (V^{2+}), during the discharge process, they undergo two single-electron reductions (Figure 2.19, Scheme 2). A radical cation of V^+ is the first reduction product, and the second reduction process is to form a neutral quinoid structure of V. N-containing active materials also permits bipolar redox reactions. *Triazines are* a family of nitrogen-containing heterocycles, which are generally polymerized together with various linkers for electrode materials [69]. The dual redox reactions work by stabilizing the extra negative charge or extra positive charge to the aromatic triazine ring with a metal ion or negative ion (Figure 2.19, Scheme 3).

2.5.5 Redox Mechanisms of Other N-containing Active Materials

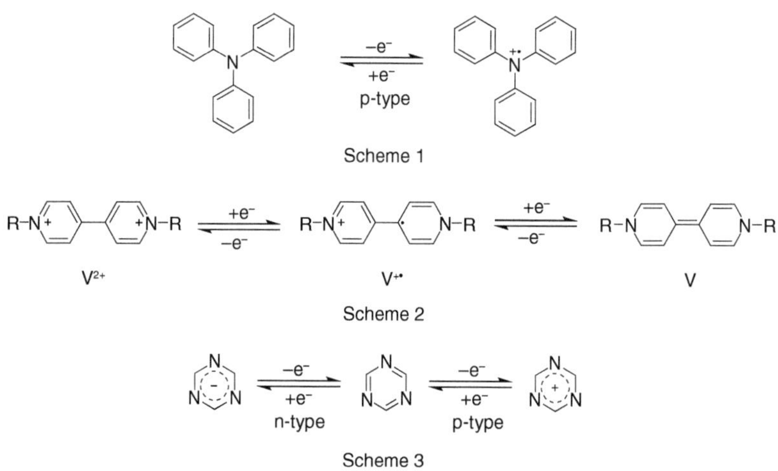

Figure 2.19 Scheme 1, the redox reaction of a representative phenylamine material, triphenylamine. Scheme 2, the redox reaction of viologens. Scheme 3, the redox reaction of the triazine.

2.5.6 Characterization Methods

Reversible reaction of 7,8-dimethylalloxazine (LC) in lithium batteries was confirmed by ex situ FTIR (Figure 2.20a) [64]. The vibrational peaks of C=N bonds gradually disappear during reduction and reappear during oxidation (Figure 2.20b) suggesting the redox center in the LC electrode is the conjugated diazabutadiene unit of –N=C–C=N–. Furthermore, the vibrational peaks of C=O bonds also show reversible disappear/appear during reduction/oxidation, indicating that the lithium insertion reactions occur at the N atoms neighboring O atoms to generate the heterocyclic structure of N–Li–O. This reversible redox behavior was also monitored by fluorescence spectroscopy (Figure 2.20c). After discharging, the absorption peak of alloxazinic fluorescence at 450 nm disappears, and the peak of isoalloxazinic fluorescence at 520 nm appears, indicating the isoalloxazinic fluorescence is reduced to isoalloxazinic fluorescence. In recharge, the signal of alloxazinic fluorescence returns to its natural state.

To prove the reaction process of 3Q in lithium batteries, ex situ ^{15}N, and ^{13}C solid-state NMR spectra were performed [61]. During cycling, the 3Q electrode involves multiple electron transfer processes, which correspond to six lithiated 3Q structures (Figure 2.21a). Ex situ solid-state NMR spectra of the 3Q electrode at various states of discharge/charge (Figure 2.21b) were shown in Figure 2.21c. Before discharging, the pristine 3Q electrode shows one strong peak (N1) at -55 ppm due to the six equivalent N atoms. After discharging to 2.4 V, there are three peaks, i.e., N2, N3, and N4, appear on the NMR spectra. Among them, N2 is the non-bonded pyrazine N located at the farthest position from Li, N3 is the non-bonded pyrazine N closest to a shared lithium-bonded bipyridine-type site, and N4 is the co-shared one Li-bonded pyrazine N. The appearance of three peaks indicates the 3Q-Li is formed. When the cell discharges to 2.0 V, these three peaks remain, and only N4

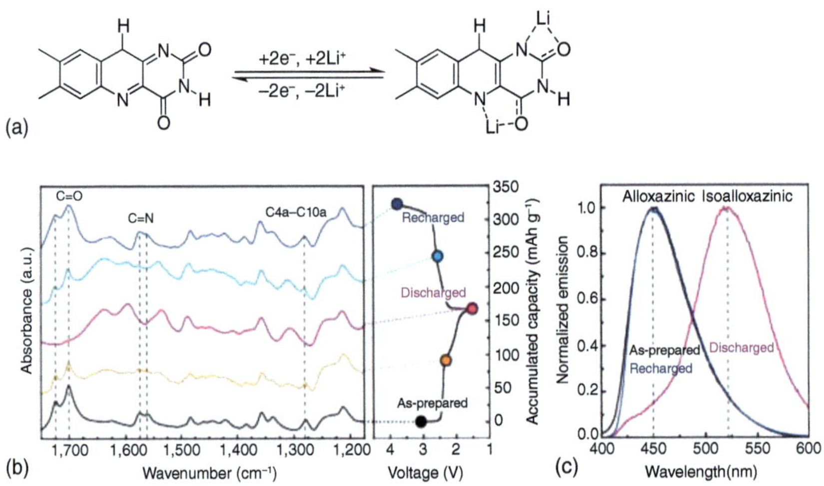

Figure 2.20 (a) The redox reaction of LC in lithium batteries. (b) Ex situ FTIR spectra of the LC electrode at various states of discharge–charge. (c) Fluorescence spectroscopy. Source: Reproduced with permission from Hong et al. [64]. Copyright 2014, Nature Publishing Group.

increases, suggesting the formation of 3Q-2Li. The peak of N4 increases in intensity at the discharge voltage of 1.82 V due to the generation of 3Q-3Li. When the cell further discharges to 1.7 V, the N4 peak decreases, and a new N5 peak appears, which belongs to the two nitrogen atoms coordinating with two lithium ions, implying that structural conversion to 3Q-4Li. With further discharging, the peak intensity of N5 increases strongly, while the N4 peak becomes weak due to the formation of 3Q-5Li. After discharging completely, the peak N5 grows, and the N4 peak disappears, indicating that all redox active sites in 3Q have accepted electrons and Li^+ to form 3Q-6Li. Furthermore, the reversible redox behavior was also proved via ex situ ^{13}C solid-state NMR spectra. The pristine 3Q electrode has three peaks of C1, C2, and C3 (Figure 2.21d). During discharging, these three peaks disappear and

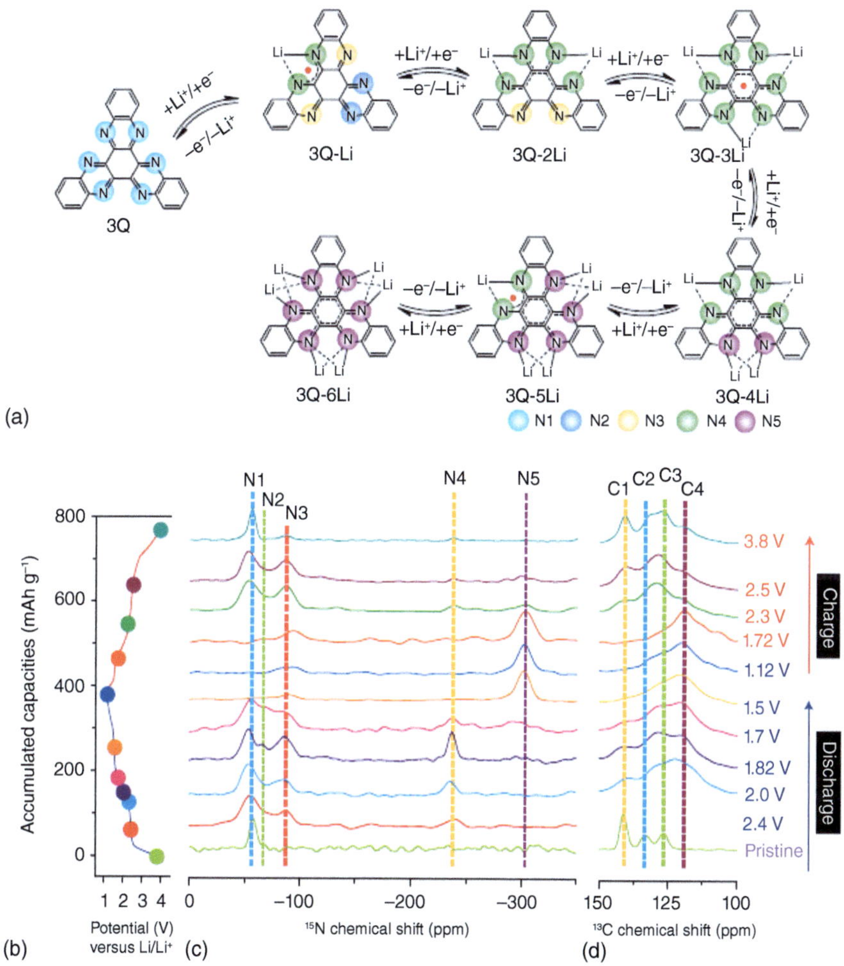

Figure 2.21 (a) Redox reaction of the 3Q electrode in lithium batteries. (b) Discharge–charge profile of a Li-3Q cell. Ex situ ^{15}N (c) and ^{13}C (d) solid-state NMR spectra of the 3Q electrode at different discharge/charge states. Source: Reproduced with permission from Che et al. [56]. Copyright 2017, Nature Publishing Group.

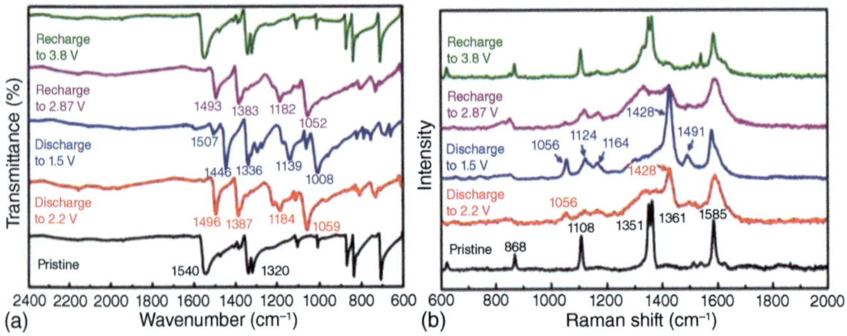

Figure 2.22 Ex situ FTIR (a) and Raman (b) spectra of the 1,4-dinitrobenzene electrode at various voltage states during cycling. Source: Reproduced with permission from Liu and Ye [66]. Copyright 2021, Wiley-VCH.

a new peak of C4 appears. By contrast, the peak of C4 disappears and the signals of the pristine 3Q electrode reappear during charging.

To study the redox reaction of the 1,4-dinitrobenzene electrode in lithium batteries (Figure 2.22), FTIR and Raman spectra were used [66]. In the FTIR spectra, the nitro group of 1,4-dinitrobenzene exhibits two characteristic stretching peaks at 1540 and 1320 cm^{-1} (Figure 2.22a). When the cell discharges to 2.2 V, these two peaks disappear completely, and some peaks of the intermediate product of nitro dianion at 1496, 1387, 1184, and 1059 cm^{-1} appear. Similarly, all characteristic peaks of 1,4-dinitrobenzene also disappear in Raman spectra except the C–C stretching peak of 1585 cm^{-1}. Meanwhile, a peak of 1428 cm^{-1} is ascribed to lithiated nitro groups emerges. After discharging, the signals of the lithiated tetraanion appear at 1446, 1336, 1139, and 1008 cm^{-1} in FTIR, and 1506, 1124, 1164, and 1428 cm^{-1} in Raman. Conversely, all the FTIR and Raman peaks return the neutral 1,4-dinitrobenzene state during charging.

2.6 Summary and Outlook

The strengths of organic electrode materials for sustainable rechargeable batteries make them very promising to replace current transition-metal electrodes. However, organic electrode materials with various redox centers have only stayed in the laboratory stage till now. Although, a certain understanding of redox reactions of these electrodes has been obtained via some advanced characterization techniques, there is still a long way to go to realize the practical application of organic electrodes. Hence, the following aspects should be considered in future research. (i) Organic materials with new redox centers. Although, the types of electrode materials have increased in recent years, organic materials with multi-electron transfer capabilities and a low molecular weight that enable high capacities of electrodes are apparently rare. (ii) Development of more advanced characterization methods. Advanced characterization methods, especially in situ should be adopted, which can monitor the reaction of organic electrodes in real time, deeply understand their redox reactions, and accurately find the reason for attenuation. In a word, the development and optimization of active molecules are inseparable from advanced characterization technologies.

Acknowledgments

The work was supported by the National Natural Science Foundation of China (Grant Nos. U2004214 and 21975225) and the National Key Research and Development Program of China (2022YFB2402200).

References

1 Song, Z. and Zhou, H. (2013). *Energy Environ. Sci.* 6: 2280–2301.
2 Lu, Y. and Chen, J. (2020). *Nat. Rev. Chem.* 4: 127–142.
3 Lu, Y., Hou, X., Miao, L. et al. (2019). *Angew. Chem.Int. Ed.* 58: 7020–7024.
4 Armand, M., Grugeon, S., Vezin, H. et al. (2009). *Nat. Mater.* 8: 120–125.
5 Häupler, B., Wild, A., and Schubert, U.S. (2015). *Adv. Energy Mater.* 5: 1402034.
6 Kye, H., Kang, Y., Jang, D. et al. (2022). *Adv. Energy Sustainability Res.* 3.
7 Kim, J.-K., Kim, Y., Park, S. et al. (2016). *Energy Environ. Sci.* 9: 1264–1269.
8 Acker, P., Rzesny, L., Marchiori, C.F.N. et al. (2019). *Adv. Funct. Mater.* 29.
9 Nguyen, T.P., Easley, A.D., Kang, N. et al. (2021). *Nature* 593: 61–66.
10 Lu, Y., Zhang, Q., Li, L. et al. (2018). *Chemistry* 4: 2786–2813.
11 Zhao, Q., Lu, Y., and Chen, J. (2017). *Adv. Energy Mater.* 7: 1601792.
12 Lee, S., Kwon, G., Ku, K. et al. (2018). *Adv. Mater.* 30: e1704682.
13 Oubaha, H., Gohy, J.F., and Melinte, S. (2019). *Chempluschem* 84: 1179–1214.
14 Yokoji, T., Kameyama, Y., Maruyama, N., and Matsubara, H. (2016). *J. Mater. Chem. A* 4: 5457–5466.
15 Han, C., Li, H., Shi, R. et al. (2019). *J. Mater. Chem. A* 7: 23378–23415.
16 Son, E.J., Kim, J.H., Kim, K., and Park, C.B. (2016). *J. Mater. Chem. A* 4: 11179–11202.
17 Wu, Y., Zeng, R., Nan, J. et al. (2017). *Adv. Energy Mater.* 7: 1700278.
18 Wang, H.G. and Zhang, X.B. (2018). *Chemistry* 24: 18235–18245.
19 Wang, H., Yao, C.-J., Nie, H.-J. et al. (2020). *J. Mater. Chem. A* 8: 11906–11922.
20 Mauger, A., Julien, C., Paolella, A. et al. (2019). *Materials* 12: 1770.
21 Renault, S., Geng, J., Dolhem, F., and Poizot, P. (2011). *Chem. Commun.* 47: 2414–2416.
22 Luo, W., Allen, M., Raju, V., and Ji, X. (2014). *Adv. Energy Mater.* 4: 1400554.
23 Han, X., Chang, C., Yuan, L. et al. (2007). *Adv. Mater.* 19: 1616–1621.
24 Han, X., Qing, G., Sun, J., and Sun, T. (2012). *Angew. Chem.Int. Ed.* 51: 5147–5151.
25 Yang, J., Xiong, P., Shi, Y. et al. (2020). *Adv. Funct. Mater.* 30: 1909597.
26 Luo, Z., Liu, L., Zhao, Q. et al. (2017). *Angew. Chem.Int. Ed.* 56: 12561–12565.
27 Li, X., Jia, K., Zhang, J. et al. (2022). *Inorg. Chem. Front.* 9: 4517–4524.

28 Zhang, H., Meng, X., Liu, X. et al. (2022). *Mater. Lett.* 319: 132250.

29 Luo, Z., Liu, L., Ning, J. et al. (2018). *Angew. Chem.Int. Ed.* 57: 9443–9446.

30 Zhang, M., Zhou, W., and Huang, W. (2021). *J. Energy Chem.* 57: 291–303.

31 Zhao, E.W., Liu, T., Jonsson, E. et al. (2020). *Nature* 579: 224–228.

32 Zhao, E.W., Jonsson, E., Jethwa, R.B. et al. (2021). *J. Am. Chem. Soc.* 143: 1885–1895.

33 Lee, M., Hong, J., Lopez, J. et al. (2017). *Nat. Energy* 2: 861–868.

34 Ma, T., Liu, L., Wang, J. et al. (2020). *Angew. Chem.Int. Ed.* 59: 11533–11539.

35 Dinnebier, R.E., Nuss, H., and Jansen, M. (2005). *Acta Crystallogr., Sect. E: Struct. Rep.* 61: m2148–m2150.

36 Tie, Z. and Niu, Z. (2020). *Angew. Chem.Int. Ed.* 59: 2–13.

37 Wang, D.Y., Guo, W., and Fu, Y. (2019). *Acc. Chem. Res.* 52: 2290–2300.

38 Speer, M.E., Kolek, M., Jassoy, J.J. et al. (2015). *Chem. Commun.* 51: 15261–15264.

39 Visco, S.J., Lutgard, C.C.M., De Jonghe, C., and Armand, M.B. (1989). *J. Electrochem. Soc.* 136: 661.

40 Weng, G.-M., Yang, B., Liu, C.-Y. et al. (2019). *Energy Environ. Sci.* 12: 2244–2252.

41 Wang, D.Y., Si, Y., Guo, W., and Fu, Y. (2020). *Adv. Sci.* 7: 1902646.

42 Tang, S., Chen, Q., Si, Y. et al. (2021). *Adv. Mater.* 33: e2100824.

43 Li, F., Si, Y., Liu, B. et al. (2019). *Adv. Funct. Mater.* 29: 1902223.

44 Daub, N., Hendriks, K.H., and Janssen, R.A.J. (2022). *Batteries Supercaps* 5: e202200386.

45 Zhang, L., Zhao, B., Zhang, C., and Yu, G. (2021). *Angew. Chem. Int. Ed.* 60: 4322–4328.

46 Preefer, M.B., Oschmann, B., Hawker, C.J. et al. (2017). *Angew. Chem. Int. Ed.* 56: 15118–15122.

47 Jahnert, T., Haupler, B., Janoschka, T. et al. (2014). *Macromol. Rapid Commun.* 35: 882–887.

48 Guo, W., Yin, Y.-X., Xin, S. et al. (2012). *Energy Environ. Sci.* 5: 5221–5225.

49 Saal, A., Friebe, C., and Schubert, U.S. (2022). *J. Power Sources* 524: 231061.

50 Matsumoto, K.I., Nakanishi, I., Zhelev, Z. et al. (2022). *Antioxid. Redox Signaling* 36: 95–121.

51 Suga, T., Ohshiro, H., Sugita, S. et al. (2009). *Adv. Mater.* 21: 1627–1630.

52 Steen, J.S., Nuismer, J.L., Eiva, V. et al. (2022). *J. Am. Chem. Soc.* 144: 5051–5058.

53 Wei, X., Xu, W., Vijayakumar, M. et al. (2014). *Adv. Mater.* 26: 7649–7653.

54 Shimizu, T., Tanifuji, N., and Yoshikawa, H. (2022). *Angew. Chem.Int. Ed.* 61: e202206093.

55 Luo, C., Borodin, O., Ji, X. et al. (2018). *Proc. Natl. Acad. Sci. U.S.A.* 115: 2004–2009.

56 Che, Y.-X., Qi, X.-N., Lin, Q. et al. (2022). *J. Mater. Chem. C* 10: 11119–11174.

57 Tian, B., Ding, Z., Ning, G.H. et al. (2017). *Chem. Commun.* 53: 2914–2917.

58 de la Cruz, C., Molina, A., Patil, N. et al. (2020). *Sustainable Energy Fuels* 4: 5513–5521.

59 Dai, G., Wang, X., Qian, Y. et al. (2019). *Energy Storage Mater.* 16: 236–242.

60 Lee, M., Hong, J., Lee, B. et al. (2017). *Green Chem.* 19: 2980–2985.

61 Peng, C., Ning, G.-H., Su, J. et al. (2017). *Nat. Energy* 2: 861–868.

62 Weng, J., Xi, Q., Zeng, X. et al. (2022). *Catal. Today* 400–401: 102–114.

63 Li, J., Huang, L., Lv, H. et al. (2022). *ACS Appl. Mater. Interfaces* 14: 38844–38853.

64 Hong, J., Lee, M., Lee, B. et al. (2014). *Nat. Commun.* 5: 5335.

65 Wang, J., Lakraychi, A.E., Liu, X. et al. (2021). *Nat. Mater.* 20: 665–673.

66 Liu, X. and Ye, Z. (2020). *Adv. Energy Mater.* 11: 2003281.

67 Liu, B., Tang, C.W., Zhang, C. et al. (2021). *Chem. Mater.* 33: 978–986.

68 Chen, Z., Su, H., Sun, P. et al. (2022). *Proc. Natl. Acad. Sci. U.S.A.* 119: e2116775119.

69 Sakaushi, K., Hosono, E., Nickerl, G. et al. (2013). *Nat. Commun.* 4: 1485.

3

Carbonyl-Based Organic Cathodes

3.1 Introduction

The crucial topic of sustainable social development involves tackling energy and environmental challenges, so developing advanced energy storage systems (devices) with high energy density emerges as a critical problem. Lithium-ion batteries (LIBs), one of the first choices for large-scale renewable energy storage, still cannot meet the market demand, although their specific energy has been around 300 Wh kg^{-1} [1]. To further improve the energy density of rechargeable batteries, developing high-capacity and low-cost cathode materials has become one of the research hotspots in recent years. Organic cathode materials have been demonstrated as promising alternatives to next-generation green batteries owing to their eco-friendliness [2], tunable chemistry, and tailorable structures [3]. What's more, organic cathode materials can support n-type, p-type, or bipolar-type redox reactions attributed to their versatile properties, which allows the designing of diverse battery configurations, such as LIBs, sodium-ion batteries (SIBs), aqueous zinc-ion batteries (ZIBs), aqueous redox flow batteries (RFBs), and other metal-ion batteries (potassium-ion and magnesium-ion) [4].

Among the various organic cathode materials, carbonyl-based ones are assigned to the most promising candidates attributed to their high specific capacity, fast reaction kinetics, outstanding electrochemical reversibility, and tailorable chemical properties [4]. Williams et al. reported the first carbonyl-based organic active material (dichloroisocyanuric acid) in 1969 [5]. Later, various carbonyl-based electrode materials have been extensively investigated, which are typically divided into quinone compounds [6–8], carboxylate salts [9, 10], imides [11, 12], and anhydrides [13]. Among them, carboxylate salts are usually used as anode electrode materials due to their low potential (approximate 1 V versus Li^+/Li) [9]. Figure 3.1 shows the redox reaction mechanism of carbonyl-based organic materials in different battery systems. The carbonyl (C=O) group in these carbonyl compounds enables an enolization reaction to form a radical monoanion ($C–O^-$) during the discharge process, and the charge process is the inverse of the discharge process.

Rechargeable Organic Batteries: Materials, Mechanisms, and Prospects, First Edition.
Yongzhu Fu, Xiang Li, Shuai Tang, and Wei Guo.

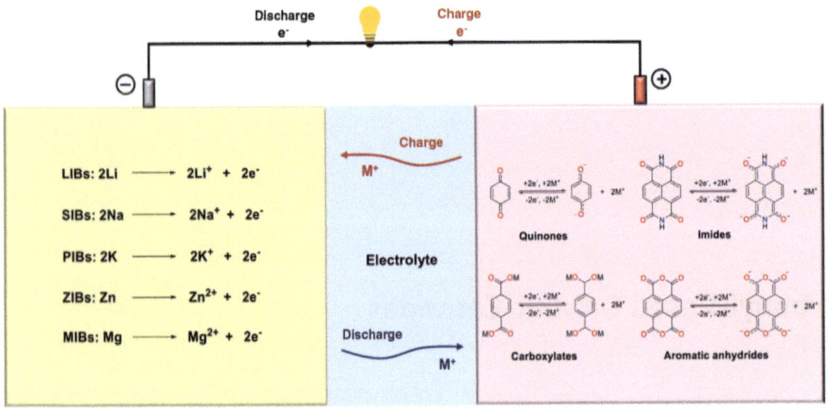

Figure 3.1 The diagram of the redox reaction mechanism of carbonyl-based organic materials for different battery systems.

In this chapter, we mainly introduce the application of carbonyl-based small molecules as cathode materials for various electrochemical energy storage devices. The progress of polymeric carbonyl materials for rechargeable batteries will be summarized in Chapter 7. Additionally, some perspectives and development strategies of carbonyl-based organic cathodes are discussed (Figure 3.1).

3.2 Quinone Compounds

3.2.1 Quinones for LIBs

To meet the growing demand for electric vehicles and likes, studying and developing high-performance electrode materials is urgently necessary. Over the past few decades, many quinone compounds have been studied as cathode materials for LIBs due to their abundance, high energy density, fast oxidation–reduction kinetics, structural diversity, and design flexibility.

However, quinone compounds' most significant challenge is their unnecessary dissolution in nonaqueous liquid electrolytes. There are two main strategies to solve the solubility of quinone compounds [14]. One utilizes chemical binding (such as amide reaction) or physical methods (such as adsorption or confinement) to immobilize quinone materials in carbon materials such as mesoporous carbon, carbon nanotubes, or graphene oxide. Another strategy is rationalizing molecular structures to lower the molecule's solubility, thereby improving the electrochemical performance. This section mainly summarizes the strategy to enhance the insolubility of small-molecule quinone compounds used in LIBs.

Chen's team synthesized many fused heteroaromatic carbonyl compounds ((BFFD, benzofuro[5,6-*b*]furan-4,8-dione), (BDTD, benzo[1,2-b:4,5-b′]dithiophene-4,8-dione) and (PID, pyrido[3,4-g]isoquinoline-5,10-dione)) and investigated their electrochemical performance as cathode materials (Figure 3.2a) [15]. These compounds, exhibiting reactions involving multiple electrons, demonstrated remarkable

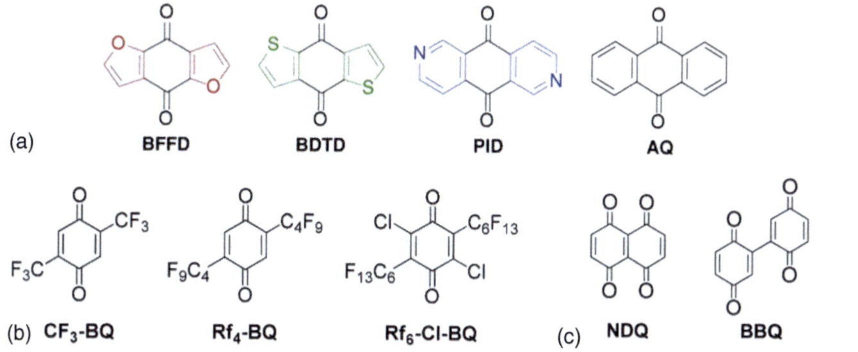

Figure 3.2 (a) The structures of the four heteroaromatic-fused carbonyl compounds. Source: Adapted with permission from Liang et al. [15], WILEY-VCH. (b) Electron-deficient benzoquinones bearing perfluoroalkyl groups. Source: Adapted with permission from Yokoji et al. [16], The Royal Society of Chemistry. (c) Structures of NDQ and BBQ derivatives. Source: Adapted with permission from Yokoji et al. [17], The Royal Society of Chemistry.

energy capacity, rate capability, and durability during sequent cycling. Matsubara's group studied several electron-deficient benzoquinone derivatives ((CF_3-BQ, 2,5-bis (trifluoromethyl)-1,4-benzoquinone), (Rf_4-BQ, 2,5-bis(perfluorobutyl)-1,4-benzoquinone), and (Rf_6-Cl-BQ, 2,5-bis(perfluorohexyl)-3,6-dichloro-1,4-benzoquinone)) as high discharge-voltage cathode materials (Figure 3.2b) [16]. Compared with batteries using electron-rich quinones, the presence of electron-withdrawing groups results in a higher voltage plateau on the charge–discharge curve. They also synthesized 1,4,5,8-Naphthodiquinone (NDQ) and 2,2′-bis-*p*-benzoquinone (BBQ, Figure 3.2c) as cathode materials for LIBs and evaluated their electrochemical properties, adding new members to the carbonyl compound family [17].

In 2016, Chen's group reported the use of sulfur-containing heterocyclic quinone (dibenzo[*b*,*i*]thianthrene-5,7,12,14-tetraone = DTT, Figure 3.3a) as cathode material for LIBs, using the conductive polymer poly(3,4-ethylenedioxythiophene):poly

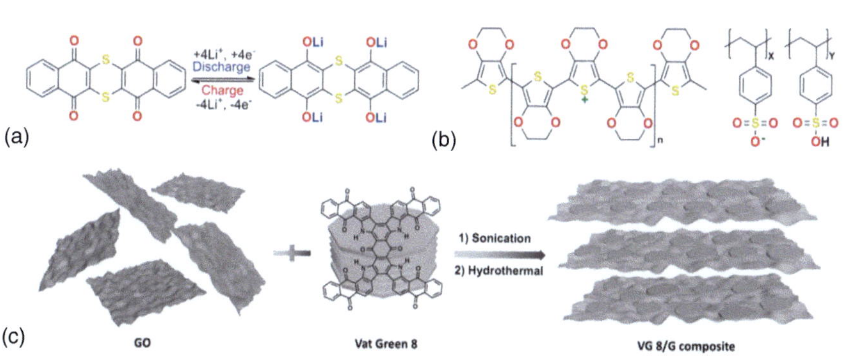

Figure 3.3 (a) Molecular structures and electrochemical mechanism of DTT. (b) The polymeric structure of PEDOT:PSS. Source: Adapted with permission from Ma et al. [18], WILEY-VCH. (c) The scheme for the synthesis of VG 8/G composite. Source: Reproduced with permission Ai et al. [19]/John Wiley & Sons.

(styrenesulfonate) = PEDOT:PSS as a binder (Figure 3.3b) [18]. The extended π-conjugated configuration of the DTT (electron donor) and cationic PEDOT (electron acceptor) leads to non-covalent interactions between them, which further helps inhibit the dissolution of DTT, thereby exhibiting sustained stability over long-cycling and exceptional performance at high charging rates in LIBs. This research provides new ideas and meaningful reference values for modifying quinone compounds' electrochemical properties from the perspective of binders.

Yu et al. synthesized a new type of organic cathode for LIBs by combining ultrasound and hydrothermal methods to synthesize a vat dye/graphene composite material (Figure 3.3c) [19]. Vat Green 8 is a green dye with an electrically active conjugated carbonyl group and an extensively condensed aromatic ring system. The π–π interaction between Vat Green 8 and graphene provides a stable composite structure, which not only effectively inhibits the dissolution of the active material but also ensures rapid electron transfer during the electrochemical reaction.

Park et al. synthesized 2,3-diamino-1,4-naphthoquinone (DANQ, Figure 3.4a) through a simple method [20]. The amino group in DANQ undergoes ring closure, forming 1*H*-naphtho[2,3-*d*]imidazole-4,9-dione (IMNQ, Figure 3.4b). This ring closure induces a 0.15 V increase in the oxidation–reduction potential, which is attributed to the electron delocalization of the heteroaromatic ring. The Li-DANQ cell has the highest initial discharge capacity of 250 mAh g^{-1} and 99% capacity retention of its initial capacity after 500 cycles at 0.2 C (Figure 3.4c). Their work demonstrates that carbonyl compounds with modulated charge/ion transport are feasible alternatives to lithium metal oxides. Afterward, in 2017, Park's team used lawsone (2-hydroxy-1,4-naphthoquinone) as an organic cathode material [21]. The lawsone molecule has a unique two-dimensional planar structure (Figure 3.4d) and an ultrahigh level of overlap between adjacent p orbitals, which can simultaneously

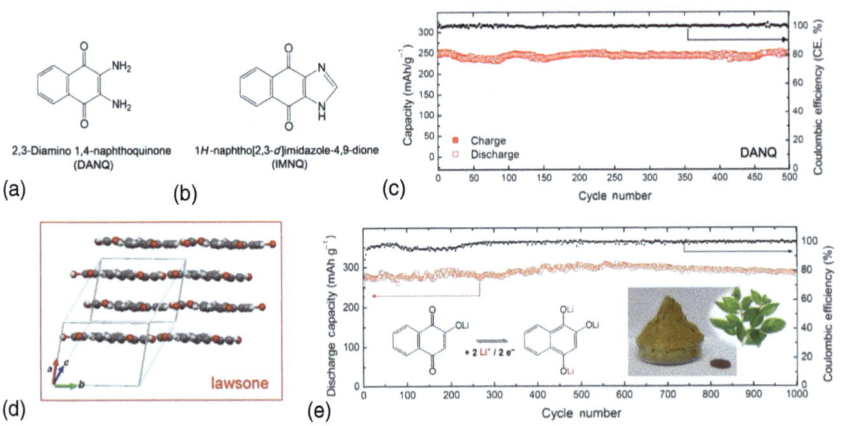

Figure 3.4 Molecular structures of (a) DANQ and (b) IMNQ. (c) Discharge/charge capacities and Coulombic efficiency of the Li-DANQ cell at 0.2 C. Source: Reproduced with permission Lee et al. [20]/American Chemical Society. (d) Images capture the molecular arrangements of lawsone. (e) Discharge/charge capacities and Coulombic efficiency of the Li-lawsone cell at 0.5 C. Source: Adapted with permission from Lee and Park [21]/John Wiley & Sons, Inc.

enhance electrical and ionic conductivities. Therefore, the LIBs based on lawsone can achieve a high initial discharge capacity of 280 mAh g^{-1} and a capacity retention rate of 99% after 1000 cycles at 0.5 C rate (Figure 3.4e).

Park et al. introduced a new molecular design strategy that uses triptycene scaffold bridged-oxidation–reduction functional units benzoquinone (BQ) to attain organic cathode materials with exceptional capacity and high energy for LIBs [22]. The three-dimensional arrangement of triptycene tribenzoquinone (TT, Figure 3.5a) facilitates the highly reversible entry of numerous redox states and increases the oxidation–reduction potential. Zhou's team first attempted to use commercially available 1,1′-Iminodianthraquinone (IDAQ, Figure 3.5b) as cathode material for LIBs, and the use of a polypropylene/Nafion/polypropylene (PNP) sandwich-type separator suppressed the shuttle effect of IDAQ [23]. This method helped IDAQ achieve excellent cycling performance (Figure 3.5c).

Chen et al. reported the application of 2,3,5,6-tetraphthalimido-1,4-benzoquinone (TPB, Figure 3.6a) for LIBs [24]. TPB has four inflexible phthalimide groups, so it is insoluble in non-proton electrolytes, which is conducive to enhancing

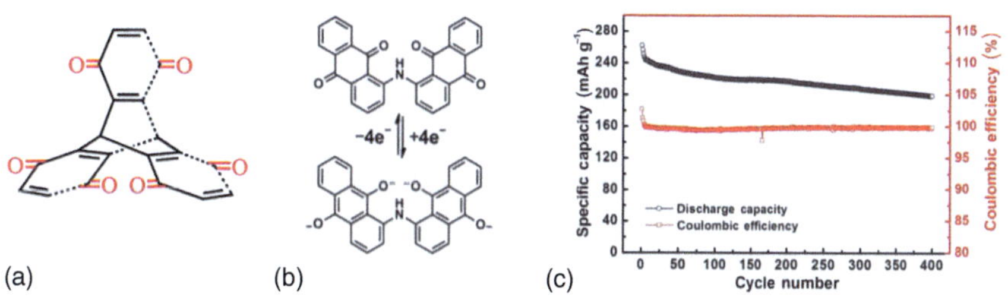

Figure 3.5 (a) Chemical structures of TT. Source: Adapted with permission from Kwon et al. [22], The Royal Society of Chemistry. (b) The electrochemical redox mechanism of IDAQ. (c) Long-term cycling of IDAQ using PNP5 separator at a current rate of 50 mA g^{-1}. Source: Reproduced with permission Song et al. [23]/John Wiley & Sons.

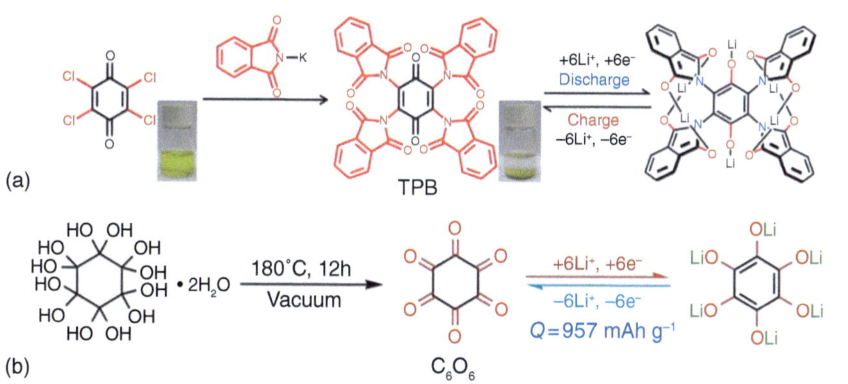

Figure 3.6 (a) The synthesis and electrochemical redox mechanism of TPB. Source: Adapted with permission from Luo et al. [24], Wiley-VCH. (b) The synthetic process and electrochemical redox mechanism of C_6O_6. Source: Adapted with permission from Lu et al. [25]/John Wiley & Sons, Inc.

electrochemical performance. They also reported the synthesis and application of cyclohexanehexone (C_6O_6, Figure 3.6b) in 2019 [25]. Because C_6O_6 provides the maximum possible number of transferred electrons, it can exhibit an ultrahigh capacity of 902 mAh g^{-1} at 20 mA g^{-1}. This study opens up new possibilities for developing sustainable LIBs by utilizing high-capacity organic cathode materials.

Zhang's group reported sulfur-rich thiazyl small molecules ((NNS, benzo[1,2-*d*:4,5-*d*′]bis ([1,2,3]thiadiazole)-4,8-dione), (NSN, dithiadiazolo-*p*-benzoquinone) and (NSS, benzoquino-bis-1,2,3-dithiazole)) as organic electrode materials (Figure 3.7a) [26]. Their research showed that sulfur-rich thiazole-based materials could be used to design small molecular-size organic electrode materials that achieve prolonged cycling stability and accelerated electrode kinetics. Huang's work developed a new type of Calix[8]quinone (C8Q, Figure 3.7b), in which the eight BQ units are linked by methylene [27]. This synthesis strategy increases the molecular weight without reducing the theoretical capacity.

Fu's group introduced a new biredox-ionic composite, ethylviologen dianthraquinone-2-sulfonate (EV-AQ$_2$, Figure 3.8a) [28], which offers multiple and reversible active sites. In the initial stages, EV-AQ$_2$ has a capacity of 199.2 mAh g^{-1} at 0.1 C, corresponding to the transfer of two electrons from the redox couple EV^{2+}/EV^0 and four electrons from two redox-active AQ^- anions. It is important to note that EV-AQ$_2$ demonstrates a significant improvement in cyclability. The capacity retention reaches 89%, and the Coulombic efficiency approaches 100% after 120 cycles at a rate of 0.5 C. These findings verify that incorporating AQ^- into the EV^{2+} scaffold has multiple redox sites and shows promise in developing high-energy density LIBs. This team also designed a novel cathode molecular structure based on naphthoquinone (NQ) in 2022 [29]. Thanks to the inherent insolubility and fast reaction kinetics of the 2,2′-(1,4-phenylenedithio) bis(1,4-naphthoquinone) (1,4-PNQ, Figure 3.7b) molecule, it presents a noteworthy initial capacity of 231 mAh g^{-1}, corresponding to 97.9% of the theoretical capacity. Notably, when employed as a cathode, the cell retains 93.5% of its initial capacity even after 500 cycles at a 5 C rate (Figure 3.8c).

Song et al. studied a new type of small-molecule organic cathode material, oxidized indanthrone (oIDT), which is simply synthesized by oxidizing indanedione (IDT) in concentrated nitric acid (Figure 3.9a) [30]. Compared with IDT, the dihydropyrazine portion is converted to pyrazine, which not only increases

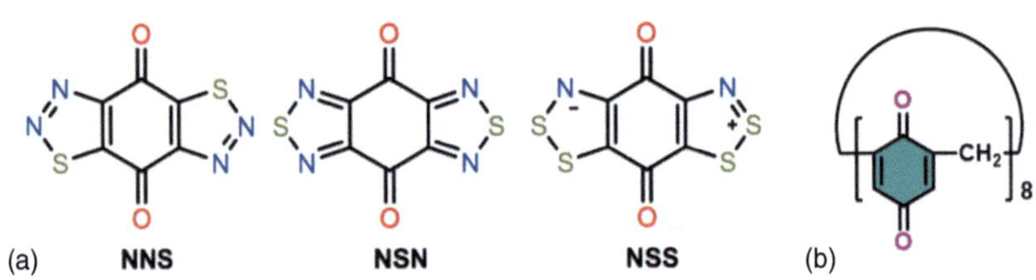

Figure 3.7 (a) Structures of NNS, NSN, and NSS. Source: Adapted with permission from Tuttle and Zhang [26], American Chemical Society. (b) The chemical structures of C8Q. Source: Adapted with permission from Huang et al. [27], John Wiley & Sons Australia, Ltd.

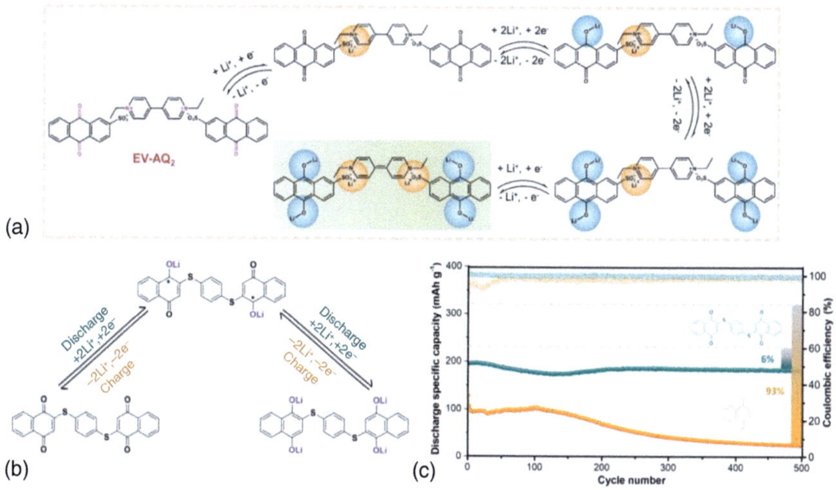

Figure 3.8 (a) The proposed electrochemical redox reactions of EV-AQ_2 cathode in Li-battery during discharging/charging process. Source: Adapted with permission from Wang et al. [28], Wiley-VCH GmbH. (b) The electrochemical redox reactions of 1,4-PNQ. (c) Cycling performances of Li/1,4-PNQ and Li/NQ at 5 C rate. Source: Reproduced with permission Zhao et al. [29]/John Wiley & Sons.

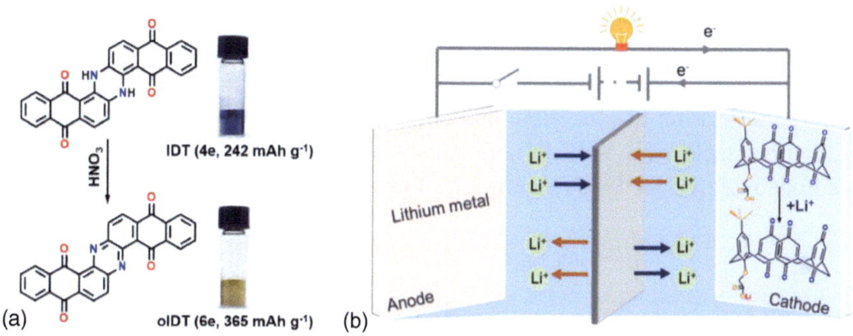

Figure 3.9 (a) Synthesis route from IDT to oIDT and the sample photographs. Source: Adapted with permission from Shi et al. [30]/Elsevier. (b) Schematic diagram of C3Q-Li application in LIBs. Source: Adapted with permission from Liu et al. [31], Elsevier B.V.

the theoretical capacity and redox potential but also improves the structural stability during the discharge–charge process. oIDT can release a high capacity of 273 mAh g^{-1} at a current density of 50 mA g^{-1} and achieve stable long-cycle performance at 500 mA g^{-1}. Huang's work used poly-carbonyl active site material calix[4]quinone (C4Q) as a structural template to design a polar molecule: methoxy acetate acid calix[3]quinone (C3Q) [31]. C3Q is converted to C3Q-Li through an *in situ* electrochemical reaction during initial discharge (Figure 3.9b). Due to the decreased solubility of C3Q-Li, the electrode exhibits excellent performance, with a capacity of 200 mAh g^{-1} after 100 cycles at 0.2 C. At a high current density of 1 C, the discharge capacity still reaches 160 mAh g^{-1} after 1000 cycles.

3.2.2 Quinones for SIBs

Rechargeable SIBs have broad application prospects in energy storage. Sodium's development and utilization are crucial to attain clean energy and sustainable development, given its abundance on the Earth. Compared with LIBs, rechargeable SIBs have relatively lower costs, which makes them spectacularly suitable for large-scale energy storage.

Using organic electrode materials has further improved the environmental friendliness, flexibility, lightweight, and low cost of sodium batteries [32]. The flexible and controllable molecular structure of organic molecules can easily achieve the insertion/extraction of Na^+, thereby realizing fast reaction kinetics and controlling changes in battery performance. Among them, quinone-type organic electrode materials have attracted significant attention due to their environmental friendliness, renewability, designability, and high-specific capacity. A variety of quinone-type electrode materials and their performance have been extensively investigated. These studies have provided critical support and impetus for developing rechargeable SIBs.

In 2015, Chen's team first prepared the disodium salt of 2,5-dihydroxy-1,4-benzoquinone (Na_2DBQ, Figure 3.10a) as a rechargeable cathode material for SIBs [33]. The Na_2DBQ electrode exhibited a high substantial discharge capacity of 265 mAh g^{-1} at 0.1 C and a long-term cycling life of 300 cycles at 1 C rate. Guo et al. used renewable Juglone biomolecules (Figure 3.10b) with redox activity for SIBs [34]. They proposed a preparation method for the Juglone/RGO hybrid electrode, which exhibits excellent electrochemical performance and guides the utilization of renewable biomaterials. In 2016, Zeng's group prepared 4,8-dihydrobenzo[1,2-*b*:4,5-*b′*] dithiophene-4,8-dione/graphene (BDT-G) composite cathode material for SIBs (Figure 3.10c) using a simple dispersion–deposition

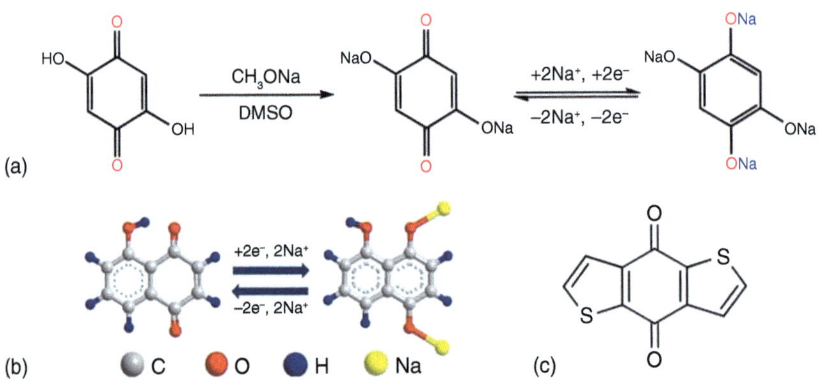

Figure 3.10 (a) The synthetic route proposed electrochemical redox mechanism of Na_2DBQ. Source: Adapted with permission from Zhu et al. [33], The Royal Society of Chemistry. (b) Electrochemical redox reaction mechanism of Juglone molecule. Source: Adapted with permission from Wang et al. [34], WILEY-VCH. (c) The chemical structure of BDT. Source: Adapted with permission from Chen et al. [35], The Royal Society of Chemistry.

method [35]. Through the infinite electron transfer of graphene's two-dimensional network, BDT-G exhibited higher reversible capacity, better rate capability, and better cycling performance than BDT, demonstrating the practicality of graphene in improving the electrochemical performance of organic electrode materials.

As one of the smallest conjugated carbonyl compounds, BQ was first studied in 1996 and has a high theoretical capacity in quinone compounds [36]. However, it suffers from serious solubility problems in electrolytes, which seriously affects the performance of batteries. Wu's team synthesized 2,6-bis(*p*-benzoquinonyl) pyridine (QPQ-1) and 2,5-bis(*p*-benzoquinonyl) pyridine (QPQ-2) (Figure 3.11a) [37]. These two molecules consist of two BQ molecules connected by an electron-withdrawing pyridine. Introducing an electron-withdrawing group can lower the lowest unoccupied molecular orbital (LUMO) energy level, which is beneficial to increase the working voltage of the electrode. Both QPQ-1 and QPQ-2 have a planar structure, and the extended molecular skeleton enhances the intermolecular interaction (π–π stacking), greatly reducing their solubility in electrolytes and improving their cycling stability. QPQ-2 has a smaller band gap and lower solubility and exhibits better electrochemical performance than QPQ-1. At 0.1 C rate, the initial discharge capacity of QPQ-2 is 214 mAh g^{-1}, and the capacity retention rate is 93% after 100 cycles. It can stably work 1,000 cycles with a capacity retention rate of 82.7% at 1 C rate (Figure 3.11b). Wu's team also synthesized 2,5-bis-(*p*-benzoquinonyl) furan (QFQ), a molecule consisting of two BQs bridged by an electron-rich furan [38]. QFQ is almost insoluble in electrolytes with a planar molecular structure. More importantly, it can serve as a high-potential P-type material to provide storage sites for PF_6^- and additional capacity due to the electron-rich character of furan (Figure 3.11c). The electrochemical performance of QFQ in half-cell was tested, and it maintained a high capacity retention rate of 84.6% after 500 cycles at 0.5 C rate (Figure 3.11d). This work demonstrates that QFQ has good stability as a bipolar electrode material and provides new ideas for designing organic electrode materials.

3.2.3 Quinones for Aqueous ZIBs

Aqueous rechargeable batteries have low costs and impressive rate capabilities due to the aqueous electrolyte's high ion conductivity. Zinc as the negative electrode has a remarkable theoretical capacity (214 mAh g^{-1}), low oxidation–reduction potential (−0.76 V versus normal hydrogen electrode [NHE]), high stability in aqueous solutions, widespread existence in nature, low toxicity, and other advantages [39]. Therefore, aqueous ZIBs have attracted widespread attention, and positive electrode materials play a vital role in these batteries. According to different redox active sites, organic electrode materials are usually divided into conductive polymers, organic sulfur compounds, and nitrogen–oxygen free radical compounds. Carbonyl compounds offer several advantages, including redox stability, structural diversity, high theoretical capacity, and fast kinetics. As a result, they have been extensively researched and utilized in aqueous ZIBs.

In 2017, Chen's group first reported that quinone could be used for aqueous ZIBs [40]. They studied a series of quinone compounds, in which calix[4]quinone

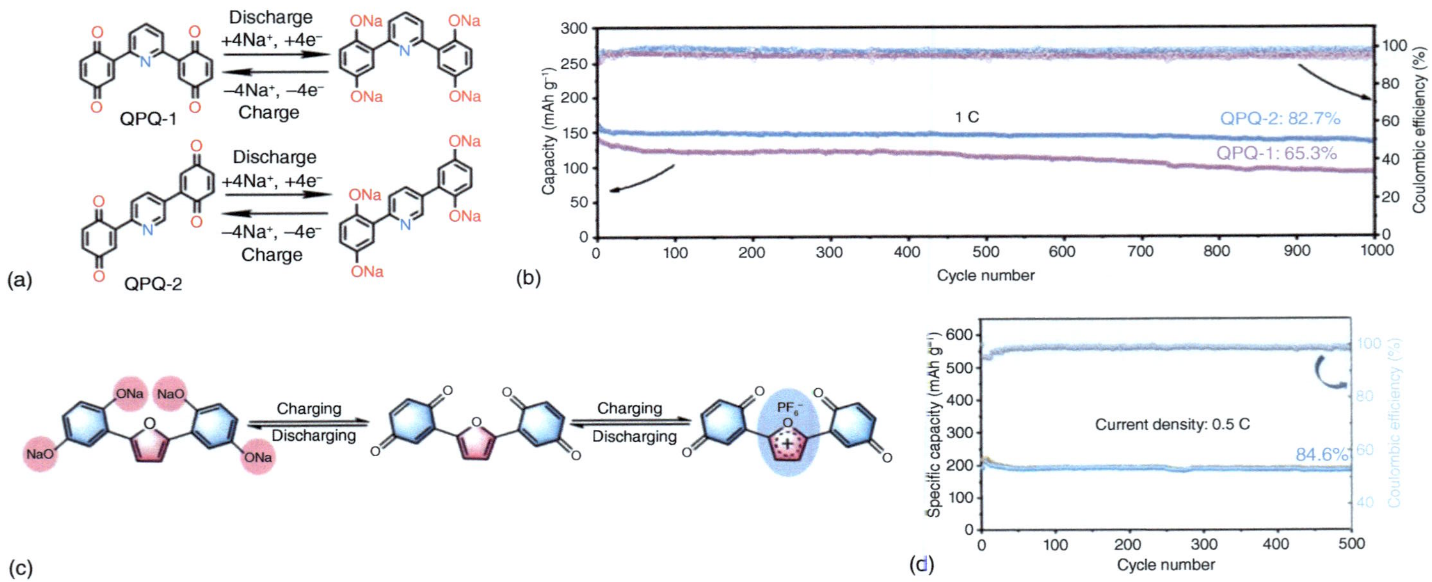

Figure 3.11 (a) Synthetic routes of QPQ-1 and QPQ-2. (b) The long-term capability of QPQ-1 and QPQ-2 at 1 C. Source: Reproduced with permission Li et al. [37]/Royal Society of Chemistry. (c) The energy storage mechanism of QFQ. (d) The long-term capability of QFQ at 0.5 C. Source: Reproduced with permission Zhang et al. [38]/Royal Society of Chemistry.

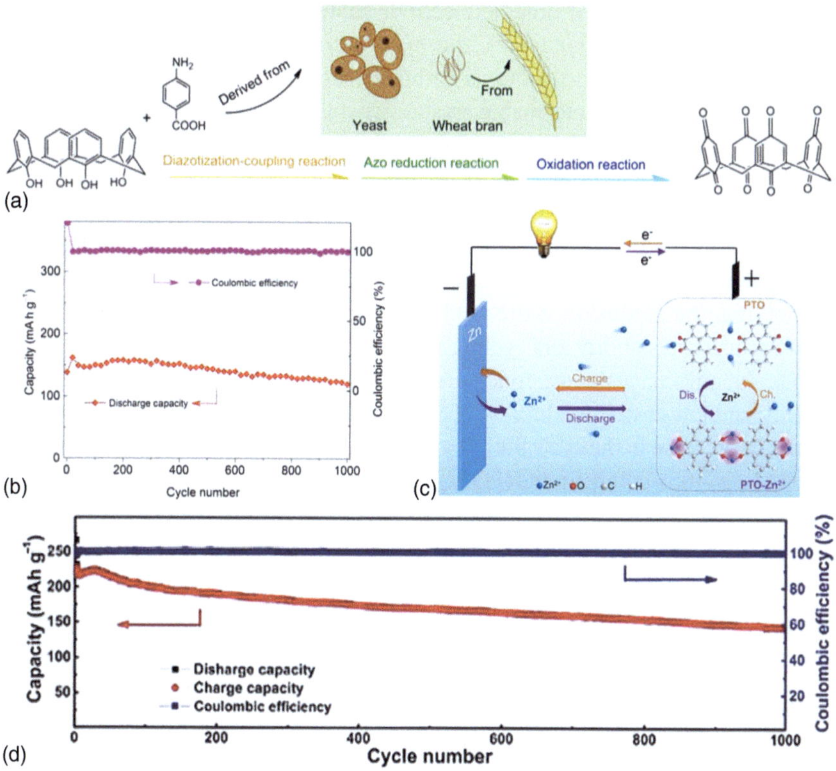

Figure 3.12 (a) Schematic diagram of preparing C4Q. (b) Cycling performance of Zn-C4Q cell and Coulombic efficiency at the current density of 500 mA g^{-1} with a Nafion separator. Source: Reproduced with permission Zhao et al. [40]/American Association for the Advancement of Science. (c) Visualization of the reversible reaction mechanism in the aqueous PTO//Zn battery. (d) The long-term capability of PTO//Zn at 3 A g^{-1}. Source: Reproduced with permission Guo et al. [41]/John Wiley & sons.

(C4Q, Figure 3.12a), in combination with a cation-selective membrane using an aqueous electrolyte, exhibits an initial capacity of 335 mAh g^{-1} at a current density of 20 mA g^{-1}, and has a long life of 1000 cycles at 500 mAh g^{-1}, with a capacity retention rate of 87% (Figure 3.12b). They also used electrostatic potential calculations to prove that the carbonyl group is the electrochemical active site, providing an excellent reference value for the application of carbonyl-based active materials in ZIBs in the future.

Subsequently, researchers began to study quinone compounds in ZIBs. Wang's work found that when pyrene-4,5,9,10-tetrone (PTO, Figure 3.12c) was used as the cathode electrode, two pairs of redox peaks appeared at 1.0 V/1.17 V and 0.63 V/0.85 V versus Zn^{2+}/Zn [41]. Similarly, two pairs of plateaus can be observed in the charge–discharge curve, showing a high capacity of 336 mAh g^{-1} and excellent reversibility. Thanks to PTO's faster reaction kinetics and higher reversibility, the PTO//Zn battery shows an energy density of 186.7 Wh kg^{-1} and an excellent cycling life of more than 1000 cycles (Figure 3.12d), which can open up the potential application of quinone compounds in wearable electronic devices.

Kundu et al. found that if organic electrode materials have a crystal structure fixed by weak intermolecular forces, the material may be ideal for storing Zn^{2+} [42]. They used tetrachloro-1,4-BQ as cathode material, demonstrating that the molecule has a plastically deformable crystal structure. The molecular column can be twisted to accommodate Zn^{2+}, limiting the volume changes during cycling and facilitating Zn^{2+} storage to be more reversible and efficient. Due to the catalytic effect of water phase transfer, the infinite growth of the discharged and charged phases decreases electrochemical reversibility. Therefore, they used the strategy of confining tetrachloro-1,4-benzoquinone in the nanopores of mesoporous carbon CMK-3 to limit the expansion of the discharged and charged phases, improving the electrochemical reversibility and cycling stability (Figure 3.13b).

Despite the advantages of quinone compounds, such as high capacity, good sustainability, and low cost, the solubility of small-molecule quinones in the electrolyte can lead to a decrease in the cycle life of batteries. To address this issue, Stoddart et al. reported a large-molecule quinone (PQ-Δ) with redox activity as cathode material for aqueous ZIBs [43]. At a current density of 150 mA g^{-1}, the initial discharge capacity was 210 mAh g^{-1}, and it could run 500 cycles without capacity decay. The consistent electrochemical performance is credited to the stable and inflexible structure of the sizable quinone molecule. In addition, divalent Zn^{2+} and water molecules are

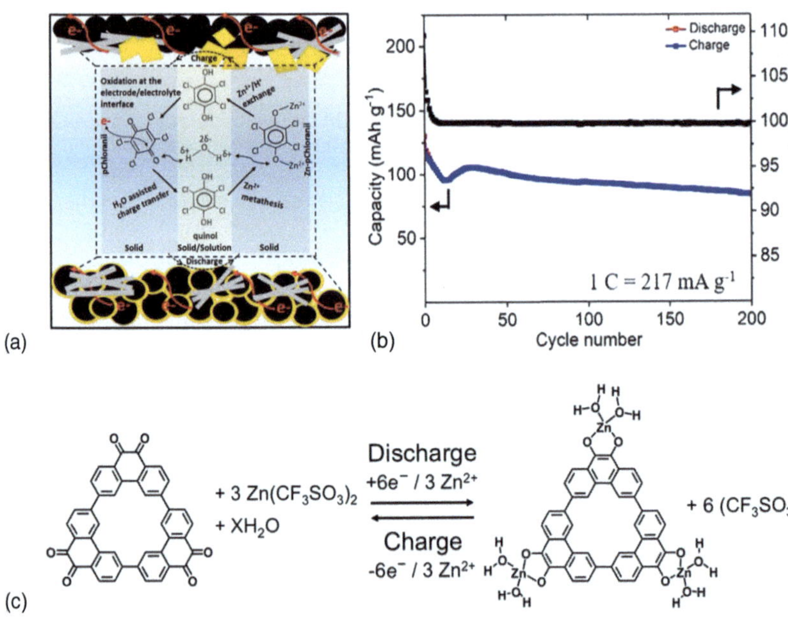

Figure 3.13 (a) Diagram illustrating the H_2O-assisted phase transfer mechanism for the formation of Zn-p-chloranil (gray fibers) and p-chloranil (yellow rhombus). (b) Galvanostatic cycling performance and Coulombic efficiency of the CMK-3 electrode infused with p-chloranil at a 1 C rate. Source: Reproduced with permission Kundu et al. [42]/American Chemical Society. (c) Electrochemical redox chemistry of PQ-Δ. Source: Adapted with permission from Nam et al. [43], American Chemical Society.

inserted into the positive electrode of PQ-Δ (Figure 3.13c), effectively reducing the interface resistance between the cathode and electrolyte and significantly improving the redox activity and cyclability of the divalent battery.

Fu's team found that modifying the chemical groups or functional moieties can influence quinone's electrochemical characteristics or behavior, such as the redox potential and cycling stability [44]. They prepared a novel carbonyl compound, 2-((4-hydroxyphenyl) amino) naphthalene-1,4-dione (APh-NQ), and combined it with carbon nanotubes as cathode electrodes (APh-NQ@CNT) for aqueous ZIBs (Figure 3.14a). Confining the active material within an interwoven carbon nanotube network helps to improve the utilization of APh-NQ. Using the same electrode preparation method, NQ composite electrode and dichloronaphthoquinone composite electrode were prepared. The performance of dichloronaphthoquinone and APh-NQ was superior to that of NQ due to the electron-withdrawing and electron-donating groups. This study not only contributes to the advancement of binder-free organic electrodes but also offers insights into the design of high-performance cathode materials.

It has been reported that highly symmetric molecules have low dipole moments and solubility in water. Therefore, Sun's group applied a symmetric

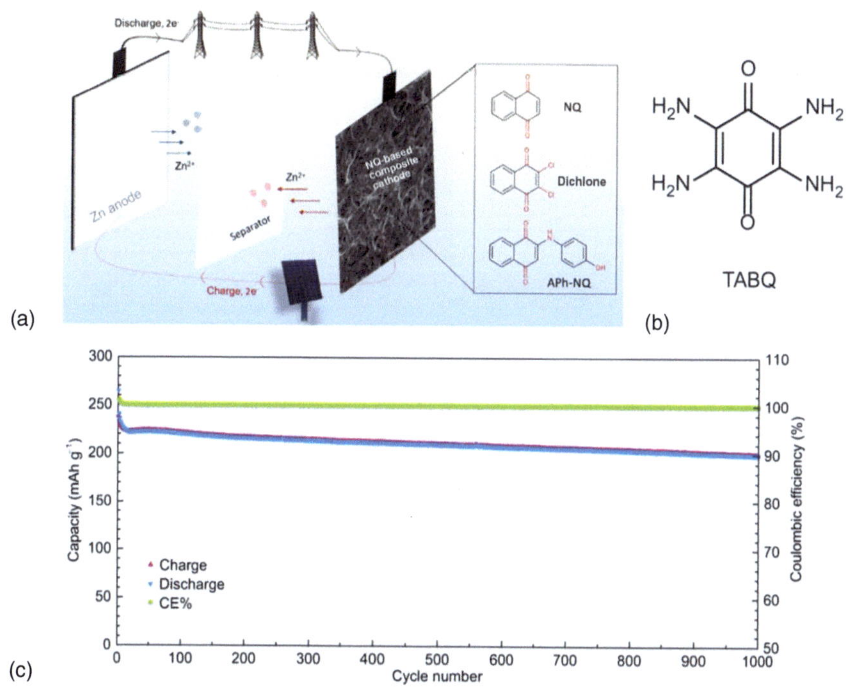

Figure 3.14 (a) Illustration of the reversible reaction mechanism of the aqueous APh-NQ//Zn battery. Source: Reproduced with permission Kumankuma-Sarpong et al. [44]/American Chemical Society. (b) The chemical structure of TABQ. (c) Capacity and Coulombic efficiency of TABQ at 5 A g^{-1} for 1000 cycles. Source: Lin et al. [45]/Springer Nature/CC BY 4.0.

tetraamino-*p*-benzoquinone (TABQ, Figure 3.14b) small molecule for aqueous ZIBs [45]. The four amino groups in the molecule can be protonated in weakly acidic electrolytes, creating a proton-active environment and initiating proton insertion. At the same time, hydrogen bonds are formed between the amino and carbonyl groups, and the breaking and rearrangement of the hydrogen bond network make proton conduction easy. This unique proton conduction mode results in activation energies for charge transfer and proton diffusion of less than 300 meV, resulting in the high capacity (303 mAh g^{-1}), excellent rate capability, and stable cycling performance of TABQ electrode (Figure 3.14c).

Song's group has reported a new type of cathode material called 5,7,12,14-tetraaza-6,13-pentacenequinone (TAPQ, Figure 3.15a), the first to integrate quinone and pyrazine structures [46]. Both C=O and C=N serve as redox-active sites. TAPQ has a high theoretical specific capacity of 515 mAh g^{-1} and an ultrahigh reversible capacity of 443 mAh g^{-1} in the voltage range of 0.1–1.6 V versus Zn^{2+}/Zn (Figure 3.15c).

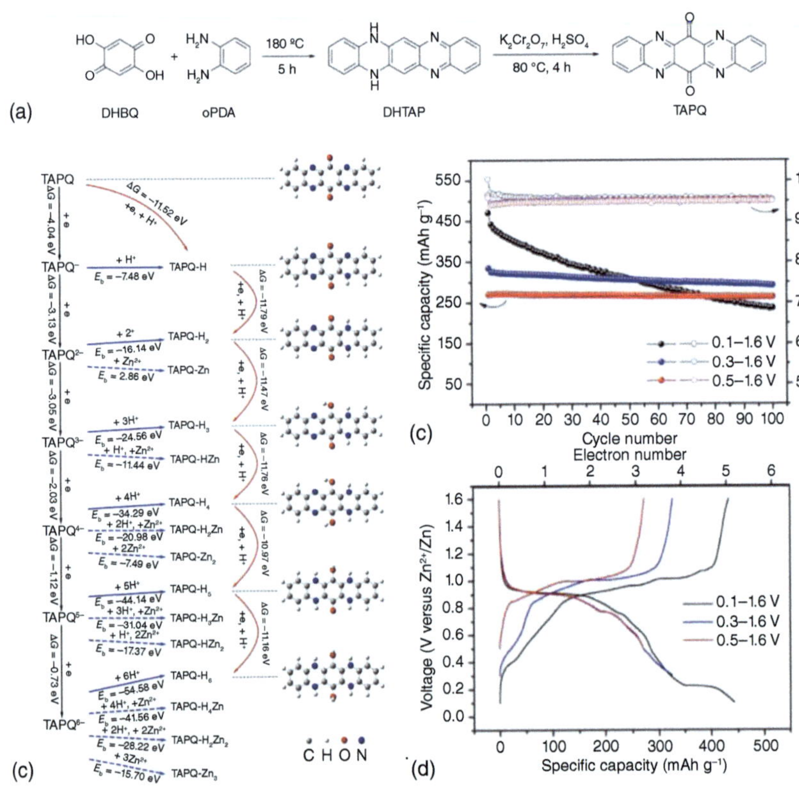

Figure 3.15 (a) The synthetic pathway for TAPQ. (b) The calculated binding energy (Eb) of TAPQH$_x$Zn$_y$, the Gibbs free energy gaps (ΔG) between TAPQ-related molecules, and proposed reaction routes of the discharging process. (c) Cycling performance and (d) discharge/charge voltage profiles of TAPQ cathode at a current rate of 50 mA g^{-1} in different voltage ranges of 0.1/0.3/0.5–1.6 V versus Zn^{2+}/Zn. Source: Reproduced from Gao et al. [46]/Elsevier.

The potential mechanism of proton and Zn^{2+} co-insertion was systematically studied by *in situ* characterization, electrochemical testing, and DFT calculations. In addition, the structural evolution mechanism of the TAPQ electrode during cycling was investigated in detail (Figure 3.15b).

3.2.4 Quinones for Other Metal-Ion Batteries

LIBs are the dominant electrochemical energy storage technology nowadays. However, increasing demand could cause a shortage of lithium resources. Research has begun to focus on exploring other metal-ion batteries. Potassium-ion batteries (PIBs), as a new type of battery, have received widespread attention [47]. Compared with LIBs, PIBs possess resource abundance and lower cost, with broad application prospects. Whereas, due to the larger ionic radius of K^+ compared to Li^+, suitable cathode materials must be found to overcome issues, such as capacity decay and structural damage.

As an organic electrode material with high redox activity, diverse structures, and low cost, carbonyl compounds have been successfully applied in PIBs. In 2016, Chen's team rationally designed nine oxocarbon salts of $M_2(CO)_n$ (M = Li, Na, K, n = 4, 5, 6) and customized their structures and substituents [48]. Different metals (Li, Na, K) as negative electrodes combined with $Li_2(CO)_n$, $Na_2(CO)_n$, and $K_2(CO)_n$ (Figure 3.16a) as cathode materials were systematically studied. It was found that the four-membered ring salts were difficult to combine with metal ions within the chosen discharge voltage range, while the five/six-membered ring salts could adsorb a specific quantity of metal ions. It is worth mentioning that each $K_2C_6O_6/K_2C_5O_5$ can bind two K^+ during the discharge process, providing a host for the rapid insertion and extraction of K^+. The $K/K_2C_6O_6$ battery could cycle 100 times at a 1 C rate with only 0.4% capacity decay per cycle (Figure 3.16b).

Xu et al. studied the electrochemical performance of anthraquinone-1,5-disulfonic acid sodium salt (AQDS, Figure 3.17a) as cathode material for PIBs [49]. Compared with anthraquinone, AQDS has two polar sodium sulfonate substitution groups, which can suppress the dissolution of organic electrolytes and improve cycling stability. After 100 cycles, the capacity remained at 78 mAh g^{-1} (Figure 3.17b).

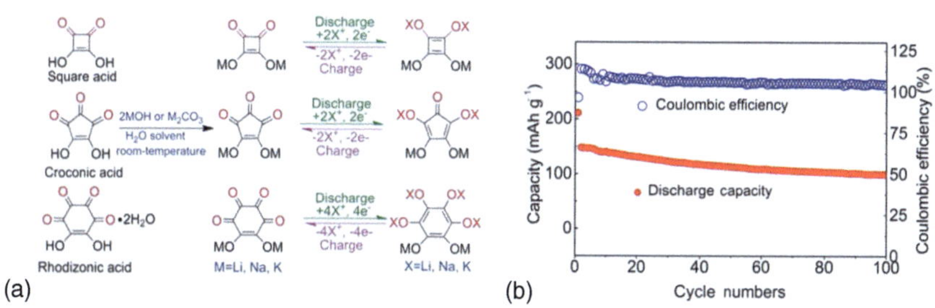

Figure 3.16 (a) Preparation, structures, and theoretic reactions of designed oxocarbon salts. (b) Cycling performance of $K/K_2C_6O_6$ battery at 1 C. Source: Adapted from Zhao et al. [48].

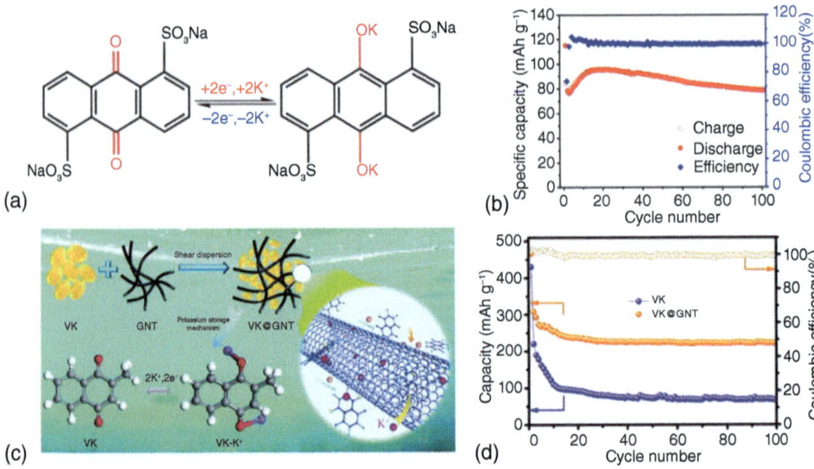

Figure 3.17 (a) Potassium storage mechanism of AQDS. (b) The cycling stability of the AQDS cathode at 0.1 C. Source: Reproduced with permission from Zhao et al. [19]/Elsevier. (c) Visualization of the hybridization process and potential potassium storage mechanism. (d) Cycling performance and Coulombic efficiency of VK and VK@GNT at 100 mA g^{-1}. Source: Reproduced with permission from Xue et al. [50], The Royal Society of Chemistry.

The research provides an effective way to develop high-performance organic electrode materials for PIBs. In recent years, due to biomolecules' renewable and degradable properties, more and more biomolecules have been applied in energy storage. Vitamin K has physiological effects, such as promoting normal blood clotting and participating in bone metabolism. Due to the existence of its carbonyl functional group, Chen's team used vitamin K for PIBs [50]. Combined with carbon nanotubes, it exhibited a remarkable reversible capacity of 300 mAh g^{-1}, which remained at 222.3 mAh g^{-1} even after 100 cycles at a current density of 100 mA g^{-1}. The Coulombic efficiency reached 99% (Figure 3.17d).

Ionic liquids have extremely low volatility, high ionic conductivity, good thermal stability, low flammability, and electrochemical stability. So, utilizing ionic liquid electrolytes will reduce organic materials' solubility and optimize batteries' safety. Huang et al. studied Pillar[5]quinone (P5Q, Figure 3.18a) as cathode material in PIBs [51], and the battery containing more conductive ionic liquid electrolyte (0.3 M KFSI-PY13FSI) demonstrated better performance. At the 2 C rate, the battery maintained a capacity of 101 mAh g^{-1} after 1000 cycles and 61 mAh g^{-1} after 5000 cycles (Figure 3.18c).

Magnesium metal has weaker reactivity with air and water than lithium metal, so magnesium-based batteries are safer than lithium batteries. As early as 2012, researchers had already used quinone as the positive electrode material for magnesium-ion batteries (MIBs). Yao's research aimed to investigate a redox-active organic crystal constructed by intermolecular interactions primarily mediated by weak forces [52]. Whether the crystal lattice can reversibly adapt to Mg^{2+} electrochemically and serve as an active material for MIBs was investigated thoroughly. 2,5-Dimethoxy-1,4-benzoquinone (DMBQ, Figure 3.19a) was selected for the MIBs.

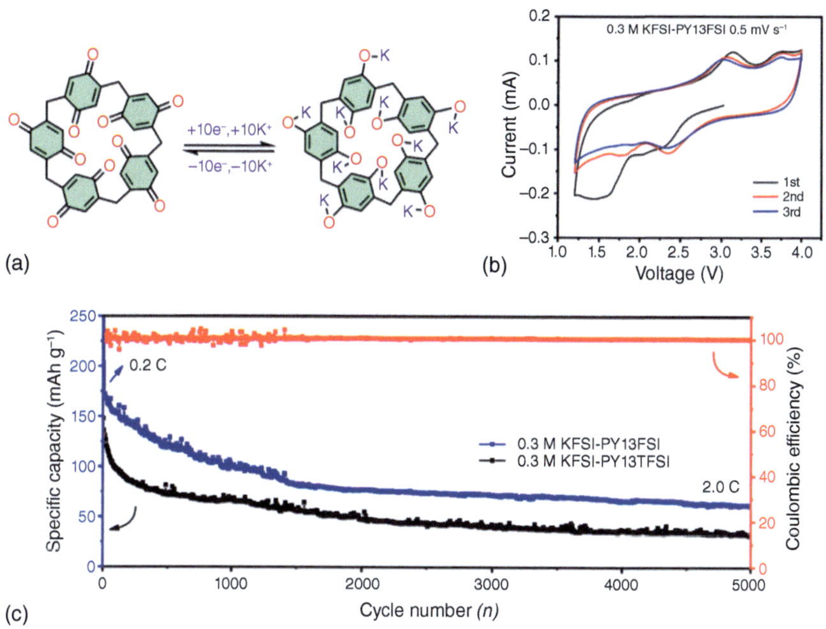

Figure 3.18 (a) Potassium storage mechanism and (b) CV curves of P5Q. (c) Cycling performance of P5Q at 2 C. Source: Reproduced with permission from Zhang et al. [51], American Chemical Society.

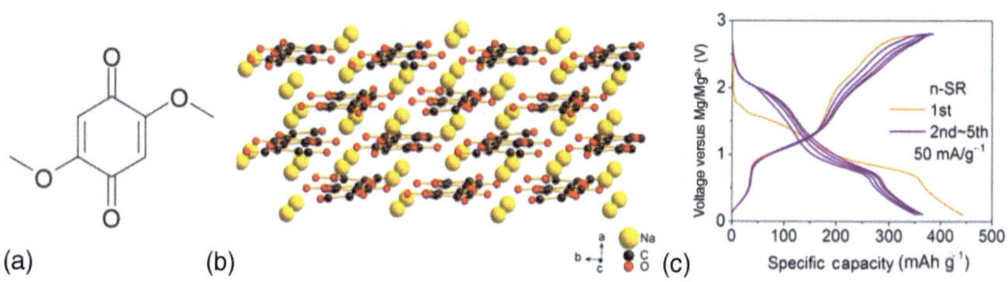

Figure 3.19 (a) Structural formula of DMBQ. Source: Adapted with permission from Sano et al. [52], The Chemical Society of Japan. (b) Crystal structure of $Na_2C_6O_6$, consisting of layers of $C_6O_6^{2-}$. Sodium ions are organized into anion layers staggered with $C_6O_6^{2-}$. (c) Galvanostatic charge/discharge curves of n-SR. Source: Reproduced with permission from Tian et al. [53], American Chemical Society.

The molecular crystal structure of DMBQ can be reversibly restored during the oxidation process, but the battery composed of DMBQ can only be cycled reversibly five times.

Li et al. developed a high-capacity Mg–Li hybrid battery activated by a magnesium–lithium double salt electrolyte [53]. The dual-salt electrolyte comprises 0.25 M all-phenyl complex and 1 M LiCl. Although the reaction is mainly driven by Li^+, the co-insertion of Mg^{2+} helps to alleviate the peeling of the C_6O_6 layer. Thanks to the construction of this dual-salt system, the Mg-$Na_2C_6O_6$ battery exhibits a high reversible capacity of 350–400 mAh g^{-1} (Figure 3.19c).

In addition to alkali metal ion and divalent metal-ion batteries (Ca^{2+} and Mg^{2+}), Al^{3+} is also an exciting choice. Al has the highest volume capacity, namely 8040 mAh cm^{-3}, four times Li's. Its weight capacity reaches 2980 mAh g^{-1} [54]. What's more, Al can be processed in the air, which brings significant advantages to battery manufacturing and dramatically improves the safety level of electrochemical storage systems. Recently, the selection of positive electrode materials for Al batteries has expanded to the family of organic molecules. Due to the weak Van der Waals interaction between molecules, enough intermolecular space is produced, which can be used for the diffusion and storage of large-volume aluminum complex carrier ions. These internal structures are also beneficial for releasing the strain generated during repeated insertion and extraction of large-volume complex ions, thereby achieving long-term cycling.

Choi et al. reported a macrocycle of tetraketone (TDK, Figure 3.20a) as a high-capacity cathode material for aluminum-ion batteries (AIBs) [55]. TDK exhibits good cyclic stability and maintains 78% capacity after 8000 cycles (Figure 3.20b). This research provides valuable insights for the design of active molecules to activate their ability to store Al^{3+}, thereby fully utilizing the advantages of multivalent batteries. Lu's team selected tetrachloro-1,4-benzoquinone (TCQ, Figure 3.20c) as cathode material for rechargeable aqueous AIBs [56]. Due to its simple structure and low molecular weight, TCQ has a high theoretical capacity (218.04 mAh g^{-1}). The existence of electron-withdrawing groups and induction effects in TCQ's extended lattice makes it easy to bond and break repeatedly with the surface Al^{3+}. In other words, TCQ's carbonyl group efficiently stores Al^{3+} through enolization oxidation–reduction reactions. The high reversibility of this reaction causes minimal volume changes and maintains the structural integrity of TCQ during charge and discharge. The Al//TCQ battery exhibited good cycling stability (capacity retention of 70.7% after 200 cycles, Figure 3.20d).

3.2.5 Quinones for RFBs

The operational principle of RFBs varies significantly from metal ion-batteries based on solid electrodes [57]. In RFBs, the redox-active materials are dissolved in external electrolyte reservoirs and transported through porous electrodes. These electrodes provide active sites for charge transfer without directly participating in electrochemical reactions. Such a design allows the separation of stored energy and output power. The amount of electrolyte determines the energy storage capacity, while the size of the cell and the number of cell stacks determine the power output. This property makes RFBs a vital energy storage system for medium- to large-scale applications, as they offer high scalability and independent control over energy and power.

Organic molecules as redox-active materials for RFBs have recently become an important topic. Among these molecules, quinone derivatives, which store charge on carbonyl groups through an ion coordination mechanism, have been one of the most extensively studied [58]. This section summarizes the recent examples of quinone molecules in RFBs.

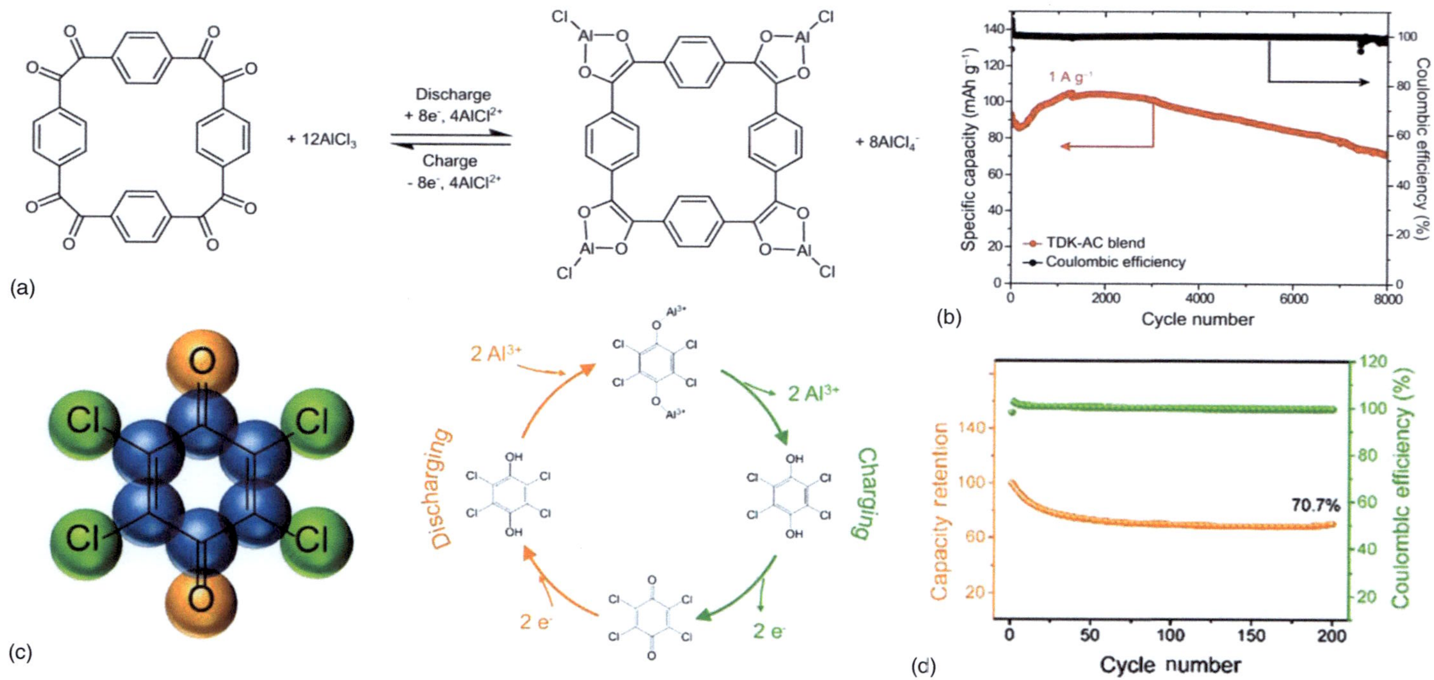

Figure 3.20 (a) Illustration of electrochemical redox mechanism of TDK. (b) Long-term cycling performance of TDK-AC blends at a current density of $1\,A\,g^{-1}$. Source: Yoo et al. [55]/Springer Nature/CC BY 4.0. (c) Illustration of the structure of TCQ and a schematic diagram depicting the structural transformation of TCQ during the electrochemical process. (d) The cycling performance of the Al//TCQ battery at a current density of $2\,A\,g^{-1}$. Source: Reproduced with permission He et al. [56]/Royal Society of Chemistry.

In 2009, Wen's group first applied quinones in RFBs [59]. Their work presented an innovative RFB that utilizes a single-flow acid Cd-chloranil battery. The electrolyte employed for the negative and positive electrodes consists of an aqueous mixture of H_2SO_4–$(NH_4)_2SO_4$–$CdSO_4$. The negative electrode is a nonreactive metal like copper foil, while the positive electrode is an insoluble organic substance, tetrachloro-*p*-benzoquinone. During the battery's operation, the electrolyte is continuously circulated through the cells using a single pump, and no membrane is required. Charging oxidizes tetrachloro-*p*-benzo-hydroquinone into chloranil at the positive electrode, while at the same time, cadmium ions are reduced to cadmium and electroplated onto the negative electrode. During discharge, the process is reversed. The experimental results showcased excellent performance, with an average Coulombic efficiency of 99% and an energy efficiency of 82% achieved over 100 cycles at a current density of $10\ mA\ cm^{-2}$.

In 2015, Aziz's team demonstrated that quinone-based batteries can be used in alkaline solutions (Figure 3.21a) [60]. The molecule (2,6-dihydroxyanthraquinone, 2,6-DHAQ) contains OH groups that donate electrons, resulting in a lower reduction potential. When placed in an alkaline electrolyte, these OH groups lose protons, allowing the molecule to dissolve at room temperature in a 1 M KOH solution. Additionally, the redox potential of this molecule decreases further in an alkaline electrolyte. The flow battery demonstrated an open circuit voltage of 1.2 V at a 50% state of charge. It could cycle continuously for 100 cycles at a current density of

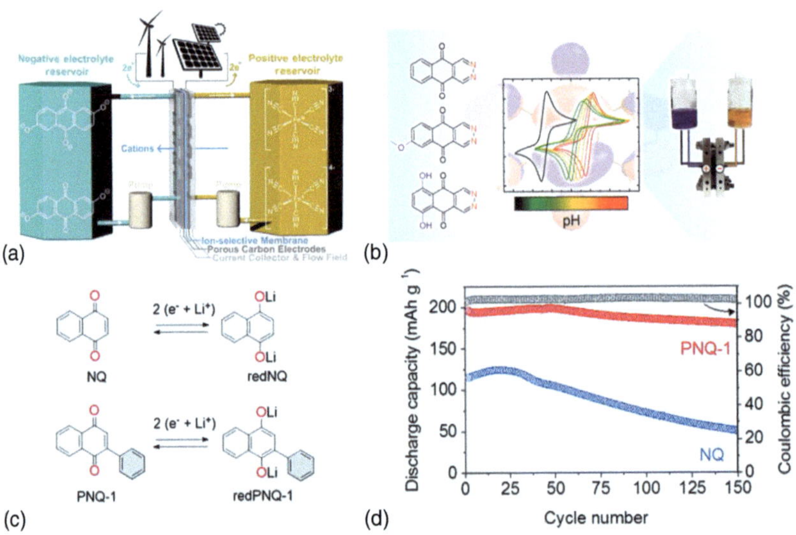

Figure 3.21 (a) Schematic of cell in charge mode. Source: Reproduced with permission Lin et al. [60]/American Association for the Advancement of Science. (b) Illustration of molecule DAD and two of its derivatives. Source: Reproduced with permission Hofmann et al. [61]/American Chemical Society. (c) Two-electron electrochemical lithiation and delithiation. (d) Cyclabilities of NQ and PNQ-1 at $100\ mA\ g^{-1}$. Source: Reproduced with permission Shin et al. [62]/Royal Society of Chemistry.

0.1 A cm^{-2}, achieving a current efficiency of over 99% and a round-trip energy efficiency of 84%.

Schröder et al. suggested a modified quinone, namely 2,3-diaza-anthracene-dione (DAD) and two of its derivatives, as a potential active material for aqueous RFBs (Figure 3.21b) [61]. Through systematic experimentation and density functional calculations, they investigated the electrochemical properties, such as redox potentials, rate constants, and diffusion coefficients, of these compounds at various pH values. A notable positive shift in the redox potential of approximately 300 mV was achieved by incorporating a diaza moiety into the anthraquinone base structure. When a methoxy group was added to the base structure of the 2,3-diaza-anthracene-dione, its electrochemical stability in aqueous acidic media was significantly enhanced, particularly at low pH. Moreover, incorporating two hydroxyl groups leads to a negative redox potential shift of 54 mV in acidic solutions and 264 mV in alkaline solutions. These findings highlighted the versatility of this new class of compounds, making them suitable as active materials in RFBs operating in both alkaline and acidic environments.

To demonstrate the concept, the 2,3-diaza-anthracene-diones were utilized as anolyte active material in a complete redox flow cell. Among them, the 2,3-diaza-anthracene-diones functionalized with a methoxy group exhibited the best cycling stability. These results can be extended to other quinone-based structures, such as NQ, enabling the development of even more effective catholyte and anolyte active materials for future organic active material-based RFBs.

Shin et al. synthesized a redox-active organic molecule called 2-phenyl-1,4-naphthoquinone (PNQ, Figure 3.21c) and utilized it in nonaqueous lithium-organic flow batteries [62]. They dissolved 0.05 M PNQ in a 1.0 M LiTFSI-TEGDME electrolyte to prepare the catholyte. The PNQ catholyte demonstrated excellent capacity retention of 92% over 150 cycles at 100 mA g^{-1}. Additionally, it achieved a high Coulombic efficiency of nearly 100% (Figure 3.21d).

In Aziz's study, 1,8-bis(2-(2-(2-hydroxyethoxy)ethoxy)-ethoxy)anthracene-9,10-dione (AQ-1,8-3E-OH) was introduced as the first water-miscible negolyte molecule for aqueous RFBs, demonstrating the highest volumetric capacity (80.4 Ah l^{-1}) among ever reported aqueous organic electrolytes (Figure 3.22a) [63]. When an electrolyte consisting of 1.5 M AQ-1,8-3E-OH as a negative component was combined with a positive electrolyte containing ferrocyanide and separated by a low-cost, non-fluorinated permselective polymer membrane at pH 7, the resulting system exhibited an open-circuit potential of 1.0 V, a volumetric capacity of 80.4 Ah l^{-1}, and an energy density of 25.2 Wh l^{-1}.

Water-soluble compounds with quinone-like charge-carrying properties have potential applications in aqueous RFBs. However, their practical use has been limited due to inadequate stability during battery operation. In Stahl's investigation, the stability of a group of quinones that are soluble in water and have redox potentials between 605 and 885 mV versus NHE was assessed under acidic conditions (1 M H_2SO_4) [64]. Four quinones (Figure 3.22b) were employed as cathodic electrolytes in an aqueous RFB, paired with anthraquinone-2,7-disulfonate as the anodic electrolyte. The RFB data and other stability tests indicated that the most stable

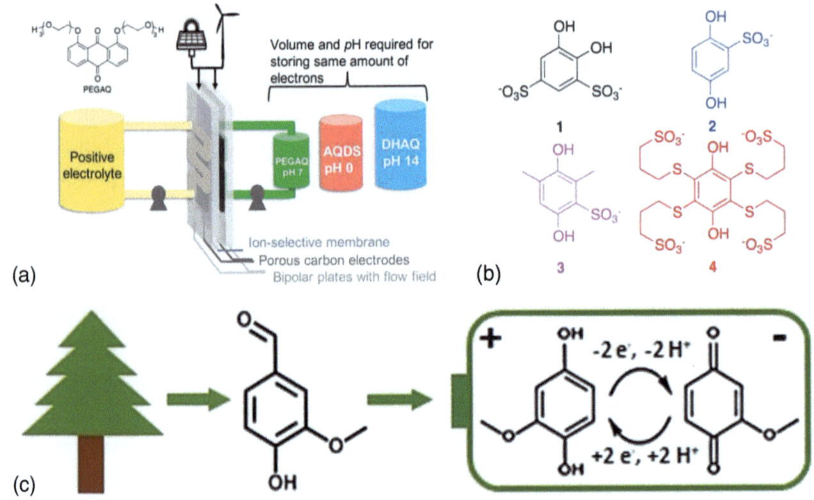

Figure 3.22 (a) Schematic flow battery with three negative electrolytes (negolytes): PEGAQ, AQDS, and DHAQ. The pH and volume required to store the same number of electrons are highlighted for each. Source: Reproduced with permission Jin et al. [63]/American Chemical Society. (b) The chemical structure of four cathode mediators. Source: Reproduced with permission Gerken et al. [64]/John Wiley & Sons. (c) Illustration of the reversible reaction mechanism of MHQ. Source: Reproduced with permission Schlemmer et al. [65]/John Wiley & Sons.

electrolyte was a quinone tetrasubstituted with four sulfonated thioether groups. These findings highlighted the significance of substituting all C–H positions of the quinone to enhance its stability. They established a foundation for advancing high-performance organic electrolytes suitable for aqueous RFBs.

Spirk et al. produced 2-methoxy-1,4-hydroquinone (MHQ, Figure 3.22c) by combining vanillin and established the groundwork for utilizing the 2-methoxy-1,4-quinone (MQ)/MHQ redox pair for RFBs [65]. These quinones deteriorated in the presence of different ions and when exposed to light. Using 0.5 M H_3PO_4 as the solvent, they stabilized the MQ/MHQ pair and obtained 97–99% Coulombic efficiency over 250 cycles in a full RFB.

The choline cation, which formed interionic hydrogen bonds with quinone sulfonate anions, was the solubilizing agent used to create the novel active organic electrolytes with high water solubility, good electrochemical reversibility, and favorable stability. 9,10-anthraquinone-2-disulfonic choline salt (AQSCH) and 1,2-dihydroxy-3,5-phenydisulfonic acid dicholine salt ($BQDS(CH)_2$) were synthesized (Figure 3.23) through simple ion exchange and acid-base neutralization, and can be used as the negative and positive electrolyte for a novel pH-neutral aqueous RFB [66]. The solubility of anthraquinone and BQ derivatives in water increased significantly due to the weaker ionic interactions and the hydrogen bonding between the cation and anion. Additionally, the choline-quinone derivatives exhibited excellent electrochemical reaction kinetics. Stable charging and discharging performance was achieved using choline chloride (CHCl) as the

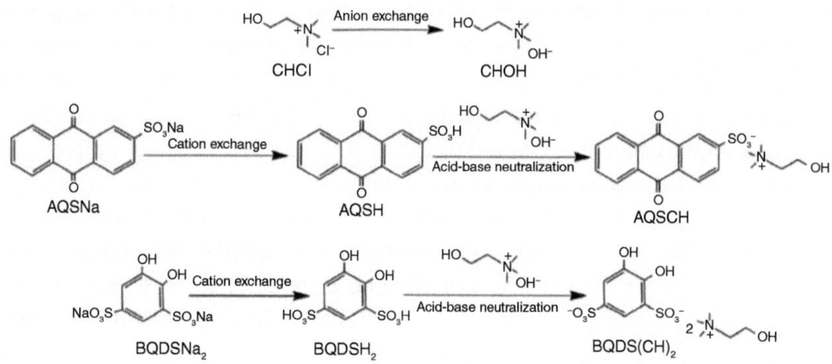

Figure 3.23 Synthesis process of AQSCH and BQDS(CH)$_2$. Source: Adapted with permission from Yang et al. [66], Elsevier Ltd.

supporting electrolyte, demonstrating the efficacy of choline-quinone derivatives as active electrolytes for novel pH-neutral aqueous RFBs with high safety, stability, and energy density.

3.3 Imides

The small-molecule imines used as positive electrode materials are generally diimines. They consist of two imine groups and a π-conjugated structure. In 2011, pyromellitic diimide dilithium salt (Figure 3.24a) was used as electrode material in LIBs by Poizot et al., which was the first example of imine as cathode material

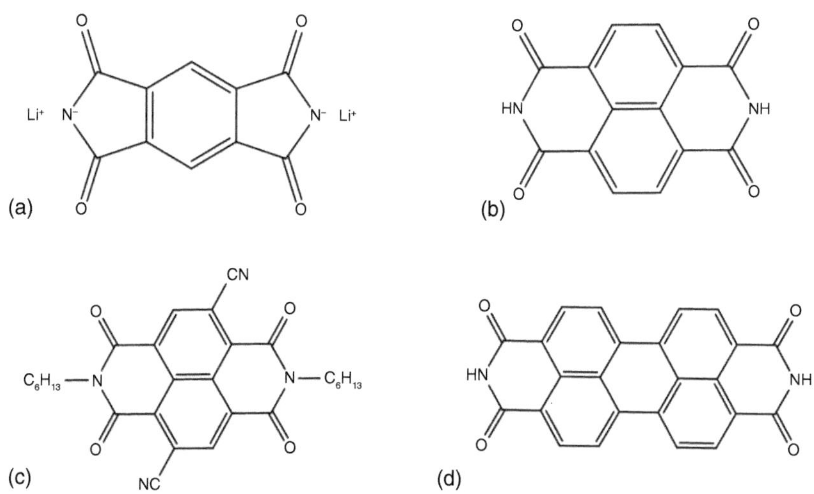

Figure 3.24 Molecular structure of some imides. (a) pyromellitic diimide dilithium salt, (b)naphthalene diimide, (c) N-hexyl-nitrile-substituted naphthalene diimide and (d) 3,4,9,10-perylene-bis(dicarboximide).

for metal-ion batteries, obtaining a reversible capacity of 200 mAh g^{-1} [67]. The electrochemical properties of amides can be regulated by regulating their molecular structure, such as by extending the aromatic system, *N*- or aryl-substitution, and designing rigid molecular structures. Dunn et al. investigated the electrochemical properties of naphthalene diimide (NDI, Figure 3.24b) and a series of derivatives, which can extend the aromatic system of the imine and adjust the solubility as well as the electronic properties of the molecule by nitrogen and aromatic substitution of the molecule, thereby affecting its electrochemical properties [68]. It was found that the N-hexyl-nitrile-substituted naphthalene diimide (Figure 3.24c) has the second-highest voltage of 2.8 V and the largest capacity utilization with an actual capacity of 100 mAh g^{-1} and a theoretical capacity of 100 mAh g^{-1}. Yang et al. further extended the aromatic structure of amide and studied a sodium ion battery with 3,4,9,10-perylene-bis(dicarboximide) (Figure 3.24d) as the cathode active material [69]. A high capacity of 140 mAh g^{-1} was obtained at a current density of 10 mA g^{-1} (Figure 3.25a). After 300 cycles at 100 mA g^{-1}, it remained at 90% of the initial capacity of 118.6 mAh g^{-1} (Figure 3.25b). A perylene-based imide derivative was used as an electrode material in zinc batteries, achieving a long cycle life of 5000 cycles at a high rate of 3000 mA g^{-1} (Figure 3.25d) [70]. The perylene-based imide derivative was synthesized by the perylene-3,4,9,10-tetracarboxylic dianhydride (Figure 3.25c), which possesses an extensive conjugation system that significantly

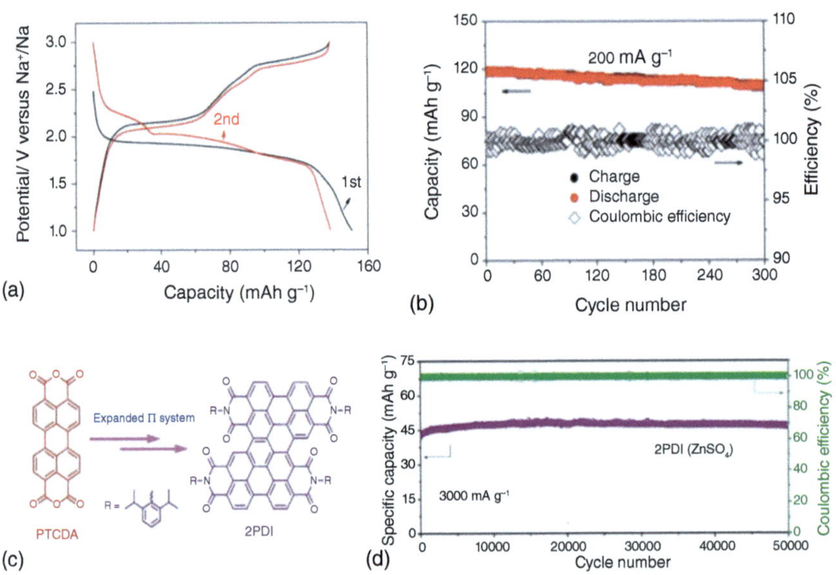

Figure 3.25 (a) Charge/discharge curves of 3,4,9,10-perylene-bis(dicarboximide) at 10 mA g^{-1}. (b) Cycle performance and Coulomb efficiency curves of 3,4,9,10-perylene-bis(dicarboximide) at 100 mA g^{-1}. Source: Reproduced with permission Deng et al. [69]/American Chemical Society. (c) The perylene-based imide derivative was synthesized from the perylene-3,4,9,10-tetracarboxylic dianhydride. (d) Cycle performance and Coulomb efficiency curves of the perylene-based imide derivative at 3000 mA g^{-1}. Source: Reproduced from Li et al. [70]/Elsevier.

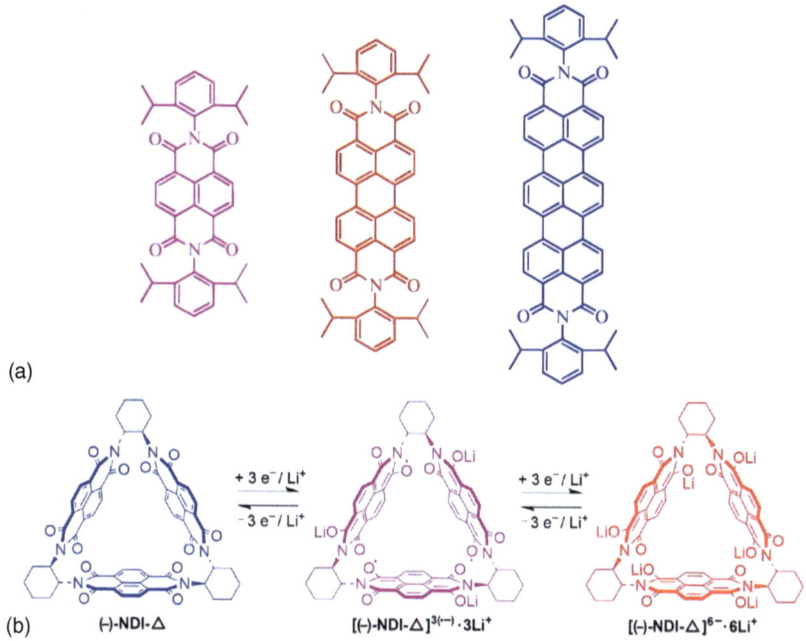

Figure 3.26 (a) Molecular structure of naphthalene diimide, perylene diimide, and terrylene diimide. Source: Adapted with permission from Li et al. [72], Elsevier B.V. (b) Schematic representation of the redox reaction of (-)-NDI-△. Source: Adapted with permission from Chen et al. [73], WILEY-VCH Verlag GmbH.

inhibited its solubility in water. Through a comparative study of several aromatic imides (Figure 3.24a, b, and d) with different aromatic ring sizes, Lu et al. found that the increase of aromatic ring size can modulate the energy of LUMO orbitals, reduce the LUMO–HOMO gap, and disperse the spin electron density, resulting in the improvement of the rate capability and cycling stability of the imide electrode [71]. Lin et al. investigated three different vertical p-extensions of imides: naphthalene diimide, perylene diimide, and terrylene diimide (Figure 3.26a), demonstrating that the vertical p-extensions produce different redox mechanisms and can improve the capacity utilization of the electrodes [72].

Stoddart et al. synthesized a substance with a rigid molecular triangular structure containing a naphthalenediimine unit (Figure 3.26b). They used it as the cathode material for LIBs, significantly reducing the molecule's solubility and improving its electrochemical properties. It obtained a capacity of about 146.4 $mAh\,g^{-1}$ at a low rate of 0.1 C and 58.1 $mAh\,g^{-1}$ at an ultrahigh rate of 100 C. Moreover, the battery held a capacity of 71.1 $mAh\,g^{-1}$ after 300 cycles at a rate of 10 C [73].

3.4 Anhydrides

Anhydrides generally possess a large conjugated structure. They can undergo two reduction reactions with an even number of carbon atoms between the two carbonyl groups for each reduction reaction. Sun et al. conducted the first study on anhydride as an organic cathode material [74]. They investigated the electrochemical performance of 3,4,9,10-perylenetetracarboxylic dianhydride (PTCDA) as an electrode material in SIBs. It was found that the initial capacity of the material was 135 mAh g^{-1} at a constant current density of 100 mA g^{-1}, but it dropped rapidly afterward, with the capacity maintenance below 50% after 90 cycles. The application of PTCDA to SIBs was also investigated by Ji et al. [75]. In SIBs, unlike in LIBs, it was found to have a higher capacity, better rate capability, and longer cycle life, with a reversible capacity of 145 mAh g^{-1} at a rate of 10 mA g^{-1}, 91 mAh g^{-1} at a constant current density of 1 A g^{-1} for charging and discharging, and a stable cycle for 200 cycles. Simultaneously, it was found that when discharged down to 0.01 V, each molecule could insert 15 sodium ions (Figure 3.27a), exhibiting a high specific capacity of 1017 mAh g^{-1}. However, this process was irreversible. In 2015, Ji et al. used PTCDA as a cathode material for PIBs. They obtained a battery with an initial capacity of 122 mAh g^{-1} at a current density of 20 mA g^{-1} and a Coulombic efficiency of 94.6% with modest rate capability and cycling performance [13]. Meanwhile, it was found that the PTCDA crystals underwent amorphization during discharge. Though the crystal structure was partly recovered during charging, the reaction was irreversible once the voltage was as low as 0.2 V. Lu et al. continued to study the application of PTCDA as a cathode material in PIBs. They proposed

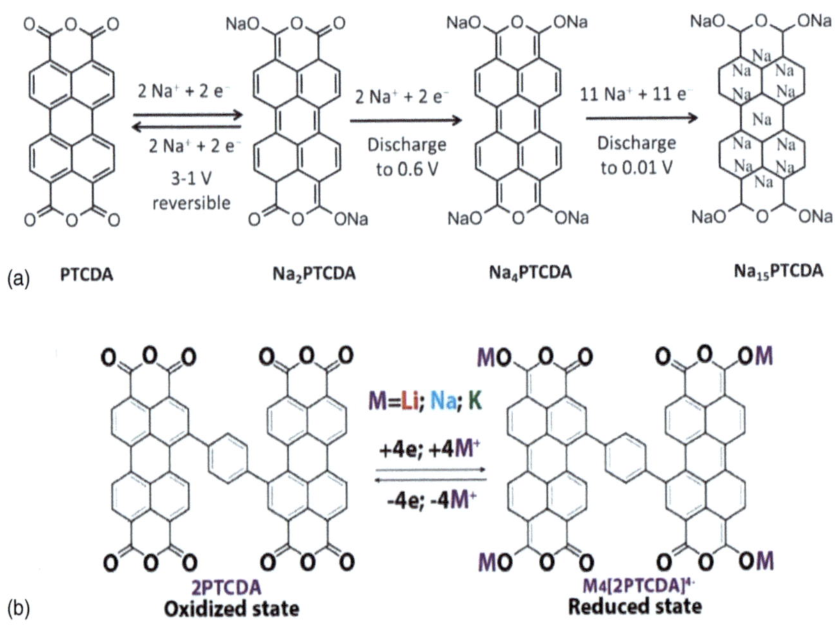

Figure 3.27 (a) Schematic representation of the REDOX reaction of PTCDA. Source: Adapted with permission from Luo et al. [75], WILEY-VCH Verlag GmbH. (b) Schematic representation of the 2PTCDA. Source: Adapted with permission from Hong et al. [76], Elsevier B.V.

that the reversible phase change of the cathode material during the charging and discharging process is a crucial factor affecting the cycling stability of the battery. Therefore, such reversible change was promoted by optimizing the electrode and electrolyte. And an organic-PIB with high power density, energy efficiency, excellent rate performance, and long cycle life could be obtained [77].

The dissolution of the anhydride during charging and discharging is one critical issue affecting the electrochemical performance. To solve this problem, Fan et al. designed an insoluble anhydride, 1,1′-(1,4-phenylene)-bis(perylene-3,4,9,10-tetracarboxylic dianhydride) (Figure 3.27b), which was applied in Li/Na/K-ion batteries and obtained a discharge capacity of 120–137 mAh g^{-1} with better cycling stability [76].

3.5 Summary and Outlook

To summarize, carbonyl-based compounds are doubtlessly promising cathode materials for next-generation rechargeable batteries thanks to their cost-effectiveness, resource abundance, high capacity, and tunable chemistry. The application attempts of carbonyl-based cathodes in diverse rechargeable batteries were presented in this chapter, including quinone compounds, carboxylate salts, imides, and anhydrides. Like many others, some challenges for carbonyl-based compounds stand before researchers, such as low electrical conductivity, low discharge voltage, and unwanted high solubility in the electrolyte. Some prospects and perspectives we praised are provided as follows:

1. Developing carbonyl-based compounds with outstanding electrical conductivity and elevated discharge voltage to promote cathodes' energy density and other electrochemical performance.
2. Reasonably designing the cathode architectures and electrolytes to prevent the dissolution of active materials. For the cathode, introducing conductive carbonaceous materials or polymerizing may effectively suppress the dissolution and shuttle of carbonyl-based active substances; for electrolytes, the solvent with weak electron donor capacity or significant difference in polarity could solve the dissolution problem.
3. Systematically elucidating the mechanism and behavior of carbonyl-based cathodes during the electrochemical process inside the rechargeable batteries via exhaustive *in situ*/*ex situ* experimental characterizations and theoretical calculation. This offers new opportunities to maximize the utilization of the carbonyl-based compounds.
4. Broadening the application of carbonyl-based compounds into all solid-state organic batteries to enhance the devices' safety and cycling performance. Replacing liquid electrolytes with solid-state electrolytes could optimize carbonyl-based small molecules' long-term cycling performance without needing additional cathode active material polymerization.

Acknowledgments

The work was supported by the National Natural Science Foundation of China (Grant Nos. U2004214 and 21975225), the National Key Research and Development Program of China (2022YFB2402200).

References

1 Liu, Y., Zhu, Y., and Cui, Y. (2019). *Nat. Energy* 4: 540–550.
2 Poizot, P., Gaubiche, J., Renault, S. et al. (2020). *Chem. Rev.* 120: 6490–6557.
3 Kim, J., Kim, Y., Yoo, J. et al. (2022). *Nat. Rev. Mater.* 8: 54–70.
4 Zhu, L., Ding, G., Xie, L. et al. (2019). *Chem. Mater.* 31: 8582–8612.
5 Williams, D.L., Byrne, J.J., and Driscoll, J.S. (1969). *J. Electrochem. Soc.* 116: 2.
6 Liang, Y., Zhang, P., and Chen, J. (2013). *Chem. Sci.* 4: 1330–1337.
7 Liang, Y., Jing, Y., Gheytani, S. et al. (2017). *Nat. Mater.* 16: 841–848.
8 Lee, M., Hong, J., Lopez, J. et al. (2017). *Nat. Energy* 2: 861–868.
9 Armand, M., Grugeon, S., Vezin, H. et al. (2009). *Nat. Mater.* 8: 120–125.
10 Jouhara, A., Dupre, N., Gaillot, A.C. et al. (2018). *Nat. Commun.* 9: 4401.
11 Geng, J., Bonnet, J.-P., Renault, S. et al. (2010). *Energy Environ. Sci.* 3: 1929–1933.
12 Song, Z., Zhan, H., and Zhou, Y. (2010). *Angew. Chem. Int. Ed.* 49: 8444–8448.
13 Xing, Z., Jian, Z., Luo, W. et al. (2016). *Energy Storage Mater.* 2: 63–68.
14 Er, S., Suh, C., Marshak, M.P., and Aspuru-Guzik, A. (2015). *Chem. Sci.* 6: 885–893.
15 Liang, Y., Zhang, P., Yang, S. et al. (2013). *Adv. Energy Mater.* 3: 600–605.
16 Yokoji, T., Matsubara, H., and Satoh, M. (2014). *J. Mater. Chem. A* 2: 19347–19354.
17 Yokoji, T., Kameyama, Y., Maruyama, N., and Matsubara, H. (2016). *J. Mater. Chem. A* 4: 5457–5466.
18 Ma, T., Zhao, Q., Wang, J. et al. (2016). *Angew. Chem. Int. Ed.* 55: 6428–6432.
19 Ai, W., Zhou, W., Du, Z. et al. (2017). *Adv. Funct. Mater.* 27: 1603603.
20 Lee, J., Kim, H., and Park, M.J. (2016). *Chem. Mater.* 28: 2408–2416.
21 Lee, J. and Park, M.J. (2017). *Adv. Energy Mater.* 7: 1602279.
22 Kwon, J.E., Hyun, C.-S., Ryu, Y.J. et al. (2018). *J. Mater. Chem. A* 6: 3134–3140.
23 Song, Z., Qian, Y., Otani, M., and Zhou, H. (2016). *Adv. Energy Mater.* 6: 1501780.
24 Luo, Z., Liu, L., Zhao, Q. et al. (2017). *Angew. Chem. Int. Ed.* 56: 12561–12565.
25 Lu, Y., Hou, X., Miao, L. et al. (2019). *Angew. Chem. Int. Ed.* 58: 7020–7024.
26 Tuttle, M.R. and Zhang, S. (2019). *Chem. Mater.* 32: 255–261.
27 Huang, W., Liu, S., Li, C. et al. (2022). *EcoMat* e12214.
28 Wang, Z., Fan, Q., Guo, W. et al. (2022). *Adv. Sci.* 9: 2103632.
29 Zhao, B., Si, Y., Guo, W., and Fu, Y. (2022). *Adv. Funct. Mater.* 32.
30 Shi, T., Li, G., Han, Y. et al. (2022). *Energy Storage Mater.* 50: 265–273.
31 Liu, S., Peng, F., Lin, Y. et al. (2023). *Chem. Eng. J.* 451: 1385–8947.
32 Xu, Y., Zhou, M., and Lei, Y. (2018). *Mater. Today* 21: 60–78.

33 Zhu, Z., Li, H., Liang, J. et al. (2015). *Chem. Commun.* 51: 1446–1448.
34 Wang, H., Hu, P., Yang, J. et al. (2015). *Adv. Mater.* 27: 2348–2354.
35 Chen, X., Wu, Y., Huang, Z. et al. (2016). *J. Mater. Chem. A* 4: 18409–18415.
36 Cao, S., Zhang, H., Zhao, Y., and Zhao, Y. (2021). *eScience* 1: 28–43.
37 Li, X., Jia, K., Zhang, J. et al. (2022). *Inorg. Chem. Front.* 9: 4517–4524.
38 Zhang, J., Jia, K., Li, X. et al. (2022). *J. Mater. Chem. A* 10: 10062–10068.
39 Xie, J. and Zhang, Q. (2019). *Small* 15: 1805061.
40 Zhao, Q., Huang, W., Luo, Z. et al. (2018). *Sci. Adv.* 4: eaao1761.
41 Guo, Z., Ma, Y., Dong, X. et al. (2018). *Angew. Chem. Int. Ed.* 57: 11737–11741.
42 Kundu, D., Oberholzer, P., Glaros, C. et al. (2018). *Chem. Mater.* 30: 3874–3881.
43 Nam, K.W., Kim, H., Beldjoudi, Y. et al. (2020). *J. Am. Chem. Soc.* 142: 2541–2548.
44 Kumankuma-Sarpong, J., Tang, S., Guo, W., and Fu, Y. (2021). *ACS Appl. Mater. Interfaces* 13: 4084–4092.
45 Lin, Z., Shi, H.Y., Lin, L. et al. (2021). *Nat. Commun.* 12: 4424.
46 Gao, Y., Li, G., Wang, F. et al. (2021). *Energy Storage Mater.* 40: 31–40.
47 Hwang, J.-Y., Myung, S.-T., and Sun, Y.-K. (2018). *Adv. Funct. Mater.* 28: 1802938.
48 Zhao, Q., Wang, J., Lu, Y. et al. (2016). *Angew. Chem. Int. Ed.* 55: 12528–12532.
49 Zhao, J., Yang, J., Sun, P., and Xu, Y. (2018). *Electrochem. Commun.* 86: 34–37.
50 Xue, Q., Li, D., Huang, Y. et al. (2018). *J. Mater. Chem. A* 6: 12559–12564.
51 Zhang, W., Tian, H., Wang, J. et al. (2022). *ACS Appl. Mater. Interfaces* 14: 38887–38894.
52 Sano, H., Senoh, H., Yao, M. et al. (2012). *Chem. Lett.* 41: 1594–1596.
53 Tian, J., Cao, D., Zhou, X. et al. (2018). *ACS Nano* 12: 3424–3435.
54 Elia, G.A., Marquardt, K., Hoeppner, K. et al. (2016). *Adv. Mater.* 28: 7564–7579.
55 Yoo, D.J., Heeney, M., Glocklhofer, F., and Choi, J.W. (2021). *Nat. Commun.* 12: 2386.
56 He, J., Shi, X., Wang, C. et al. (2021). *ChemComm.* 57: 6931–6934.
57 Han, C., Li, H., Shi, R. et al. (2019). *J. Mater. Chem. A* 7: 23378–23415.
58 Zhu, F., Guo, W., and Fu, Y. (2023). *Chem. Asian J.* 18: e202201098.
59 Xu, Y., Wen, Y., Cheng, J. et al. (2009). *Electrochem. Commun.* 11: 1422–1424.
60 Lin, K., Chen, Q., Gerhardt, M.R. et al. (2015). *Science* 349: 1529–1532.
61 Hofmann, J.D., Pfanschilling, F.L., Krawczyk, N. et al. (2018). *Chem. Mater.* 30: 762–774.
62 Shin, D.-S., Park, M., Ryu, J. et al. (2018). *J. Mater. Chem. A* 6: 14761–14768.
63 Jin, S., Jing, Y., Kwabi, D.G. et al. (2019). *ACS Energy Lett.* 4: 1342–1348.
64 Gerken, J.B., Anson, C.W., Preger, Y. et al. (2020). *Adv. Energy Mater.* 10: 2000340.
65 Schlemmer, W., Nothdurft, P., Petzold, A. et al. (2020). *Angew. Chem. Int. Ed.* 59: 22943–22946.
66 Yang, Y., Wang, D., Zheng, J. et al. (2023). *Electrochim. Acta* 452: 142292.
67 Renault, S., Geng, J., Dolhem, F., and Poizot, P. (2011). *Chem. Commun.* 47: 2414–2416.
68 Vadehra, G.S., Maloney, R.P., Garcia-Garibay, M.A., and Dunn, B. (2014). *Chem. Mater.* 26: 7151–7157.

69 Deng, W., Shen, Y., Qian, J. et al. (2015). *ACS Appl. Mater. Interfaces* 7: 21095–21099.

70 Li, L., Wang, Y., Gong, W. et al. (2023). *Chem. Eng. J.* 465: 142824.

71 Chen, J., Gu, S., Hao, R. et al. (2022). *ACS Appl. Mater. Interfaces* 14: 44330–44337.

72 Li, L., Wang, J., Chen, M. et al. (2019). *Chin. Chem. Lett.* 30: 2254–2258.

73 Chen, D., Avestro, A.J., Chen, Z. et al. (2015). *Adv. Mater.* 27: 2907–2912.

74 Han, X., Chang, C., Yuan, L. et al. (2007). *Adv. Mater.* 19: 1616–1621.

75 Luo, W., Allen, M., Raju, V., and Ji, X. (2014). *Adv. Energy Mater.* 1400554.

76 Hong, Y., Hu, J., Tang, W. et al. (2022). *Energy Storage Mater.* 52: 61–68.

77 Fan, L., Ma, R., Wang, J. et al. (2018). *Adv. Mater.* 30: 1805486.

4

Sulfur-Containing Organic Cathodes

4.1 Introduction

Organosulfur compounds are environmentally amicable and rich in resources; they are considered as promising electrode materials because of their tunable structures, precise active sites, and high theoretical specific capacities [1, 2]. As early as 1988, organosulfides were used in battery systems and showed good electrochemical activity [3], and since then, researchers have not stopped mining and exploring organosulfur electrodes (Figure 4.1). The most attractive advantage of organosulfur is controllable structure, which is not only limited to the growth of sulfur chains and the simple modification of molecular structures but also in the regulation of functional groups. Making full use of this characteristic is of great significance for the development and application of organosulfur-based batteries.

The transmission of electrons and ions in most organosulfur-based batteries depends on the breaking and reformation of S—S bonds in organosulfides [13]. Due to the limited active site, the specific capacity and energy density of organodisulfides (R_1–S_2–R_2, R: organic group) are quite low. Based on the controllable structures of organosulfides, organic polysulfides (R_1–S_n–R_2, $n > 2$) with multiple electrochemical reaction sites were designed and synthesized and showed high specific capacity in rechargeable batteries. Organic functional groups play an important role in regulating the electrochemical properties of organosulfur, organosulfides-containing heteroatoms (e.g. N, Se, O, etc.) are introduced. The addition of heteroatoms makes organosulfur-based batteries show unique redox behavior. In addition, most organosulfides show slow kinetics in batteries because of their low conductivity, and the addition of inorganic materials with high conductivity can make up for this defect. Organosulfur-inorganic hybrid cathode materials were displayed, and they exhibited fast kinetics and enhanced electrochemical performance in rechargeable batteries. Finally, some organosulfides with unique electrochemical properties are introduced.

Rechargeable Organic Batteries: Materials, Mechanisms, and Prospects, First Edition.
Yongzhu Fu, Xiang Li, Shuai Tang, and Wei Guo.

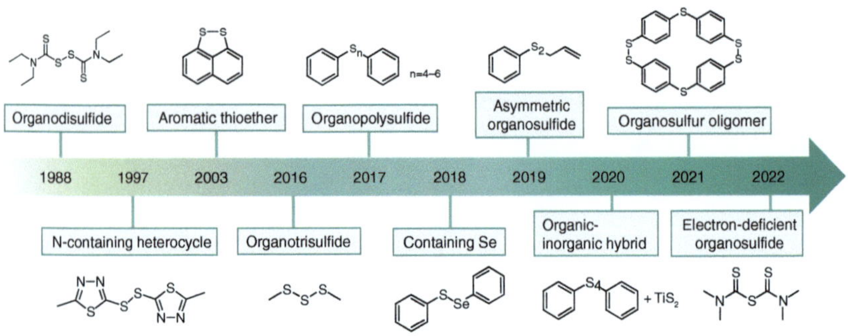

Figure 4.1 The development of small-molecule organosulfides for batteries. The dates correspond to the first published developments featuring each compound [3–12].

4.2 Organodisulfide

Organosulfides were first introduced into battery systems by Visco in 1988, where tetraethyl thiuram disulfide (TETD) was used as cathode in lithium batteries and exhibited electrochemical reversibility [3]. Afterward, Visco et al. explored the electrochemical performance of a series of organodisulfide cathodes (Figure 4.2), i.e. tetramethylthiuram disulfide (TMTD), dimethyl disulfide (DMDS), diphenyl disulfide (DPDS), di-fluorophenyl disulfide (DFPDS), etc. [14], further confirming the potential application of organosulfur compounds in energy storage systems. Although these organodisulfides exhibit certain electrochemical reversibility, they are limited by their poor cycling performances and low energy densities, which make it difficult to compete with traditional inorganic battery systems. With the in-depth study of the battery systems, organosulfur-based batteries with high performance have been realized by regulating the structure of the organosulfides and optimizing other components in the battery systems.

DPDS may lead to the shuttle effect due to their solubility in ether-based electrolytes, resulting in low Coulombic efficiency and poor cycling stability of batteries.

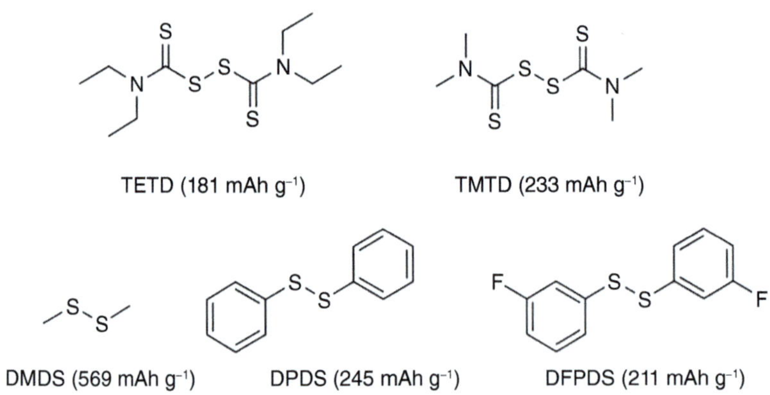

Figure 4.2 Organodisulfides and their theoretical capacities.

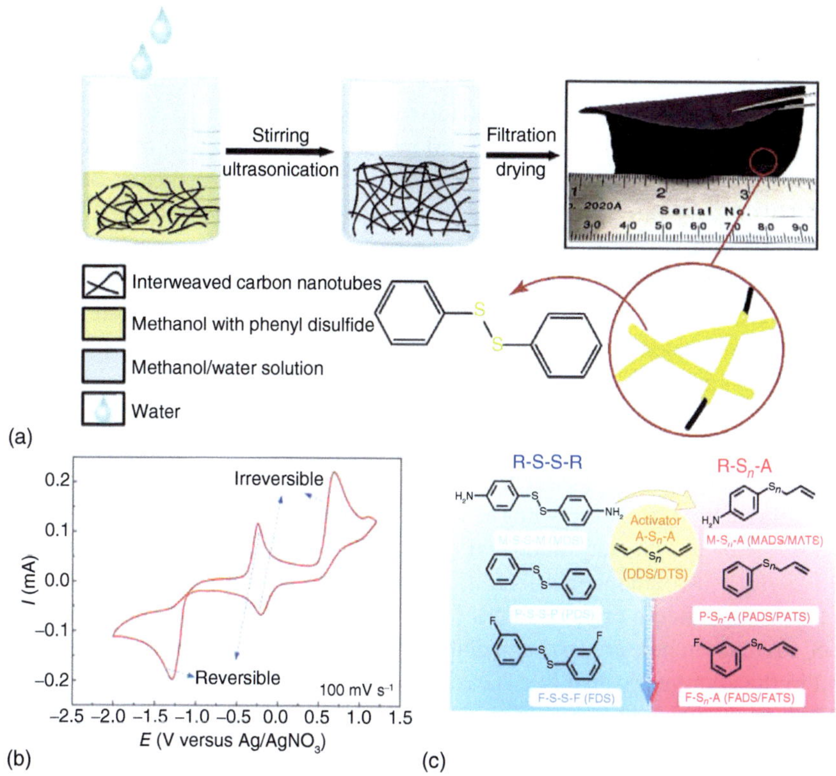

Figure 4.3 (a) Schematic illustration of the DPDS@CNT cathode preparation. Source: Adapted with permission from Bhargav et al. [15]/Royal Society of Chemistry. (b) Cyclic voltammogram (CV) of 5 mM TETD on the glassy carbon working electrode. Source: Reproduced with permission from Zhang et al. [16] John Wiley & Sons. (c) Schematic of asymmetric organosulfide synthesis activated by allyl group. Source: Adapted with permission from Weng et al. [9], Royal Society of Chemistry.

To avoid this problem, Bhargav et al. chose carbon nanotube (CNT) as the conductive carbon matrix for DPDS and prepared DPDS@CNT composite cathode by a "phase-extraction" technique (Figure 4.3) [15], which was achieved by exploiting the difference in solubility of DPDS in water and ethanol. DPDS@CNT is directly used as a cathode for rechargeable lithium batteries without additional current collectors or binders, such as a cathode, which can reduce the mass of inactive materials and achieve a high overall energy density. DPDS@CNT cathode was able to meet the full infiltration of electrolytes and thus improve the utilization of active materials. The Li/DPDS cell achieves a theoretical specific capacity of 89% and a Coulomb efficiency of 99% at 1 C rate. This work demonstrates the great significance of optimizing organosulfur-based cathodes for the improvement of electrochemical performance.

In addition to showing considerable performance in lithium coin cells, organosulfides have also demonstrated certain feasibility in flow batteries with great application prospects. Zhang et al. applied low-cost TETD and TMTD as models to explore the electrochemical performance of organosulfides in flow batteries [16].

Comparing TMTD and TETD, it is found that TETD has a larger molecular weight and polarization, which facilitates the regeneration of disulfide bonds and achieves fast reaction kinetics. Therefore, TETD is chosen as the object in the subsequent exploration. The reversible cleavage and formation of S—S bonds led to the excellent electrochemical reversibility of Li/TETD flow cell. Besides, electron transport was also reflected on N heteroatoms, which belongs to irreversible redox behavior (Figure 4.3b), and this reaction mechanism is synergistically confirmed by CV, ultraviolet and visible spectrum (UV), and nuclear magnetic resonance (NMR). The Li/TETD flow cell can cycle stably for 100 cycles at 10 mA cm^{-2} and retain a capacity retention rate of 94%. The successful application of organodisulfide in flow batteries shows that organosulfides have great development potential in large-scale energy storage systems.

Conventional organodisulfides tend to exhibit slow kinetics and low achievable capacity. Diallyl disulfide (DDS, A–S–S–A) easily forms allyl radical ($A^{\cdot}$) due to low C—S bond energy, but its electrochemical activity is poor due to self-decomposition and recombination reactions. Weng et al. used a pre-synthesis method to make allyl replace one of the organic functional groups (R) in symmetric organodisulfides (R–S–S–R, R = aminophenyl [MDS], phenyl [DPDS], fluorophenyl [FDS]) and then form asymmetric organosulfides ($A{-}S_n{-}R$) [9], and the allyl organosulfides ($A{-}S_n{-}A$) chosen here are DDS and diallyl trisulfide (DTS) (Figure 4.3c). The A—S bond in the asymmetric disulfide is easily broken to form $A^{\cdot}$• under electrochemical reduction, which is favorable to achieve fast kinetic properties. Subsequently, R–S–$S^{\cdot}$ is further reduced to lithium sulfide (Li_2S) and R-SLi, and such a reaction mechanism drives the A–S–S–A to achieve the $3e^-$ reaction and thus exhibits a high specific capacity. Among them, FDS+DDS exhibits a high specific capacity of 202 Ah l^{-1}. In addition, the Li/FDS+DDS flow cell shows a higher average discharge voltage (2.13 V), lower voltage polarization, and more stable cycle performance compared to the precursors (FDS and DDS).

4.3 Organopolysulfides

4.3.1 Basic Organopolysulfides

Although organodisulfides exhibit stable cycling performance, their low energy density makes them difficult to gain a foothold in commercial energy storage technologies. Based on the structural tunability of organosulfur, researchers have developed many strategies to modulate its molecular structure to optimize its electrochemical performance, including the introduction of more active sites.

As early as 2003, Inamasu et al. demonstrated the superior specific capacity of organopolysulfides to disulfides, naptho[1,8-*cd*:4,5-*c′d′*]bis[1,2]dithiol and naphtho [1,8-*cd*][1,2]dithiol has a similar structure [5], but the former has one more S—S bond than the latter, which nearly doubles its specific capacity (Figure 4.4). Although such naphthalene ring molecules exhibit a certain electrochemical reversibility and high capacity, their cyclic stability is still lacking. Therefore, the design and

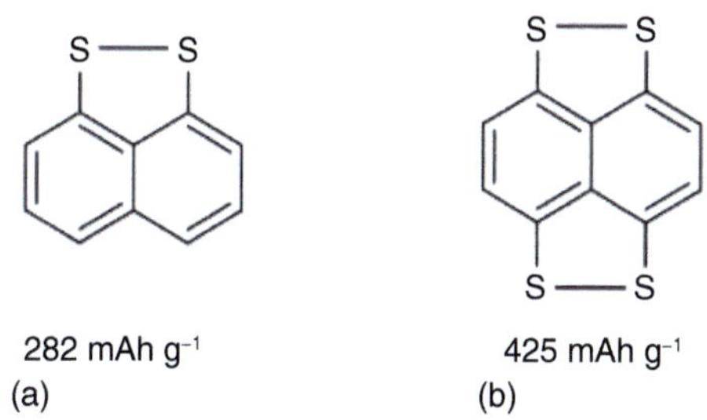

Figure 4.4 Structural formula and theoretical capacities of (a) naphtho[1,8-*cd*][1,2]dithiol and (b) naptho[1,8-*cd*:4,5-*c′d′*]bis[1,2]dithiol.

optimization of organosulfur molecules also need to consider the unity of molecular structure and electrochemical performance.

Based on the design perspective of the high capacity and stable structure of organosulfur, Hu et al. synthesized tetrathiotetracene (TTT) and hexathiapentacene (HTP) with multiple reaction sites using a solvent-free zone-melting chemical-vapor-transport (ZM-CVT) [17]. The synthesized TTT and HTP were successfully verified by matrix-assisted laser-desorption ionization time-of-flight (MALDI-TOF) mass spectra. HTP and TTT can transfer 4 and 6 electrons, respectively, exhibiting high theoretical specific capacities (TTT: 281 mAh g^{-1}, HTP: 317 mAh g^{-1}). The crystal structure of HTP exhibits closely spaced channel-like features, which are favorable for the transport of lithium ions. In addition, the density functional theory (DFT) results show that the bond energy of the trisulfide bond in HTP is lower than that of the disulfide bond in TTT, which results in a lower reduction potential (E_{red}) of HTP in lithium batteries (Figure 4.5a). The Li/HTP cell can release a specific capacity of 265 mAh g^{-1} at 0.1 C, and its voltage profile shows good consistency as the cycle continues, reflecting good electrochemical reversibility (81% capacity retention after 60 cycles). In contrast, the Li/TTT cell released 257 mAh g^{-1} in the first cycle, which rapidly decayed to 51 mAh g^{-1} after 30 cycles. DFT revealed the lithiation mechanism of HTP and TTT, where lithium ions tend to react with sulfur atoms on the same side to form Li· · ·S bonds and then

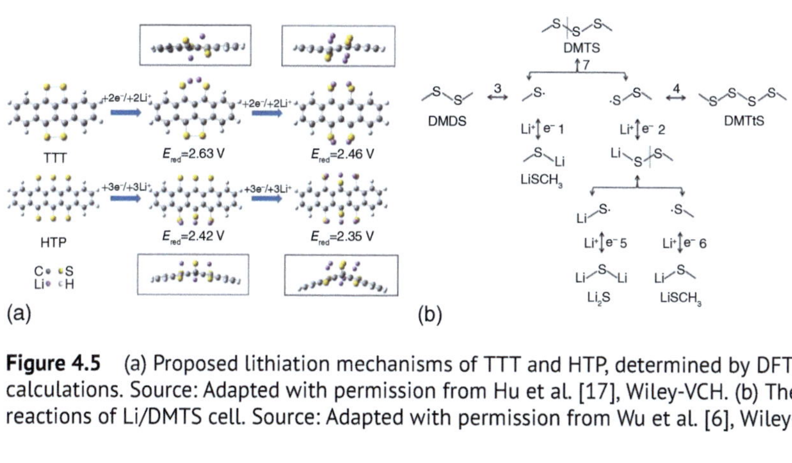

Figure 4.5 (a) Proposed lithiation mechanisms of TTT and HTP, determined by DFT calculations. Source: Adapted with permission from Hu et al. [17], Wiley-VCH. (b) The redox reactions of Li/DMTS cell. Source: Adapted with permission from Wu et al. [6], Wiley-VCH.

react with sulfur groups on the other side, where lithiation of Li/HTP cell follows a two-step and three-electron reaction mechanism.

Organotrisulfides (RSSSR) have three sulfur atoms as active sites, compared to organodisulfide, one more sulfur atom can not only increase the theoretical capacity but also reduce the dissociation energy of S—S bonds. RSSSR as cathodes and matched with suitable anodes are expected to realize high-energy density organosulfur-based batteries. Wu et al. chose dimethyl trisulfide (DMTS) as the research model to explore the reaction mechanism and electrochemical behavior of RSSSR in rechargeable lithium batteries [6]. The dissociation energy of the S—S bonds in DMTS is smaller than that of DMDS, which means that the S—S bonds in DMTS are easier to break and the reaction kinetics are faster. In the process of lithiation, DMTS first undergoes homolytic cleavage to form $^{\cdot}SCH_3$ and $^{\cdot}SSCH_3$ (Figure 4.5b). Most of these radicals are lithiated into $LiSCH_3$ and $LiSSCH_3$, but some of them recombine to form DMDS and dimethyl tetrasulfide (DMTtS), which are rapidly reduced to form $LiSCH_3$ and $LiSSCH_3$. After that, $LiSSCH_3$ continues to be lithiated, in which the S—S bond is broken, and Li_2S and CH_3SLi are generated. This reaction mechanism was synergistically confirmed by X-ray diffraction (XRD), X-ray photoelectron spectroscopy (XPS), and gas chromatography-mass spectrometry (GC-MS). Based on the $4e^-$ reaction of DMTS, its theoretical specific capacity can be calculated to be 849 mAh g^{-1}. Li/DMTS cell exhibits a specific capacity as high as 720 mAh g^{-1} in the first cycle at 0.1 C, and it can retain 82% of its initial capacity after 50 cycles with a Coulombic efficiency of 98%.

DMTS may cause the instability of batteries due to its strong volatility. The structure of the organosulfur molecule is tunable, which allows to select different R groups in the RSSSR to realize the performance regulation of the organosulfur molecule. Recently, Wu et al. reported a novel organosulfur cathode (DPTS), which is a yellow liquid. The synthesis of DPTS catholyte relies on the coupling reaction of DPDS and sulfur (Figure 4.6a) [18]. Compared with DMTS, the phenyl group of DPTS is electron-absorbing group, which contributes to higher discharge voltage and better stability in the batteries. Li/DPTS cell exhibits a lithiation mechanism similar to that of Li/DMTS cell, where one molecule of DPTS can combine 4 Li^+ to

(a) (b) (c)

Figure 4.6 (a) Schematic illustration of the synthesis of diphenyl trisulfide (DPTS) in liquid electrolyte at elevated temperature (i.e. 70 °C). Source: Reproduced with permission Wu et al. [18]/American Chemical Society. (b) Digital photos of the mixture of DPTtS and electrolytes. (c) Schematic illustration of the interface. Source: Adapted with permission from Lv et al. [19]/John Wiley & Sons, Inc.

produce two C_6H_5SLi and one Li_2S. In the discharge process, DPTS is first lithiated to generate PhSSLi ($C_6H_5SSSC_6H_5 + 4Li^+ + 4e^- \leftrightarrow 2C_6H_5SLi + Li_2S$), which corresponds to the discharge plateau of 2.4 V. It is worth noting that the discharge voltage plateau of DPTS (2.4 V) is higher than that of DMTS (2.1 V), which is because two phenyl groups in DPTS are electron-withdrawing groups, which reduces the energy level of the lowest unoccupied molecular orbital (LUMO). After that, PhSSLi was completely lithiated into PhSLi and Li_2S ($C_6H_5SSLi + 2Li^+ + 2e^- \leftrightarrow C_6H_5SLi + Li_2S$), resulting in the voltage plateau of 2.0 V in Li/DPTS cell. This cell with a loading of 3.9 mg cm^{-2} exhibits a specific energy of up to 751 Wh kg^{-1} at 0.5 C and can maintain an initial specific energy of 91% after 50 cycles. Even at DMTS loadings as high as 7.7 mg cm^{-2}, the Li/DPTS cell can release a specific energy of 686 Wh kg^{-1}.

DPTtS is a liquid organosulfide with four sulfur atoms that can undergo six electron transfer reactions. DPTtS can achieve a theoretical specific capacity as high as 570 mAh g^{-1} in lithium batteries. Lv et al. combined DPTtS with localized high-concentration electrolyte (LHCE) to construct a new type of liquid–liquid interface between cathode and electrolyte (Figure 4.6b,c) [19], which is more conducive to rapid interface kinetic transformation than the solid–liquid interface formed by inorganic sulfur and electrolyte. And DPTtS is insoluble in LHCE, which can effectively inhibit the shuttle effect of active substances. PhSLi will be produced during the discharge process of Li/DPTtS cell, and a small amount of PhSLi builds a bridge of ionic transport at the interface between DPTtS and electrolyte, thus promoting ionic conductivity and improving reaction kinetics. When Li/DPTtS cell is matched with TTFE80 (the 1,2-dimethoxyethane (DME)/1,1,2,2-tetrafluoroethyl-2,2,2-tri-fluoroethyl ether (TTFE) volume ratio is 2 : 8) electrolyte, it can form LiF-rich solid-electrolyte interphase (SEI) at lithium anode, which can protect the anode well and realize the stable long cycle of the batteries. The Li/DPTtS cell based on TTFE80 electrolyte shows excellent rate performance and cycle stability (the capacity retention rate was 64.9% after 750 cycles). The capacity of Li/DPTtS pouch cell with DPTtS loading of 3.1 g can reach 997.3 mAh, and the capacity retention rate is 82.1% after 50 cycles at a current of 100 mA.

Organopolysulfides exhibit high theoretical specific capacities, but due to the long sulfur chains often involved in the formation of Li_2S, this multiphase transformation leads to a decrease in electrochemical stability and an increase in voltage polarization. A typical case is lenthionine, which has a theoretical specific capacity as high as 853.6 mAh g^{-1} due to multiple sulfur atoms as active sites. However, during lithiation, the S—S—S bond of lenthionine is broken to form Li_2S (Figure 4.7a) [20], which will be converted into polysulfides during charging. The shuttle of polysulfides will cause continuous loss of active materials, which is reflected in the continuous attenuation of capacity. Gao et al. introduced DPDS as a receptor into Li/lenthionine cell. During the charging process, PhS$^{\cdot}$ was able to capture the radical fragments $^{\cdot}$S$^{\cdot}$ and $^{\cdot}$S–C–S$^{\cdot}$ of lenthionine to form phenyl polysulfides (Figure 4.7b). Phenyl polysulfide replaces lenthionine as the main reactant during the subsequent battery cycling process. The Li/lenthionine cell with DPDS can cycle for more than 500 cycles and still retain 81.1% of the initial capacity. Under

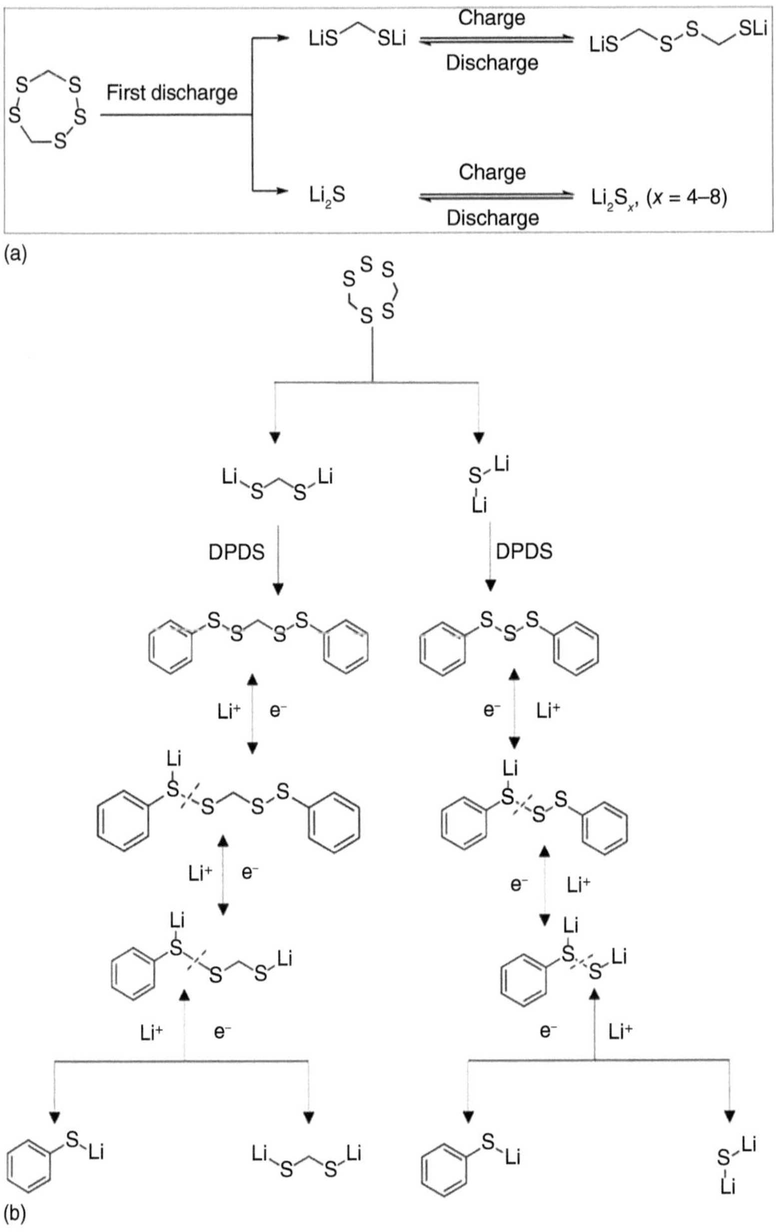

Figure 4.7 The redox reactions of (a) lenthionine and (b) lenthionine with DPDS cathode in a lithium cell. Source: Adapted with permission from Gao et al. [20], Wiley-VCH.

the same conditions, the Li/lenthionine cell only has 35.6% of the initial capacity remaining. The introduction of suitable radical receptors plays an important role in regulating the redox reaction process of organosulfur, which is another method to improve the electrochemical performance of organosulfur-based batteries.

4.3.2 Thiol-derived Organopolysulfides

Although organopolysulfides have high theoretical specific capacities, they suffer from many problems, such as structural instability, caused by long sulfur chains. As a member of small-molecule organosulfur family, organothiol compounds can undergo energy exchange according to a reversible reaction process (2SH ↔ S–S), and their theoretical specific energy far exceeds that of conventional battery materials. In recent years, based on the structural tunability of organic compounds, many organothiol derivatives have been designed, synthesized, and studied. Most of the organothiol derivatives have both abundant active sites and unique structures, which can not only ensure the realization of high energy density but also consider the cycling stability of batteries.

The method of increasing reactive sites by introducing additional sulfur atoms is applied to organothiol molecules. Bhargav et al. synthesized a series of phenyl polysulfides ($C_6H_5S_xC_6H_5$, $4 \leq x \leq 6$) using benzenethiol (PhSH) with sulfur powder as precursors in a solvent-free and catalyst-free method (Figure 4.8a) [7]. In which PhSH and sulfur are mixed and stirred in appropriate molar ratios, the high acidity of –SH induces ring-opening reactions of cyclo-octasulfur and thus generates

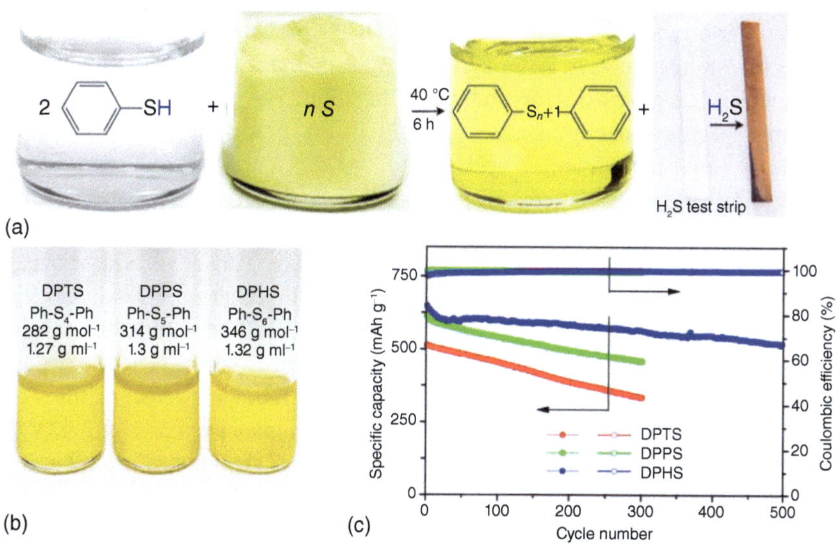

Figure 4.8 (a) Equation along with a visual representation of the phenyl polysulfide synthesis process. A lead acetate-based test strip was used to confirm the H_2S evolution. (b) Optical image of the liquid polysulfides along with their properties. (c) Long-term cycling performance of the polysulfides at 1 C. Source: Adapted with permission from Bhargav et al. [7]/American Chemical Society.

phenyl polysulfanes (PhS_xH), while the unstable high-order phenyl polysulfanes undergo intermolecular condensation reactions and release H_2S, prompting the formation of phenyl polysulfides (Figure 4.8b). The longer sulfur chain contributes to the realization of higher theoretical specific capacity and energy density in lithium batteries, among which diphenyl hexasulfide (DPHS) can release a specific capacity of 650 $mAh\,g^{-1}$ in the first cycle at 1 C and retain 80% of the specific capacity after more than 500 cycles (Figure 4.8c). Moreover, Li/DPHS cells can achieve a specific energy of up to 1302 $Wh\,kg^{-1}$ at an electrolyte/DPHS ratio of 3 $\mu l\,mg^{-1}$, which promotes the development of high-energy density organosulfur-based batteries.

To explore the effects of different thiol substitution positions on the electrochemical behavior and performance of organosulfur-based batteries, Li et al. selected 1,2-, 1,3-, and 1,4-benzenedithiols (BDTs) as a model in a series of electrochemical characterizations in lithium battery systems [21]. The three BDTs were lithiated into lithium benzenedithiolates (LBDTs), and the experimental results showed that the order of their electrochemical performance was ortho > para > meta. Theoretical calculations show that the –SLi groups in 1,2- and 1,4-LBDTs are more like electron-donating groups, which can activate the benzene ring, and the resulting conjugation effect can accelerate the lithiation reaction. 1,2-, 1,3-, and 1,4-LBDT generate free radicals during charging, which then recombine to form cyclic dimers, trimers, and tetramers, respectively (Figure 4.9a–c). Among them, the cyclic dimer of 1,2-LBDT exhibits the lowest polarization potential and fast kinetics, and the Li/1,2-LBDT cell can release a specific capacity of 340 $mAh\,g^{-1}$ in the first cycle at 0.5 C and maintain 84.1% of the initial capacity after 100 cycles.

Subsequently, Li et al. used 1,3-BDT and sulfur powder as precursors to synthesize three intertermolecular cyclic polysulfides (ICPSs) (Figure 4.10a) [22]. Among them, 2,3,4,6,7,8-hexathia-1,5(1,3)-dibenzenacyclooctaphane (HDBCO) is the main product, which has a crown structure similar to sulfur. The Li/ICPSs cell exhibits a discharge-specific capacity of 625.0 $mAh\,g^{-1}$ in the first cycle and can retain 445.8 $mAh\,g^{-1}$ after more than 100 cycles.

Organothiols are capable of self-polymerization under appropriate conditions to produce oligomers with low solubility, contributing to the realization of organosulfur-based batteries with long-cycle stability. Pan et al. synthesized oligomeric cyclic organodisulfides (macrocyclic 2,5-difluorobenzene-1,4-disulfide, F-MCDS) in dimethyl sulfoxide by a one-step oxidation method with 2,5-difluorobenzene-1,4-dithiol (2F-BDT) as a precursor [23], which is white–yellow powder (Figure 4.10b). Benefiting from the strong electron-withdrawing effect of the –F group, the Li/MCDS cell exhibits high discharge voltage plateaus (2.45 and 2.2 V). F-MCDS includes dimer, trimer, and tetramer of 2F-BDT, which are insoluble in ether electrolytes, enabling F-MCDS to achieve a stable cycle of more than 1000 cycles in lithium batteries. Subsequently, Ren et al. used a similar oxidation method to synthesize 2,4,5,7,9,10-hexathia-1,3,6,8 (1,4)-tetrabenzenacyclodecaphane (HTBCO) with low solubility in ether electrolytes using TBBT as the precursor (Figure 4.10c) [24]. When HTBCO is matched with a lithiated carbon paper (Li-CP) anode to assemble a full cell operation, the discharge products of HTBCO are dissolved in the electrolyte during the initial cycling process to participate

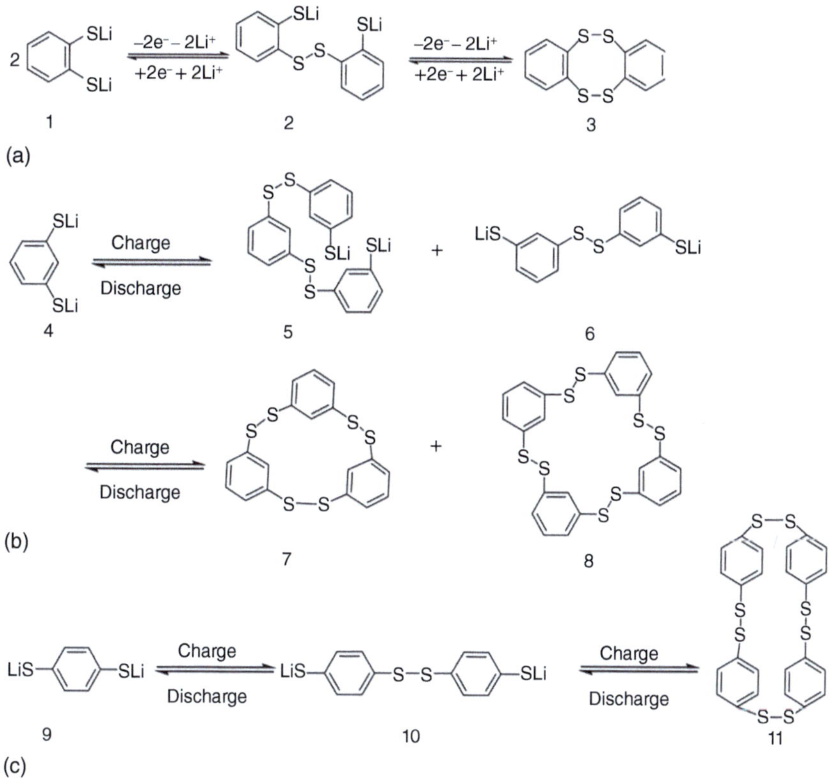

Figure 4.9 Redox reactions of (a) 1,2-LBDT, (b) 1,3-LBDT, and (c) 1,4-LBDT during battery cycling. Source: Adapted with permission from Li et al. [21], Wiley-VCH.

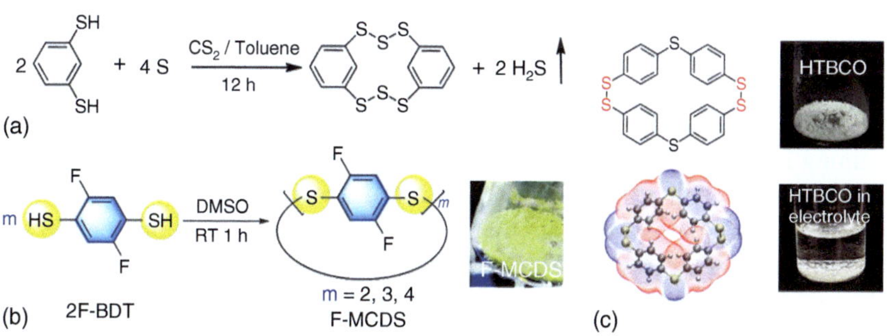

Figure 4.10 (a) Synthesis of HDBCO with 1,3-benzenedithiol and elemental sulfur precursors. Source: Adapted with permission from Li et al. [22], Royal Society of Chemistry. (b) The synthesis process of F-MCDS from 2F-BDT in DMSO. Source: Adapted with permission from Pan et al. [23]/Royal Society of Chemistry. (c) The chemical structure and electrostatic potential (ESP) maps of the HTBCO (left), the photographs of raw HTBCO and HTBCO in the electrolyte for 24 hours in vials (right). Source: Adapted with permission from Ren et al. [24]/Royal Society of Chemistry.

in the formation of SEI on the Li-CP surface. Experimental characterization shows that organic/inorganic composite SEI will be formed on the Li-CP surface, which well suppresses the appearance of solvent co-intercalation phenomenon, prompting HTBCO to demonstrate good compatibility with Li-CP anode, and the Li-CP/HTBCO cell achieves stable long-cycle performance. The Li-CP/HTBCO full cell achieves stable long-cycle performance. Li-CP as the anode can improve the safety of the battery compared to lithium anode, and the excellent electrochemical performance of Li-CP/HTBCO full cell shows the promising development of organosulfur batteries.

Due to the abundance and low price of sodium (Na), Na batteries have become one of the next-generation promising energy storage technologies. However, due to the severe shuttle effect and the strong activity of metallic Na, the development and application of Na batteries have been greatly hindered. Tang et al. selected 4,4′-thiobisbenzenethiol (TBBT), 1,4-BDT, and DPDS as models to explore the effect of molecular size of organosulfur on the electrochemical performance of Na-organosulfur battery [11]. TBBT and 1,4-BDT will form dimer and trimer during charging, respectively. Theoretical calculation results show that the order of *van der Waals* sizes of these three molecules during charging is TBBT (13.16 Å) > 1,4-BDT (11.46 Å) > DPDS (11.18 Å). The Na/TBBT cell matched with Nafion membrane can stably cycle for 300 cycles (≈2420 hours) and still maintain 77% of theoretical specific capacity. In contrast, 1,4-BDT and DPDS with smaller molecular sizes show poor cycle stability, which is caused by the shuttle effect of organosulfur. Due to the large molecular size of TBBT, it can be well limited by Nafion membrane, and the shuttle effect in Na/TBBT cell is greatly suppressed. This work highlights the importance of molecular structure and membrane size effects for achieving stable battery cycling, providing design inspiration for optimizing battery systems (Figure 4.11a).

Based on the feasibility of organothiols in sodium battery applications, Li et al. screened several organothiols with different conjugated structures by DFT simulations, namely 1,2-BDT, 2,3,6,7-tetramercapto naphthalene (NTT) and 2,3,6,7,10,11-triphenylenehexathiol (THT) [25], where 1,2-BDT, NTT, and THT correspond to 6, 10, and 18π electrons, respectively. The DFT results show that the THT containing 18π electrons has the smallest energy gap between the highest occupied molecular orbital and LUMO (HOMO–LUMO), which can ensure fast charge transport. In addition, the thiol group in THT has the lowest binding energy to ether solvent, which can reduce the solubility of organosulfur molecules in the electrolyte. In addition, THT monomers easily polymerize into oligomers, which makes the THT-based redox mediator easily confined by the nanochannels of the Nafion membrane, thus well suppressing the shuttle effect (Figure 4.11b). Large conjugated organic compounds can exhibit attractive electrochemical behavior, which highlights the importance of conjugated structures in molecular design.

Differences in the number of thiol groups may have an effect on the polymerization products of organothiols, which in turn affects their electrochemical properties. Jiao et al. conducted a series of explorations and characterizations in Na batteries using TBBT, 1,3-BDT, and 1,3,5-benzenetrithiol (BTT) with different numbers of

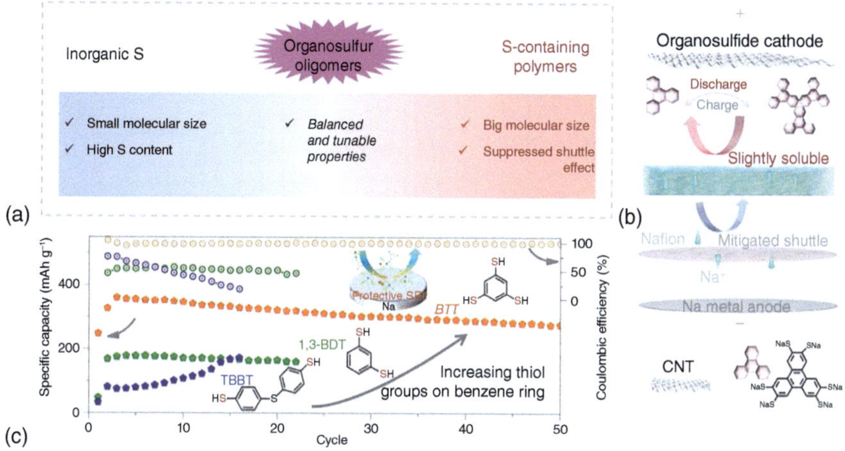

Figure 4.11 (a) The comparison of the organosulfur compounds with inorganic S and S-containing polymers as cathodes of rechargeable batteries. Source: Reproduced with permission from Tang et al. [11], John Wiley & Sons. (b) Schematic illustration of the Na-THT cell. Source: Reproduced with permission from Li et al. [25], Royal Society of Chemistry. (c) Cycling performance of the Na/TBBT, Na/1,3-BDT, and Na/BTT cells. Source: Reproduced with permission from Jiao et al. [26], Elsevier.

thiol groups as research models and found that the electrochemical performance of organothiol compounds in batteries is not only related to the size of the polymers they form under electrochemical conditions [26]. Theoretical calculations and experimental results show that the thiol groups of BTT tend to combine with Na^+ in the electrolyte and participate in the solvation sheath. Also, BTT and concentrated $NaPF_6$ salts can synergistically participate in the formation of SEI, and the uniform and dense SEI layer well protects the Na anode from repetitive exposure of fresh Na to the electrolyte in long-cycle tests, and the Na/BTT cell can stably cycle for more than 200 cycles and maintain an average of 98.9% (Figure 4.11c). In contrast, the cycling performance of Na/TBBT and Na/1,3-BDT cells was slightly inferior. To prove that the excellent performance of the Na/BTT cell is related to the involvement of BTT in the formation of SEI, the author removed the Na anode from the cycled Na/BTT cell and installed it into the Na/TBBT cell for electrochemical testing, and the cycling performance of the Na/TBBT cell was significantly improved, which further confirmed the speculation that BTT is involved in the formation of SEI to stabilize the cell cycling. This work provides new design and research ideas for the construction of high-performance Na-organosulfur batteries.

Polymers can effectively suppress dissolution and shuttling problems due to their low solubility in organic electrolyte. Among them, organothiol-derived polymers are easy to synthesize, possess high theoretical specific capacity and stable structure, making them become promising cathode material candidates. Bhargav et al. designed and synthesized polyethylene hexasulfide (PEHS) with a sulfur content of up to 87% (Figure 4.12a) [27]. Based on a 10-electron redox reaction, Li/PEHS cell has a theoretical specific energy of up to 2600 Wh kg^{-1}. Although PEHS has

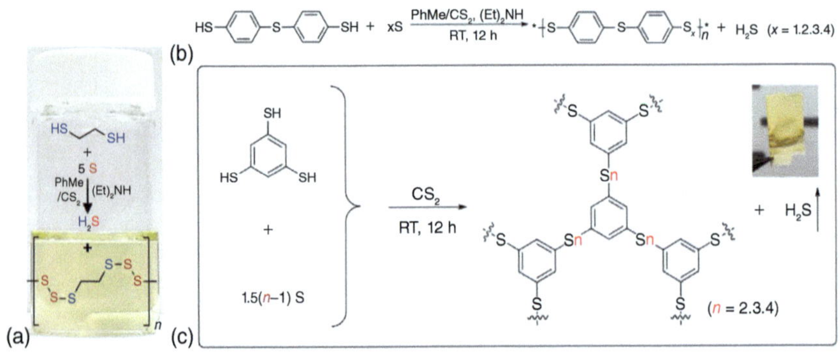

Figure 4.12 (a) Optical image of the polyethylene hexasulfide polymer in 1:1 v/v mixture of CS_2/toluene. Source: Adapted with permission from Bhargav et al. [27]/American Chemical Society. (b) Condensation polymerization between TBBT and sulfur. Source: Adapted with permission from Sang et al. [28]/Royal Society of Chemistry. (c) Scheme of the synthesis of organosulfur polymers (BTTP) with different sulfur contents. Source: Adapted with permission from Sang et al. [29], Elsevier.

a high content of sulfur, it does not contain long polysulfide chains, which can reduce the formation of high-order polysulfides during cycling, thereby improving cycling stability. The Li/PEHS cell releases a specific capacity of 774 mAh g^{-1} in the first cycle and remains a capacity retention rate of 71% after 350 cycles. Subsequently, Sang et al. adopted a similar polymerization method, successively using TBBT (Figure 4.12b) and BTT (Figure 4.12c) as reaction substrates to react with sulfur powder to synthesize various polymers [28, 29]. All of them exhibit superior electrochemical reversibility and stability in lithium batteries. This again confirms the feasibility and general applicability of this type of polymerization, providing a practical route to develop more sulfur-based polymers.

4.4 Heteroatom-Containing Organosulfides

The competitive advantage of organosulfides as electrode materials is their designability. Functional groups and heteroatoms, as components in organic molecules, have a significant impact on the redox properties of organosulfides. Therefore, the regulation of the electrochemical behavior of organosulfides is expected to be achieved by introducing suitable functional groups and heteroatom groups.

4.4.1 Organosulfides-Containing N-Heterocycles

As early as 1991, Naoi et al. introduced a series of nitrogen-containing heterocyclic organosulfur compounds (Figure 4.13) [4]. Although they showed some electrochemical activity in lithium batteries, the limited cycle life did not allow them to further expand their development. In addition, the role of nitrogen in organosulfur in electrochemical reactions has not been clearly described.

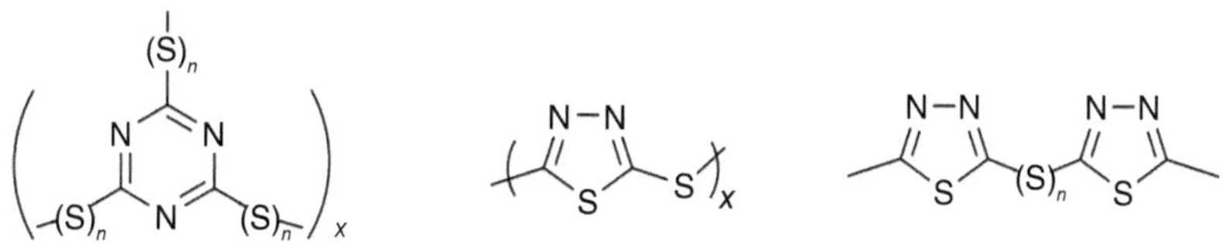

Figure 4.13 The molecular structure of tricyanuric acid (TTCA), polymers of 2,5-dimercapto-1,3,4-thiadiazole (DMcT), dimers of disulfide ($n = 2$), trisulfide ($n = 3$), tetrasulfide ($n = 4$) of 5-methyl-1,3,4-thiadiazole-2-thiol (MTT).

Bhargav et al. chose low-priced dipentamethylenethiuram tetrasulfide (PMTT) as the research model to explore the unique redox behavior of this nitrogen-containing heterocyclic organosulfur in lithium batteries (Figure 4.14a) [30]. The CV of Li/PMTT cell shows three reversible redox peaks, corresponding to three charging and discharging plateaus (Figure 4.14b). During the initial lithiation stage, S—S bond cleavage in PMTT generates $C_5H_{10}NCSSLi$ (LiPMDTC **2**) and radical **3** (Figure 4.14c). LiPMDTC **2** did not undergo the cleavage of bond during further lithiation. Instead, due to the N-electrons flow from nitrogen to sulfur via planar delocalized π orbitals, LiPMDTC **2** can exist as an intermediate (Li_2PMDTC^+ **7**) with a positive nitrogen center in the presence of excess Li^+. Owing to the strong Coulomb interaction between Li_2PMDTC^+ **7** and polysulfides, the polysulfides can be retained on the cathode side, which reduces the occurrence of the shuttle effect and maintains the stable cycle of the batteries. In the subsequent lithiation process, the radical **3** reacts with Li^+ to generate **4**. Then, the S—S bond in **4** is broken, which promotes the formation of LiPMDTC **2** and lithium persulfide radical **5**, and the radical **5** is finally lithiated into Li_2S. Dipentamethylenethiuram hexasulfide (PMTH) has a similar reaction mechanism to PMTT, but it involves a longer sulfur chain, and polysulfide is easy to produce. Benefiting from the adsorption of polysulfide by Li_2PMDTC^+ **2**, both Li/PMTT and Li/PMTH cells showed excellent electrochemical performance, and their capacity retention rate was not less than 85% after 100 cycles. Li/PMTH cell can release specific energy of 1156 Wh kg^{-1} even when E/PMTH ratio is as low as 3 μl mg^{-1}.

Magnesium (Mg) is rich in resources and low cost, it can provide high volumetric capacity (3833 mAh cm^{-3}), and it is not easy to generate dendrites during the reaction process, which ensures the safety of the batteries. However, it is difficult to find suitable electrolytes and cathodes to work with Mg anodes, which limits the development of Mg batteries. Considering the failure mechanism of Mg–S batteries due to the severe shuttle effect of polysulfides, Callander et al. introduced PMTT as cathode into Mg battery systems [31]. Due to the special structural characteristics of nitrogen heterocycles, PMTT combines with Mg ions to generate magnesium pentamethylene dithiocarbamate ($Mg(PMDTC)_2$) and MgS_2 during the oxidation process. This process is similar to the reaction mechanism of Li/PMTT cells and is confirmed by DFT calculations (Figure 4.14d). The Mg/PMTT cell achieves a specific capacity as high as 295 mAh g^{-1} in the first cycle. Subsequently, the PMTT-derived sulfur/mesoporous carbon composite was synthesized by a mild heat treatment, which achieved a stable

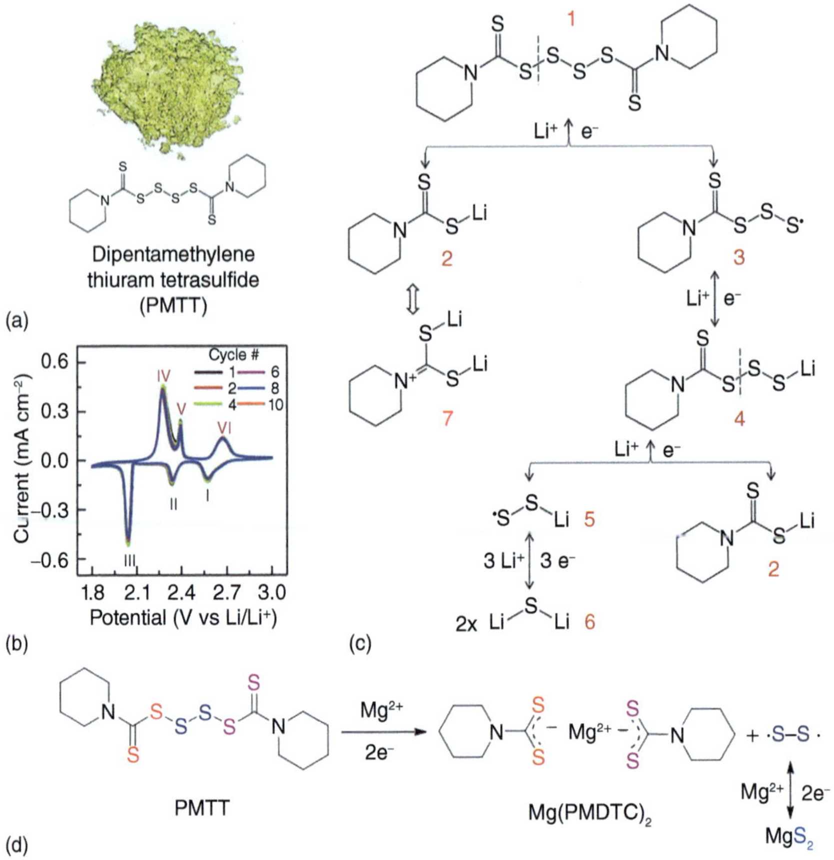

Figure 4.14 (a) Optical image of commercial PMTT powder along with its chemical structure; (b) CV of PMTT-CNT cathode performed at 0.05 mV s^{-1}; (c) Redox reactions of PMTT during battery cycling. Source: Adapted with permission from Bhargav et al. [30]/Royal Society of Chemistry. (d) Possible reaction mechanism of PMTT with Mg based on DFT calculations. Source: Adapted with permission from Kaland et al. [31], American Chemical Society.

cycle of more than 100 cycles in Mg battery (the capacity retention rate was 77% after 100 cycles). The successful application of PMTT and its derivative cathode provides an important reference for the selection of cathode materials for Mg batteries.

In order to further explore the electrochemical behavior induced by nitrogen heterocycles, Wang et al. selected DPDS and a series of pyridine disulfides as research models, including 2,2′-dipyridyl disulfide (2,2′-DpyDS), 4,4′-dipyridyl disulfide (4,4′-DpyDS), and 2,2′-dipyridyl disulfide-*N′N′*-dioxide (DpyDSDO) (Figure 4.15a) [32]. Although the electrochemical reaction sites of these four organosulfur compounds are all S—S bonds, they show different discharge voltage plateaus in lithium half-cells, and the discharge voltages of three nitrogen-containing organosulfur compounds are significantly higher than those of DPDS. Theoretical calculation results of LUMO and HOMO of three nitrogen-containing organosulfur compounds

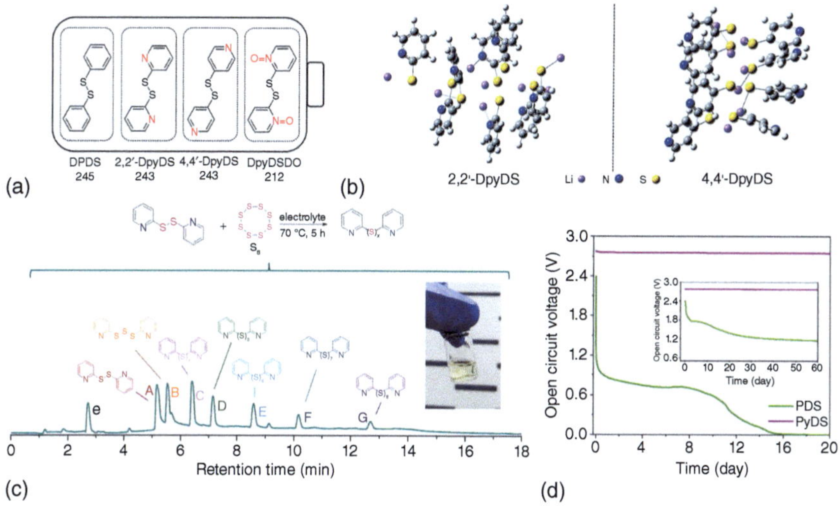

Figure 4.15 (a) Chemical structures and theoretical specific capacities of DPDS, 2,2′-DpyDS, 4,4′-DpyDS, and DpyDSDO. (b) Cluster structures of the discharge products of 2,2′-DpyDS and 4,4′-DpyDS. Source: Adapted with permission from Wang et al. [32], Royal Society of Chemistry. (c) The synthesis of a spectrum of Py_2S_x in electrolyte; the inset is a photograph of the prepared Py_2S_x catholyte. Source: Adapted with permission from Wang et al. [33]/John Wiley & Sons, Inc./CC BY 4.0. (d) Self-discharge of the Li/PyDS and Li/DPDS cells; the anode is Li-CP. Source: Adapted with permission from Fan et al. [34], American Chemical Society.

show that DpyDSDO with the strongest electron-withdrawing group has the lowest LUMO (−1.60 eV), which corresponds to the highest discharge voltage plateau (2.8 V) of Li/DpyDSDO cell.

The discharge product of 2,2′-DpyDS, lithium pyridine-2-thiolate, can form tight clusters in the form of N···Li···S bridge bonds (Figure 4.15b), thereby reducing its solubility and reducing the shuttle to ensure the long-cycle stability of the battery. The Li/2,2′-DpyDS cell has a capacity retention rate of 69% after 500 cycles. In contrast, Li/4,4′-DpyDS cell exhibits similar poor cycling performance to DPDS (the capacity retention rate was only 57% after 100 cycles), which was caused by the fact that the distance between N and S in 4,4′-DpyDS was far from being able to form clusters similar to lithium pyridine-2-thiolate. This work reveals the unique redox mechanism of nitrogen-containing heterocyclic organosulfur and points out that the improvement of discharge voltage can be achieved through functional group modulation.

Although pyridine disulfides exhibit unique electrochemical behavior and stable cycle performance, their low specific energy makes their application prospects very slim. Wang et al. synthesized pyridine polysulfide (Py_2S_x, $3 \leq x \leq 8$) with high theoretical energy density by one-pot synthesis method (Figure 4.15c) [33]. The pyridine polysulfide synthesized here is a mixture of several organosulfides with different numbers of sulfur atoms, which has been verified by ultrahigh-performance liquid chromatography quadrupole time-of-flight mass spectrometry (UPLC-QTOF-MS). Li/Py_2S_x cell exhibits three discharge plateaus, the first of which corresponds to the break of S—S bond connected to pyridine ring, which promotes the generation of lithium pyrithione and polysulfide radicals, and the next two platforms correspond to the transformation of polysulfide radicals into lithium polysulfide and lithium sulfide. In the process of lithiation, lithium pyrithione can form a variety of complexes with lithium polysulfide and lithium sulfide in the form of N···S···Li, so that the dissolution shuttle effect of polysulfide and lithium pyrithione can be suppressed, and the stable cycle performance of Py_2S_x in lithium batteries can be ensured. Li/Py_2S_x cell releases a specific capacity of 388.4 $mAh\,g^{-1}$ at 1 C and maintains a capacity retention rate of 70.5% after 1200 cycles. When the cut-off voltage is set to 2.0 V, the generation of Li_2S is avoided, and the Li/ Py_2S_x cell can be cycled for 1000 cycles at 5 C, and still maintain 62.5% of the initial capacity.

The different chemical reactivity of 2,2′-PyDS and DPDS with Li^0 is also one of the reasons for the difference in their electrochemical properties in lithium batteries [34]. The electronegativity of pyridine nitrogen in 2,2′-PyDS is higher than that of carbon in benzene in DPDS, nitrogen tends to absorb electrons from the pyridine ring to form an electron-deficient pyridine ring. The electron on Li^0 can temporarily stay at the top of the pyridine ring due to the electrostatic interaction with the electron-deficient pyridine ring; meanwhile, the S—S bond in 2,2′-PyDS remains unchanged, indicating that 2,2′-PyDS does not react with Li^0 in liquid electrolyte, which is consistent with the stable open circuit voltage of Li/2,2′-PyDS cell. On the contrary, the open-circuit voltage of Li/DPDS cell decays rapidly (Figure 4.15d). 2,2′-PyDS and DPDS were applied to the membrane-free redox flow battery, respectively, the Li/PyDS flow cell can be stably cycled for more than 230 cycles with a high Coulombic efficiency (>99%), while the Li/DPDS flow cell cannot work under the same conditions.

4.4.2 Organosulfides-Containing Selenium

Chalcogenide of selenium has an electrical conductivity of up to $1\times10^{-3}\ S\,m^{-1}$, and the high conductivity contributes to the fast kinetic properties of batteries. However, compared with sulfur, selenium has a relatively low theoretical specific capacity due to its large atomic mass. The generation of complexes of S and Se is expected to achieve both the high capacity of sulfur and the fast kinetic of selenium. The easily tunable structure of organosulfur allows us to synthesize the target substances by some simple strategies, which include the design and synthesis of phenyl selenosulfides.

To investigate the unique redox behavior of S—Se bonds, Guo et al. synthesized phenyl selenosulfide (PhS-SePh) with a single S—Se bond through a simple exchange reaction using DPDS and diphenyl diselenide (DPDSe) as precursors (Figure 4.16a) [8], which have compromised structural properties compared to the two raw materials. The discharge product of PhS–SePh is a mixture of PhSLi and PhSeLi, which has higher entropy and lower Gibbs free energy, resulting in Li/PhS-SePh cell exhibiting a high discharge voltage. Moreover, the mixture adheres to CNTs, together with the high electrical conductivity of selenium, which drives

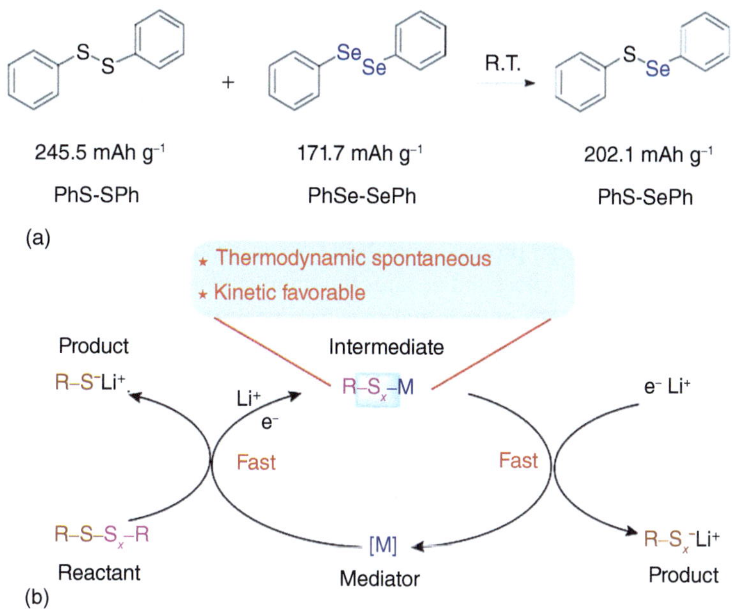

Figure 4.16 (a) Scheme of reaction of DPDS and DPDSe to form PhS-SePh along with their specific capacities. Source: Adapted with permission from Guo et al. [8], Royal Society of Chemistry. (b) Proposed catalytic cycle of Se radical in organosulfide cathodes to enhance reaction kinetics. Source: Adapted with permission from Zhang et al. [35], Elsevier.

the Li/PhS-SePh cell to achieve low voltage polarization and stable electrochemical performance.

Later, Xie et al. added a small amount of DPDSe as a mediator to the DPDS cathode and used this system as a research model to investigate the reaction mechanism of the organosulfur-based batteries [35]. By studying and analyzing the reaction thermodynamics and charging/discharging processes of Se-mediated DPDS cathodes (Figure 4.16b), it was found that the reaction pathway involves a catalytic cycle mediated by Se radicals, and it combines radical exchange and electrochemical redox, which can greatly improve the reaction kinetic of organosulfur-based batteries. This work deepens our knowledge and understanding of the reaction mechanism of organosulfides, and provides valuable research ideas that allow us to rationalize the design of organosulfur-based batteries with fast kinetics from a reaction mechanism perspective.

Although PhS-SePh exhibits interesting electrochemical behavior and improved cycling performance in lithium batteries, the limited specific capacity and energy density do not allow it to be competitive among many energy storage technologies. Cui et al. introduced sulfur into DPDSe to obtain phenyl selenosulfides with more reactive sites, of which phenyl selenosulfide (DPDSe-S) and phenyl selenodisulfide (DPDSe-S_2) have theoretical specific capacities up to 311.4 and 427.4 mAh g^{-1}, respectively [36]. Li/DPDSe-S and Li/DPDSe-S_2 cells both show three discharge plateaus. The first discharge plateau in DPDSe-S is around 2.4 V, which corresponds to the process that Li^+ and e^- attacks the sulfur atom in DPDSe-S to generate intermediate product **2** (Figure 4.17a). Then, the S—Se bond breaks, resulting in the formation of PhSeLi and PhSSeLi, which promote the appearance of the second discharge plateau. The final long plateau corresponds to the generation of Li_2S. The lithiation mechanism of DPDSe-S_2 is similar to that of DPDSe-S. Li^+ and e^- attack the sulfur atom first, and the intermediate **2′** is generated here (Figure 4.17b), followed by the formation of PhSeLi and Li_2S. DPDSe-S and DPDSe-S_2 both generate insoluble Li_2S in the discharge process, and the Li_2S can adsorb PhSeLi, which effectively inhibits the shuttle effect and ensures the stability of cycle performance. The Li/DPDSe-S cell retains 77% of its initial specific capacity (252 mAh g^{-1}) after 200 cycles, and the Li/DPDSe-S_2 cell can release 330 mAh g^{-1} of specific capacity in the first cycle, and the capacity decay rate is only 0.135% per cycle over 200 cycles.

The high electrical conductivity of selenium makes it very competitive as cathode, but the shuttle effect caused by polyselenide is severe, which makes it difficult

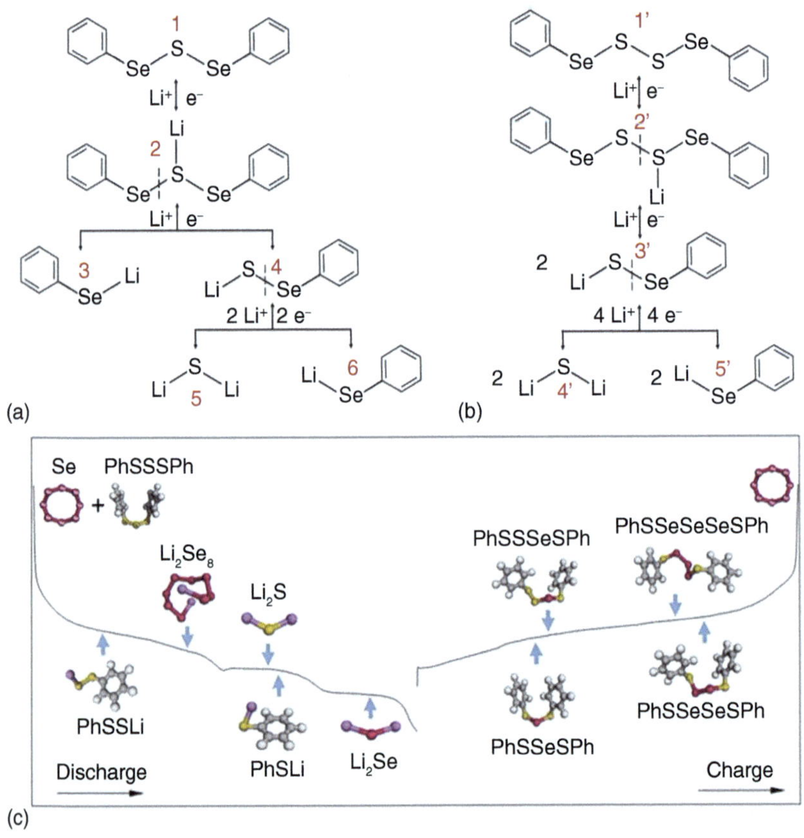

Figure 4.17 Redox reactions of (a) DPDSe-S and (b) DPDSe-S_2 in rechargeable lithium batteries. Source: Adapted with permission from Cui et al. [36], Wiley-VCH. (c) The redox reactions of the DPTS-Se cathode in a lithium cell. Source: Adapted with permission from Zhao et al. [37], Wiley-VCH.

to guarantee the stability of the batteries. The mixture of DPTS and selenium (Se) is expected to generate dynamic S—Se bonds to enhance the cycle stability of the rechargeable batteries, Zhao et al. prepared a DPTS-Se hybrid electrode by a two-step method and applied it in lithium battery [37]. Li/DPTS-Se cell shows different discharge curves from Li/Se and Li/DPTS cell. In order to explore the reaction mechanism, UPLC-QToF-MS was used to detect the charged products of Li/DPTS-Se cell, and several kinds of phenyl selenosulfides were detected, which included PhSSeSPh, PhSSeSSPh, PhSSeSeSPh, and PhSSeSeSeSPh. The formation of phenyl selenosulfides is related to the radical recombination of Li/DPTS-Se cells during oxidation, and the detailed reaction process is shown in Figure 4.17c. In the process of lithiation, firstly, the S—S bond in DPTS is broken to generate PhSS•, which is lithiated into PhSSLi, while Se is converted into Li_2Se_8. Then, PhSSLi was further lithiated into PhSLi and Li_2S, and Li_2Se_8 was lithiated into Li_2Se. Upon the oxidation process, PhSLi, Li_2S, and Li_2Se are delithiated to form PhS$^\cdot$, $^\cdot$Se$^\cdot$ and $^\cdot$S$^\cdot$, and the combination of these three radicals leads to the formation of phenyl selenosulfides. Phenyl selenosulfides have high electrochemical stability, which effectively inhibits the capacity attenuation caused by the dissolution shuttle of polyselenides.

Selenium nanowires in DPTS-Se provide high conductivity for composite cathode, and DPTS provides acceptor for selenium radical. This inference is confirmed by experimental characterizations and theoretical calculations, and it is also reflected in the electrochemical performance of batteries. The Li/Se cell releases an initial specific capacity of 820.7 mAh g^{-1} at 0.1 C and decays to 128.3 mAh g^{-1} after 200 cycles. While the Li/DPTS-Se cell achieves a theoretical specific capacity of 96.5% in the first cycle and maintains a capacity rate of 69.2% after 250 cycles, and its Coulomb efficiency is consistently over 99%. In the DPTS-Se hybrid cathode, the immobilization of selenium atoms by dynamic covalent bonding of S—Se ensures the cycling stability of Li/DPTS-Se cell, which further deepens our understanding of the unique properties of S—Se bonds.

The introduction of Se heteroatoms has an obvious effect on the improvement of the electrochemical performance of organosulfur-based batteries, mainly in terms of kinetic performance and cycling stability. The rational design of organosulfur cathodes with S—Se bonds is promising for achieving high energy density, fast kinetics, and stable cycling performance of organosulfur-based batteries.

4.4.3 Organosulfides-Containing Other Heteroatom

The functional groups of organic molecules can affect their physical and chemical properties, such as solubility, density, and electrical conductivity. The control of the electrochemical behavior of organic batteries is expected to be achieved by modulating the structure and functional groups of molecules. The diversity of structures and functional groups of organic molecules allows us to design and synthesize target molecules according to different needs, which provides a rich resource for the development of electrode materials.

Tellurium is the most metallic nonmetallic element with an electrical conductivity of up to $2 \times 10^2\ S\ m^{-1}$. Limited by the large molecular mass, tellurium-based cells mostly exhibit a low specific capacity. In recent years, a variety of Te_xS_y (x and y provide atomic ratios) cathodes have been developed and applied in battery systems, but the redox processes involved have not been clearly described. Chen et al. designed and synthesized phenyl tellurosulfide (PhS-TePh) and investigated the redox processes involved in S—Te bond in lithium battery systems (Figure 4.18a) [38]. Experimental characterization and theoretical calculations jointly reveal the unique electrochemical behavior of PhS-TePh, which involves the formation of multi-centered electron bonds during lithiation, which not only contributes to the higher voltage plateau exhibited by Li/PhS-TePh cell but also achieves stable long-cycle performance in dilute electrolyte systems (the concentration of Li salt < 0.2 M). This work is a guideline for understanding the reaction mechanism of Te_xS_y composite cathode and provides a feasible research reference for the mechanistic study of other battery systems.

DPTtS as cathode can undergo a six-electron reaction and have a high theoretical specific capacity. Guo et al. introduced DPTtS with electron-withdrawing groups (CF_3) and electron-donating groups (CH_3O) into lithium batteries as cathode and investigated the effect of different functional groups by comparing their electrochemical properties (Figure 4.18b) [39]. These three tetrasulfides all present two discharge plateau regions, where the medium voltage values in the high voltage discharge region (>2.1 V) show the order of CH_3ODPTtS

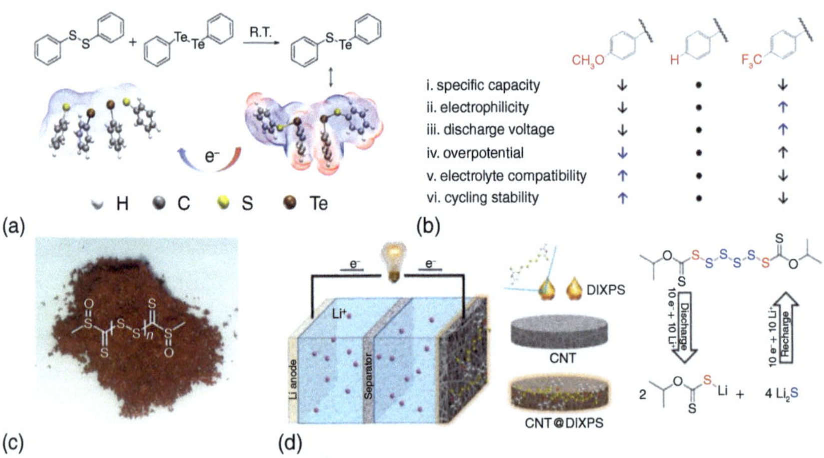

Figure 4.18 (a) A schematic diagram of the reaction of DPDS with PhTe-TePh to form PhS-TePh, ESP maps of PhS-TePh, color codes for atoms: gray, C; orange, Te; yellow, S; red, O; grey, C; and white, H. Source: Adapted with permission from Chen et al. [38], Royal Society of Chemistry. (b) Effects of functional groups on phenyl rings on the properties and electrochemical characteristics of bis(aryl) tetrasulfides; positive effects are highlighted in blue. Source: Adapted with permission from Guo et al. [39], Wiley-VCH. (c) The optical image of SOPS powder. Source: Adapted with permission from Zeng et al. [40]/Elsevier. (d) Schematic structure and reaction mechanism of the Li/DIXPS cell. Source: Adapted with permission from Bhargav et al. [41], Wiley-VCH.

(2.18 V) < DPTtS (2.22 V) < CF_3DPTtS (2.30 V), which is consistent with the order of electron-accepting ability of OCH_3, H, and CF_3 groups. Theoretical calculations showed that the LUMO of the three organosulfides is in the order of CF_3DPTtS < DPTtS < CH_3ODPTtS, which prompted Li/CF_3DPTtS cell to exhibit the highest reduction potential. In addition, the functional groups have an influence on the compatibility of the tetrasulfides with the electrolyte. Taking DPTtS as a reference, the addition of CH_3O groups increases while CF_3 decreases the compatibility of the tetrasulfide with the electrolyte, which leads to the achievement of low overpotential and stable cycling performance for Li/CH_3ODPTtS cell (retain 81% of the initial capacities over 100 cycles), while Li/CF_3DPTtS cell demonstrates large overpotential and poor cycling stability (retain only 43% of the initial capacities over 100 cycles). The modulation of the electrochemical performance of organosulfur-based batteries was achieved by introducing different functional groups, which provides an example and guidance for the development of more optimized organosulfur cathodes.

Organosulfur-containing oxygen heteroatoms are a promising class of electrode materials. Zeng et al. synthesized solid polysulfides (SOPS) with electron-withdrawing sulfoxide by a radical coupling method (Figure 4.18c) [40]. The long sulfur chain structure of SOPS contains multiple intermediate sulfur atoms with low dissociation energy, which contributes to fast kinetics. In addition, the presence of sulfoxide leads to the moderate polarity of SOPS, resulting in its low solubility in DOL/DME, which greatly reduces the losses caused by the dissolution and shuttling of active materials in the electrolyte. Moreover, due to the sulfoxide group, Li/SOPS cell tends to produce low-order lithium polysulfide ($Li_2S_{3.87}$) with low solubility during lithiation, which also reduces the occurrence of shuttle effect. It is due to these unique properties of SOPS that Li/SOPS cell achieves a specific capacity of 1166 $mAh\,g^{-1}$ at 750 $mA\,g^{-1}$ and releases 659 $mAh\,g^{-1}$ even at current densities up to 3000 $mA\,g^{-1}$.

Diisopropyl xanthogen polysulfide (DIXPS) is also a member of organosulfides-containing oxygen heteroatoms [41]. In order to explore the electrochemical performance of DIXPS in lithium batteries, DIXPS catholyte was injected into the free-standing CNT paper as the cathode, which was assembled into a lithium half-cell and characterized by a series of electrochemical methods (Figure 4.18d). DIXPS can undergo a transfer reaction of 10 e^-, and its theoretical specific capacity can reach 672 $mAh\,g^{-1}$. Li/diisopropyl xanthogen disulfide (DIXDS) cell has a single discharge plateau, which corresponds to the break of S—S bond. The discharge process of Li/DIXPS cell is similar to that of Li/DIXDS cell, but it has two discharge plateaus, 2.30 V and 2.08 V, respectively. The voltage plateau of 2.08 V corresponds to the generation of Li_2S, and the Li_2S generated here is nanoscale, which facilitates contact with lithium xanthate and thus promotes the charging reaction. Li/DIXPS cell involves the formation of polysulfide in the cycle process, which may cause a shuttle effect and lead to the loss of active materials. The mesomorphic form of lithium xanthate has the potential to combine with lithium polysulfide, which can promote the conversion of polysulfide and inhibit the shuttle effect. The Li/DIXPS cell can cycle over 1000 cycles at 4 C with only 0.026% capacity decay per cycle.

Inorganic phosphate sulfides have been shown to be electrode materials with good electrochemical activity, where phosphorus is an interesting heteroatom with a trivalent state, and the introduction of suitable phosphorus-containing groups in organosulfur is expected to achieve attractive electrochemical properties. Wang et al. synthesized 1,4-bis(diphenylphosphine)tetrasulfide (BDPPTS) by electrochemical oxidation (Figure 4.19a,b) [42]. The synthesis method was highly selective and yielded up to 96% of the target product. BDPPTS was used as the cathode and assembled into a lithium half-cell. CNTs are used as the cathode matrix in Li/BDPPTS cells, and due to the π–π stacking between the aromatic groups in BDPPTS and CNTs, the shuttle effect of soluble active materials is limited. The strong electron absorption of diphenyl phosphine prompted the Li/BDPPTS cell to exhibit a discharge voltage of up to 2.9 V, which contributed to high specific energy. Figure 4.19c shows the lithiation process of the Li/BDPPTS cell, Li^+ first attacked the S_α atom in BDPPTS, which made the Li/BDPPTS cell show a discharge plateau of 2.9 V. After that, both S_α—S_β bonds were broken, resulting in the generation of $^{\cdot}$S–S$^{\cdot}$ radicals, and the reduction reaction of $^{\cdot}$S–S$^{\cdot}$ radicals led to the appearance of multiple discharge plateaus. The last discharge plateau of 2.1 V corresponds to the generation of Li_2S. The Li/BDPPTS cell achieves a specific capacity of 311.6 mAh g^{-1} in the first cycle at 0.2 C, and it exhibits a capacity retention rate of 57% after 200 cycles. In order to suppress the loss of active materials caused by polysulfides, the cut-off voltage of the Li/BDPPTS cell is adjusted to 2.5 V. Currently, the Li/BDPPTS cell shows only one discharge plateau at 2.9 V, and the sulfur-like voltage profile disappears. After 500 cycles, the charge–discharge voltage profiles of Li/BDPPTS cells are consistent with those of the initial cycle, the overpotential is not obviously increased, and the capacity retention rate is up to 74.8%.

4.5 Organosulfur–Inorganic Hybrid Cathodes

Organosulfur is a highly competitive class of electrode materials because of its precise active site and easily tunable molecular structure, and its redox behavior can be well regulated by introducing suitable functional groups. However, most organosulfides suffer from low electrical conductivity, leading to their slow kinetics in rechargeable batteries. In addition, organosulfur-based batteries tend to generate soluble products (*i.e.* RSLi and LiPSs) during discharge, which may trigger dissolution and shuttle effects similar to those involved in Li–S batteries, resulting in loss of active materials and thus capacity degradation of the batteries.

Graphene is a class of carbon materials with good electrical conductivity and is often used as a cathode matrix. Hu et al. prepared polyacrylic acid-modified porous graphene oxide (GO-PAA) (Figure 4.20a) [43], and the mesopores in GO-PAA facilitate the diffusion of lithium ions and electrons. Subsequently, two-dimensional organic polysulfone (PS) with unique structure was synthesized on GO-PAA by direct coupling reaction, and the resulting composite cathode (pGPS) has both high sulfur content and stable chemical structure. The pGPS can be used as a free-standing electrode in lithium batteries, which allows the rapid transmission

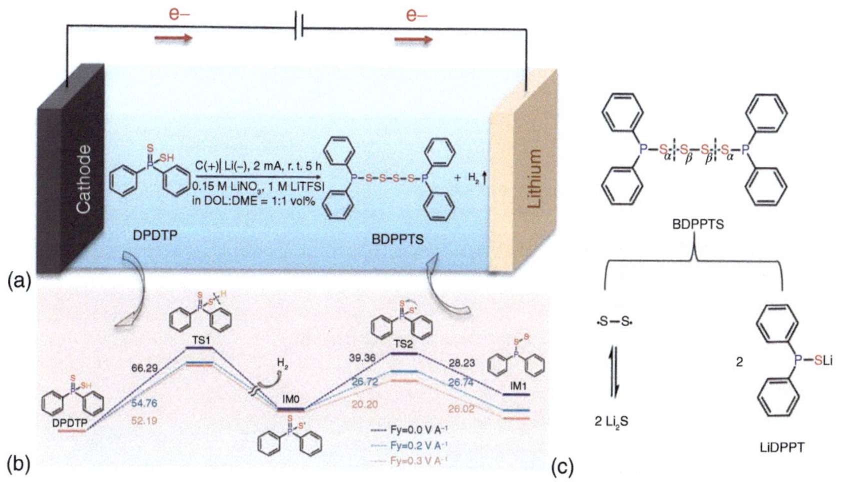

Figure 4.19 (a) Schematic of the synthesis of BDPPTS. (b) Proposed reaction process under external electric field: energy profiles (in kcal mol^{-1}) for the electrochemical oxidation of diphenyl dithiophosphinic acid (DPDTP) leading to the formation of linear organopolysulfides BDPPTS (Fy = 0.0 V Å^{-1}, dark blue lines; Fy = 0.2 V Å^{-1}, blue lines; Fy = 0.3 V Å^{-1}, red lines). (c) Proposed electrochemical redox mechanism of Li/BDPPTS cell. Source: Wang et al. [42]/Springer Nature/CC BY 4.0.

of Li^+. The Li/pGPS cell achieves high specific capacity (1045 mAh g^{-1} at 1 C) and stable cycling performance at a high rate (95.3% capacity retention for 1000 cycles at 5 C). Not only that, Li/pGPS cell can cycle for 800 cycles and retain the specific capacity of 652 mAh g^{-1} at 5 C when the sulfur loading is as high as 20.3 mg cm^{-2}.

The modification of substrate materials is not limited to the enhancement of conductivity, but the introduction of suitable adsorbent materials plays an important role in the improvement of battery performance. DMTS has a high theoretical specific capacity (849 mAh g^{-1}), but the dissolution of DMTS and its discharge products in ether-based electrolyte may cause a shuttle effect, thus leading to the loss of active materials, it is of great significance to introduce matrix materials with adsorption properties to solve this problem. He et al. prepared a three-dimensional nitrogen-doped graphene sponge with Fe_3O_4 nanoparticles (3DFNG) as the matrix of DMTS (Figure 4.20b) [44], in which the N is derived from pyrrolic N and pyridinic N with Lewis base properties, which have a significant adsorption effect on sulfides. Fe_3O_4 is not only highly conductive but also adsorbs DMTS and its discharge products due to the polar nature of Fe_3O_4. The Li/DMTS cell with 3DFNG can release a specific capacity of 822 mAh g^{-1} at 0.1 C, and it can be stably cycled for more than 500 cycles at 1 C, corresponding to a capacity attenuation rate as low as 0.09% per cycle.

Transition metal sulfides are attractive inorganic materials with high electrical conductivity and catalytic activity, among which titanium disulfide nanosheets (TiS_2 NSs) have both high lithium diffusivity and electrical conductivity (5.24×10^3 S m^{-1}), Fan et al. constructed cathode substrate by embedding TiS_2 NSs in a multi-walled

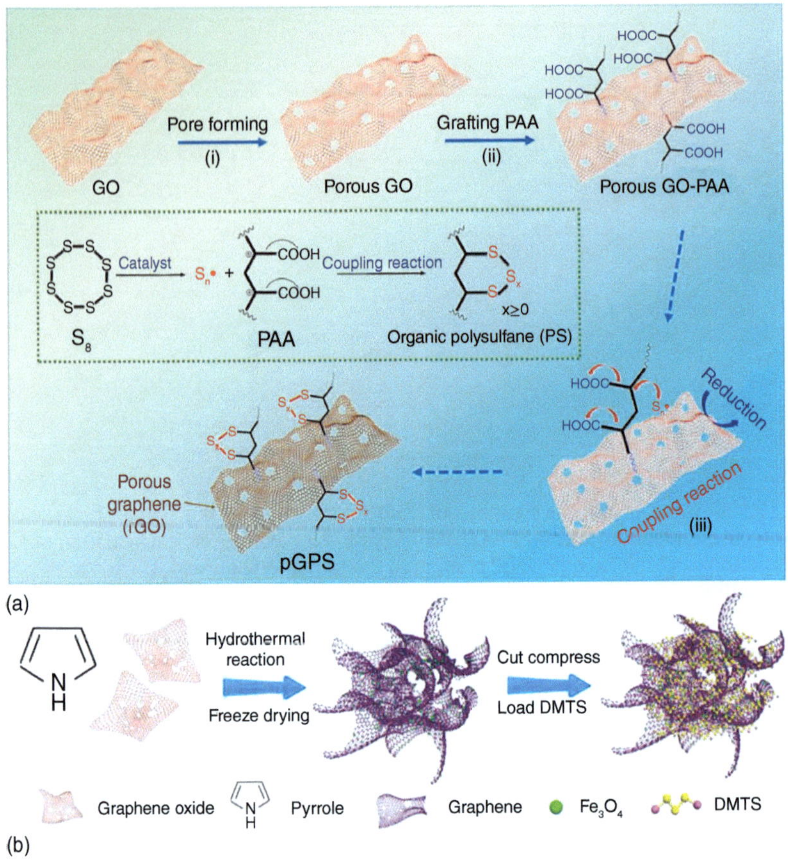

Figure 4.20 (a) Schematic illustration showing the formation mechanism of polysulfane directly grafted on pGPS. Source: Reproduced from Hu et al. [43]/Elsevier. (b) Synthesis of the 3DFNG composite. Source: Reproduced from He et al. [44]/Elsevier.

CNT (TiS_2 NSs@MWCNT) network and applied it to Li/DPTtS battery system [10]. The catalytic effect of TiS_2 can promote the electrochemical conversion process of DPTtS, which leads to a low voltage polarization and fast kinetics of the Li/DPTtS cell (Figure 4.21a). The S—S bonds of Li/DPTtS cell are broken during discharge, which leads to the formation of PhSLi and polysulfide. The dissolution and shuttle effect of PhSLi and polysulfide in ether-based electrolytes may cause the loss of active materials (Figure 4.21b, c). Benefiting from the adsorption of PhSLi and lithium polysulfide by TiS_2, the shuttle effect is effectively limited. The discharge capacity of DPTtS/TiS_2 NSs cathode after 200 cycles at 0.5 C is 467.6 mAh g^{-1}, accounting for 81.9% of the capacity of the fifth cycle, while that of pure DPTtS cathode is only 66.1%. Even when the electrolyte/DPTtS ratio is as low as 3.8 μl mg^{-1}, the DPTtS/TiS_2 NSs cathode can achieve a specific capacity of up to 444 mAh g^{-1} in Li half-cell.

MoS_2 has received much attention due to its strong chemisorption and catalytic effect on lithium polysulfide. For Li/DPTtS batteries with shuttle effect, a composite

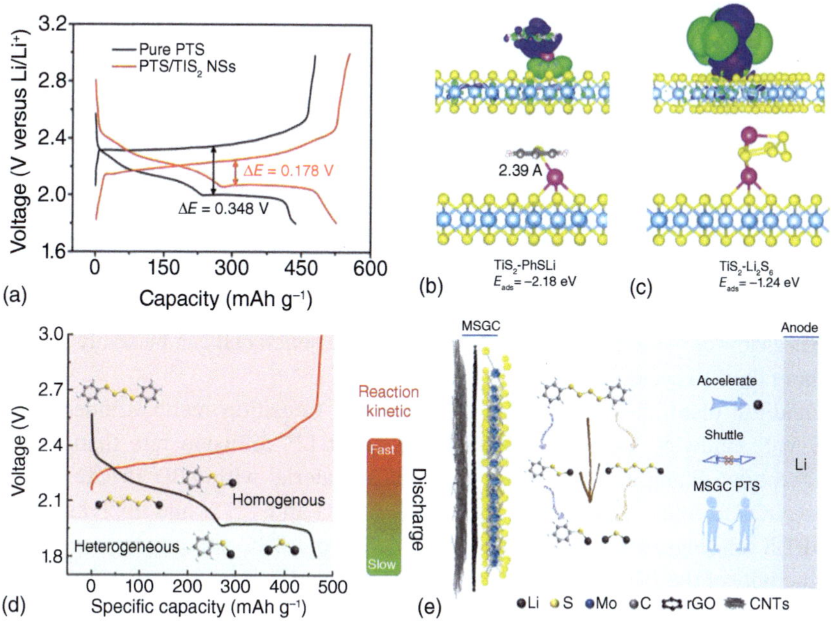

Figure 4.21 (a) Voltage profiles of pure DPTtS and DPTtS/TiS$_2$ NSs electrodes of the 10th cycle at 0.2 C rate. The fully optimized adsorption configurations of (b) PhSLi and (c) Li$_2$S$_6$ on TiS$_2$ with 3D differential charge contour surfaces. The carbon, hydrogen, lithium, sulfur, and titanium elements are shown in gray, white, purple, yellow, and blue, respectively. The electronic attachment and detachment densities are shown in blue and green surfaces, respectively. Source: Reproduced with permission Fan et al. [10]/John Wiley & Sons. (d) Discharge reactions are homogenous above 2.0 V, and discharge reactions are heterogeneous below 2.0 V. (e) Illustration of Li/DPTS battery performance enhancement by the MSGC substrate. Source: Reproduced with permission Lv et al. [45]/John Wiley & Sons.

cathode matrix (MSGC) composed of 1T MoS$_2$, graphene oxide, and CNTs was designed by Lv et al. [45] The electrochemical reaction process of Li/DPTtS cell involves the liquid–solid phase conversions (Figure 4.21d). Before discharging to 2.0 V, DPTtS and intermediates can be dissolved in an ether-based electrolyte, and this discharging process keeps homogeneous electrochemical reaction. After 2.0 V, the intermediates gradually transformed into insoluble PhSLi and Li$_2$S, which was a heterogeneous reaction. The reaction process is shown in Figure 4.21e, in which the MSGC matrix provides abundant active sites and pores so that it can be closely combined with the charge and discharge products of DPTtS, ensuring sufficient catalytic activity. Moreover, the high electronic conductivity and hydrophilicity of MSGC are also conducive to rapid electron transfer and ion diffusion.

In Li/DPTtS@MSGC cell, the instability of 1T-MoS$_2$ would induce a spontaneous transition to the stable 2H-MoS$_2$, and the transition of 2H and 1T phases is a dynamic process. During the discharge process, the continuous attack of Li$^+$ leads to the change of Mo-d orbitals, which stabilizes the octahedral coordination of Mo in the

1T phase and is the junction between the formation of 2H and 1T phases. When Li^+ accumulates to the critical concentration, MoS_2 will change from 2H to 1T phase. MoS_2 returns to 1T when the discharge reaches 1.8 V. During the charging process, 1T-MoS_2 will spontaneously transform into stable 2H-MoS_2, and the formation of 1T-2H mixed-phase MoS_2 was observed when charging to 2.5 V. This dynamic 1T-2H mixed-phase conversion facilitates electron transport and accelerates the electrochemical reaction. The mixed phase has a lower adsorption energy than the single 1T and 2H phases, which can reduce the activation energy of the reaction and improve the catalytic effect. Li/DPTtS@MSGC cell achieves a reversible specific capacity of 524 mAh g^{-1} at 0.5 C and retains 69.8% after 950 cycles. Even when the E/DPTtS ratio is as low as 0.7 μl mg^{-1}, the Li/DPTtS@MSGC pouch cell can be stably cycled for more than 10 cycles.

Vanadium disulfide (VS_2) is also a member of transition metal sulfides with an S–V–S sandwich structure, which has a faster Li^+ diffusion rate than MoS_2 and graphite. Zhang et al. prepared a substrate material with a 3D structure (3D VS_2-rGOCNTs, VSGC) by combining VS_2 with rGO and CNTs and applying it to Li/DPTtS cell (Figure 4.22a) [46]. The introduction of VSGC not only improved the conductivity of the DPTtS cathode and enhanced the utilization of active materials, but also improved the kinetic performance of the cell by promoting the electrochemical conversion process through the catalytic effect. The Li/DPTtS@VSGC can be stably cycled for 500 cycles at 0.5 C with a capacity decay rate of only 0.01% per cycle. In addition, the Li/DPTtS@VSGC pouch cell with DPTtS loading of 5.3 mg cm^{-2} and E/DPTtS = 6.2 μl mg^{-1} achieves a specific energy of 392.6 Wh kg^{-1} in the first cycle and is capable of 70 stable cycles (Figure 4.22b).

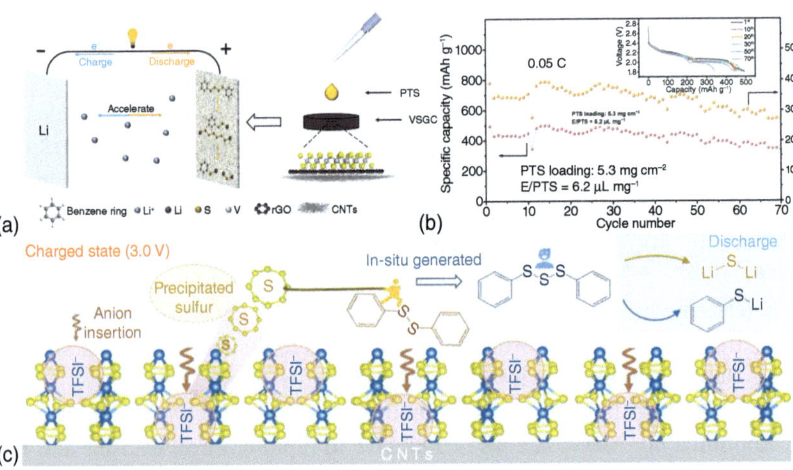

Figure 4.22 (a) Schematic representation of the structure of a Li/DPTtS@VSGC cell and construction of DPTtS@VSGC cathode. (b) Cycling performance of Li/DPTtS pouch cell. Source: Reproduced with permission Zhang et al. [46]/John Wiley & Sons. (c) Schematic diagram of the reaction mechanism for CNTs/VS_4/DPDS. Source: Reproduced with permission Wang et al. [47]/John Wiley & Sons.

Transition metal polysulfides (MS_x, $x > 2$) are a promising class of sulfur-derived cathode materials with high theoretical specific capacity. VS_4 was first discovered in 1906 [48], and it is oxidized to form metallic V and Li_2S in lithium batteries, with a theoretical specific capacity as high as $1196\,mAh\,g^{-1}$. Wang et al. assembled VS_4/CNTs as cathode into lithium half-cell [47], and found that bis(trifluoromethanesulfonyl) imide ($TFSI^-$) anions would be intercalated into VS_4 when charged to high potential (3.0 V), resulting in sulfur evolution (Figure 4.22c). Considering that this sulfur may undergo undesired reactions leading to severe shuttle effects, DPDS is introduced as a sulfur acceptor to capture these sulfur atoms. In the charging process, DPDS in CNTs/VS_4/DPDS electrode combines with S escaping from VS_4 to generate DPTS, which avoids the conversion of sulfur to polysulfide and inhibits the loss of active materials caused by the dissolution and shuttle of polysulfide. The lithium half-cell assembled with CNTs/VS_4/DPDS cathode exhibits high discharge capacity and stable cycle performance (the specific capacity of $130.7\,mAh\,g^{-1}$ is maintained even after 300 cycles at $240\,mA\,g^{-1}$). Such an organic–inorganic composite cathode design can improve the utilization of active materials and the capacity retention of the battery, providing a novel research idea for the application of transition metal sulfides and the limitation of polysulfides.

4.6 Other Organosulfur Cathodes

The diversity of organosulfur molecules has contributed to the booming development of organic batteries. Most organosulfur has high solubility in organic electrolyte, and the high solubility may trigger shuttle effects leading to loss of active material. Tetrathiafulvalenes (TTFs), a member of the organosulfur molecule, was originally used as a donor for molecular conductors. It has been shown that the solubility of TTFs derivatives in electrolytes decreases with the increase in the number of units of TTFs. Therefore, the design of organosulfur molecules with multiple TTFs units can effectively overcome their solubility problems in electrolytes, and the introduction of suitable active sites in the design of TTFs derivatives is important for their electrochemical performance improvement.

Kato et al. designed and synthesized pentakis-fused TTF derivatives possessing two cyclohexene-1,4-diylidene-inserted TTF units by retro-Diels–Alder reaction (Figure 4.23a) [49]. The solubility of tetrakis(methylthio) derivative (TTMD) in liquid electrolyte is low because of the extension of the TTF units. TTMD is allowed to undergo 10 e^- transfer reaction in lithium batteries, and the theoretical specific capacity of Li/TTMD cell can reach 196 mAh g^{-1}. Benefiting from the high conductivity of pentane-fused TTF derivatives, Li/TTMD cell shows excellent high-rate performance. In addition, multi-fused TTF systems, including TTMD, are multi-electron donors, the maximum oxidation potentials of donors shift to a more positive voltage region with the increase of the number of electrons. The average discharge voltage of Li/TTMD cell can reach 3.56 V, which leads to an energy density as high as $700\,mWh\,g^{-1}$. Due to the low solubility of TTMD in the

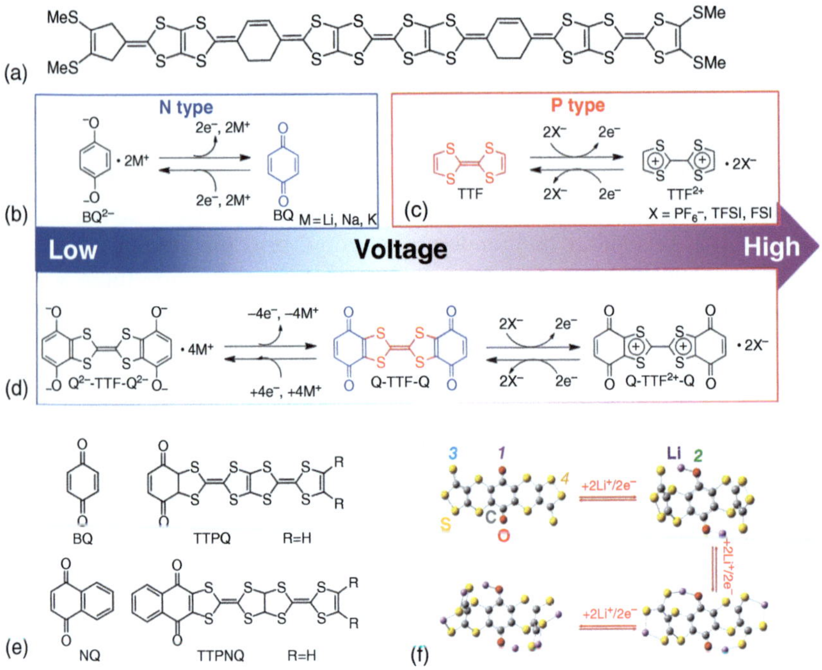

Figure 4.23 (a) The structure of TTMD. Source: Adapted with permission from Kato et al. [49], Royal Society of Chemistry. Structures and redox schemes of (b) BQ, (c) TTF, and (d) Q-TTF-Q. Source: Adapted with permission from Kato et al. [50], Royal Society of Chemistry. (e) Chemical structures of BQ, NQ, TTPQ, and TTPNQ. Source: Adapted with permission from Misaki et al. [51], Wiley-VCH. (f) The structural evolution of TPQD electrodes during the lithiation process. Source: Adapted with permission from Shadike et al. [52], Wiley-VCH.

electrolyte, the capacity attenuation caused by the dissolution and shuttle of active substances is effectively inhibited, and the Li/TTMD cell can still maintain 72% of the initial capacity after 30 cycles.

In fact, the design of TTFs derivatives is not limited to the extension of TTF units. Kato et al. introduced carbonyl groups into TTFs derivatives and designed organosulfides with both n-type and p-type material characteristics (Q-TTF-Q) (Figure 4.23b–d) [50]. Q-TTF-Q can accept both cations and anions during redox and exhibits a theoretical specific capacity of up to 441 mAh g^{-1}, and it has achieved good applications in Li^+, Na^+, and K^+ batteries. Among them, Li/Q-TTF-Q cell exhibits high discharge capacity (236 mAh g^{-1}) and average voltage (2.8 V) with energy density up to 661 mWh g^{-1}. Such TTF derivatives with both n-type and p-type material characteristics also attracted the attention of other researchers. Noda et al. substituted benzoquinone (BQ) for a terminal TTF unit in 2,2′-bi[5-(1,3-dihthiol-2-ylidene)-1,3,4,6-tetrathiapentanylidene] (TTPY) to synthesize the fused TTF-BQ triad (TTPQ), subsequently, the naphthoquinone analog of TTF-TTF-BQ triad (TTPNQ) was synthesized by introducing naphthoquinone (NQ) into TTF (Figure 4.23e) [51]. The addition of carbonyl groups facilitated the achievement of high discharge voltage and specific capacity, where Li/TTPQ cells could achieve a specific capacity of 299 mAh g^{-1} and energy density of up to 855 mWh g^{-1}.

The strategy of rational introduction of carbonyl groups is not only reflected in TTFs derivatives, Shadike et al. synthesized 2,3,4,6,8,9,10,12-octathia biscyclopenta [*b*,*c*]-5,11-anthraquinone-1,7-dithione (TPQD) using dithiane and benzoquinone as precursors [52]. The electrochemical active sites of TPQD consist of carbonyl and disulfide bonds, and the Li/TPQD cell can undergo a six-electron reaction (Figure 4.23f), which provides an initial capacity of 251.7 mAh g^{-1} at 0.1 C. In addition, the extended π-conjugated structure of TPQD facilitates excellent rate performance and cycling stability, and the Li/TPQD cell can cycle more than 200 cycles at 5 C with only 0.07% capacity degradation per cycle. This study demonstrates the design feasibility and practicality of organosulfur molecules that can help realize high-performance organic batteries.

Tetramethyl thiuram monosulfide (TMTM) is a cheap and easily available organosulfide, but it has never been used in battery systems because it has no electrochemical activity. Chen et al. promoted TMTM *in situ* conversion into an electron-deficient organosulfide (ED-OS) cathode material with electrochemical activity through an electrochemical oxidation strategy [12]. TMTM underwent configuration transformation during electrochemical oxidation (Figure 4.24a), which was revealed by MS, NMR, XPS, and DFT. In the electrochemical oxidation process, the electrons of nitrogen atoms in TMTM are detached, and nitrogen positive ions (N^+) are formed, which attract anion $TFSI^-$ in the electrolyte to form N^+–$TFSI^-$. It will stretch the electron cloud on the adjacent S atom to form S radical, which leads to the C—S bond on the other thiuram group to break and generate another S radical. Both S radicals can independently form S—S bonds, and these bonds have low energy barriers, which are easy to break and quickly combine with lithium-ion. Figure 4.24b shows the several ED-OS formed by TMTM after electrochemical oxidation.

Based on the electrochemical oxidation mechanism shown above, Li/TMTM cell can obtain reversible electrochemical behavior through simple first-cycle charge oxidation. Due to the low energy barriers of S—S bonds formed by TMTM during electrochemical oxidation, the Li/TMTM cell shows fast reaction kinetics. In order to evaluate the influence of *in situ* electrochemical oxidation degree on the electrochemical behavior of TMTM-EO, the oxidation degree of TMTM-EO was controlled by controlling the specific capacity. When the electrochemical oxidation degree is 90 mAh g^{-1}, the obtained organosulfur electrode (TMTM-90) exhibits faster reaction kinetics and better cycle stability, which is related to the kinetic variables, mass transfer, and surface effect of the electrode–electrolyte interface. Li/TMTM-90 cell can be cycled 8000 cycles at 10 C and the capacity retention rate can reach 70% (Figure 4.24c). In addition, Li/TMTM-90 cell also shows excellent electrochemical behavior in low-temperature environment (−50 °C), it can be cycled for more than 1000 hours and retain 75% of its initial capacity, which shows its practical application potential in low-temperature environment.

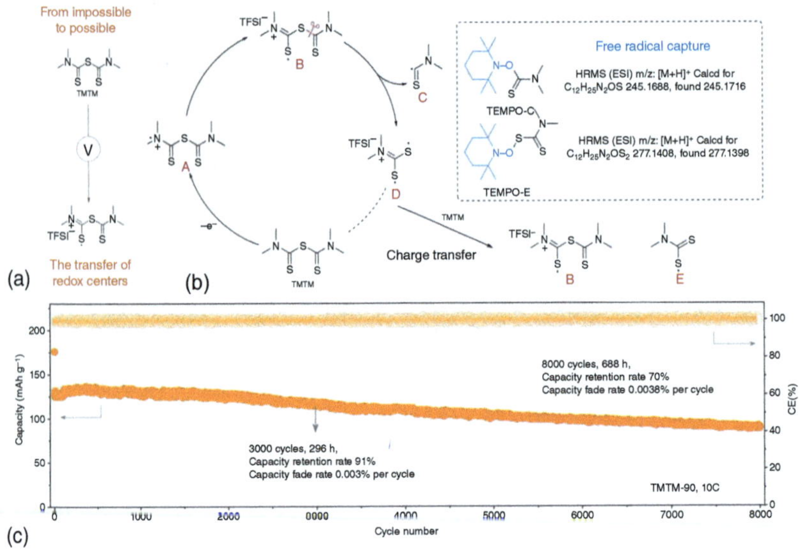

Figure 4.24 (a) Molecular transformation mechanism of TMTM. (b) Proposed molecular structure transformation mechanism of TMTM via electrochemical oxidation. (c) Cycling performance of TMTM-90 at 10 C. Source: Reproduced with permission Chen et al. [12]/American Chemical Society.

4.7 Summary and Outlooks

The controllable structure of organosulfides makes them attractive as cathode materials. Organosulfides with high specific capacity can be obtained by regulating the length of sulfur chains in molecules, and the electrochemical behavior of organosulfur-based batteries can be controlled by regulating functional groups, e.g. organosulfides-containing nitrogen heterocycles have higher discharge voltage. In addition, the introduction of suitable inorganic materials can make organosulfides exhibit fast kinetics in rechargeable batteries. The diversity of organosulfur compounds, coupled with their tunable structures, make organosulfur cathode materials increase fresh development blood and flourish. Although organosulfides have shown unique advantages as cathode materials, their future development still faces many challenges, and there is still much room for their commercial application. The future research direction should focus on the following aspects (Figure 4.25): (i) design and synthesize organosulfides with stable structure and high specific capacity; (ii) rationally regulate the molecular structure of existing organosulfides, so as to optimize their electrochemical properties; (iii) develop advanced *in situ* characterization techniques to monitor the reaction processes in the batteries, and combine theoretical calculation and simulation to show the reaction mechanisms of organosulfur-based batteries; (iv) introduce inorganic materials with high conductivity and adsorption into organosulfur cathodes to improve the

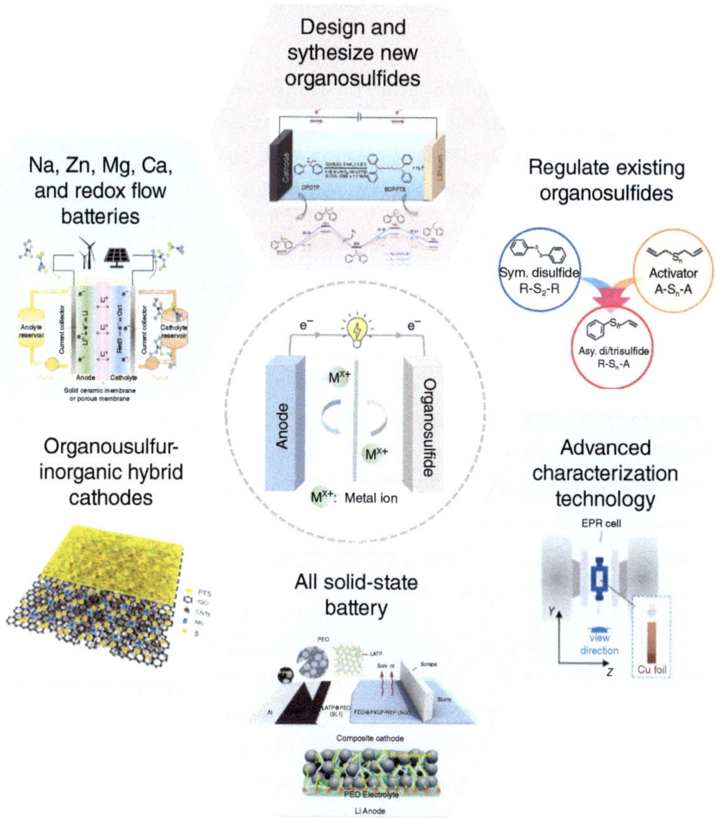

Figure 4.25 Prospects of organosulfur-based batteries. Source: Wang et al. [42]/Springer Nature/CC BY 4.0. Reproduced with permission Zhang et al. [16]/John Wiley & Sons. Reproduced with permission Weng et al. [9]/Royal Society of Chemistry. Reproduced with permission Lv et al. [45]/John Wiley & Sons. Reproduced with permission Geng et al. [53]/American Chemical Society. Reproduced with permission Ling et al. [54]/American Chemical Society. Reproduced with permission Ma et al. [55]/Royal Society of Chemistry.

dynamics and cycle stability of batteries; (v) develop organosulfur-based batteries with solid-state electrolytes to improve battery safety; (vi) expand the application of organosulfur compounds in other metal (e.g. Na, Zn, Ca, or Mg) batteries and redox flow batteries.

Acknowledgments

The work was supported by the National Natural Science Foundation of China (Grant Nos. U2004214 and 21975225) and the National Key Research and Development Program of China (2022YFB2402200).

References

1 Poizot, P., Gaubicher, J., Renault, S. et al. (2020). *Chem. Rev.* 120: 6490–6557.
2 Wang, D.Y., Guo, W., and Fu, Y. (2019). *Acc. Chem. Res.* 52: 2290–2300.
3 Visco, S.J. and Jonghe, L.C.D. (1988). *J. Electrochem. Soc.* 135: 2905–2909.
4 Naoi, K., Kawase, K.-I., and Inoue, Y. (1997). *J. Electrochem. Soc.* 144: L170–L172.
5 Inamasu, T., Yoshitoku, D., Sumi-otorii, Y. et al. (2003). *J. Electrochem. Soc.* 150: A128–A132.
6 Wu, M., Cui, Y., Bhargav, A. et al. (2016). *Angew. Chem. Int. Ed.* 55: 10027–10031.
7 Bhargav, A., Bell, M.E., Karty, J. et al. (2018). *ACS Appl. Mater. Interfaces* 10: 21084–21090.
8 Guo, W., Bhargav, A., Ackerson, J.D. et al. (2018). *Chem. Commun.* 54: 8873–8876.
9 Weng, G., Yang, B., Liu, C. et al. (2019). *Energy Environ. Sci.* 12: 2244–2252.
10 Fan, Q., Guo, W., Si, Y. et al. (2020). *Adv. Funct. Mater.* 31: 2001493.
11 Tang, S., Chen, Q., Si, Y. et al. (2021). *Adv. Mater.* 33: 2100824.
12 Chen, Q., Li, L., Wang, W. et al. (2022). *J. Am. Chem. Soc.* 144: 18918–18926.
13 Guo, W., Wang, D.Y., Chen, Q., and Fu, Y. (2022). *Adv. Sci.* 9: 2103989.
14 Liu, M., Visco, S.J., and Jonghe, L.C.D. (1990). *J. Electrochem. Soc.* 137: 750–759.
15 Bhargav, A., Patil, S.V., and Fu, Y. (2017). *Sustainable Energy Fuels* 1: 1007–1012.
16 Zhang, L., Zhao, B., Zhang, C., and Yu, G. (2021). *Angew. Chem. Int. Ed.* 60: 4322–4328.
17 Hu, P., He, X., Ng, M.F. et al. (2019). *Angew. Chem. Int. Ed.* 58: 13513–13521.
18 Wu, M., Bhargav, A., Cui, Y. et al. (2016). *ACS Energy Lett.* 1: 1221–1226.
19 Lv, X., Yang, Q., Zhang, X. et al. (2022). *Angew. Chem. Int. Ed.* 61: e202213160.
20 Gao, M., Lan, J., Fu, Y., and Guo, W. (2022). *ChemSusChem* 15: e202200423.
21 Li, F., Si, Y., Liu, B. et al. (2019). *Adv. Funct. Mater.* 29: 1902223.
22 Li, F., Si, Y., Li, Z. et al. (2020). *J. Mater. Chem. A* 8: 87–90.
23 Pan, Q., Lan, J., Si, Y. et al. (2022). *Chem. Commun.* 58: 5602–5605.
24 Ren, Y., Si, Y., Guo, W., and Fu, Y. (2023). *Chem. Commun.* 59: 3289–3292.
25 Li, X., Si, Y., Tang, S., and Fu, Y. (2023). *J. Mater. Chem. A* 11: 8694–8699.
26 Jiao, C., Si, Y., Wang, Z. et al. (2023). *Chem. Eng. J.* 455: 140562.
27 Bhargav, A., Chang, C.H., Fu, Y., and Manthiram, A. (2019). *ACS Appl. Mater. Interfaces* 11: 6136–6142.
28 Sang, P., Si, Y., and Fu, Y. (2019). *Chem. Commun.* 55: 4857–4860.
29 Sang, P., Song, J., Guo, W., and Fu, Y. (2021). *Chem. Eng. J.* 415: 129043.
30 Bhargav, A., Ma, Y., Shashikala, K. et al. (2017). *J. Mater. Chem. A* 5: 25005–25013.
31 Kaland, H., Hadler-Jacobsen, J., Fagerli, F.H. et al. (2020). *ACS Appl. Energy Mater.* 3: 10600–10610.
32 Wang, D.-Y., Si, Y., Li, J., and Fu, Y. (2019). *J. Mater. Chem. A* 7: 7423–7429.
33 Wang, D.Y., Si, Y., Guo, W., and Fu, Y. (2020). *Adv. Sci.* 7: 1902646.
34 Fan, Q., Si, Y., Guo, W., and Fu, Y. (2021). *J. Phys. Chem. Lett.* 12: 900–906.
35 Zhang, W., Ma, F., Guo, S. et al. (2022). *J. Energy Chem.* 66: 440–447.

36 Cui, Y., Ackerson, J.D., Ma, Y. et al. (2018). *Adv. Funct. Mater.* 28: 1801791.
37 Zhao, J., Si, Y., Han, Z. et al. (2020). *Angew. Chem. Int. Ed.* 59: 2654–2658.
38 Chen, Q., Si, Y., Guo, W., and Fu, Y. (2022). *Chem. Commun.* 58: 10993–10996.
39 Guo, W., Wawrzyniakowski, Z.D., Cerda, M.M. et al. (2017). *Chem. Eur. J.* 23: 16941–16947.
40 Zeng, Q., Li, Y., Wu, K.-H. et al. (2018). *Energy Storage Mater.* 12: 30–36.
41 Bhargav, A. and Manthiram, A. (2020). *Adv. Energy Mater.* 10: 2001658.
42 Wang, D.Y., Si, Y., Guo, W., and Fu, Y. (2021). *Nat. Commun.* 12: 3220.
43 Hu, H., Hu, Y., Cheng, H. et al. (2021). *J. Power Sources* 491: 229617.
44 He, J., Bhargav, A., and Manthiram, A. (2019). *Energy Storage Mater.* 23: 88–94.
45 Lv, X., Guo, W., Song, J., and Fu, Y. (2022). *Small* 18: 2105071.
46 Zhang, P., Fu, Y., and Guo, W. (2023). *Small* 19: 2207047.
47 Wang, Z., Li, X., Guo, W., and Fu, Y. (2021). *Adv. Funct. Mater.* 31: 2009875.
48 Hillebrand, W.F. (1907). *J. Am. Chem. Soc.* 29: 1019–1029.
49 Kato, M., Senoo, K., Yao, M., and Misaki, Y. (2014). *J. Mater. Chem. A* 2: 6747–6754.
50 Kato, M., Masese, T., Yao, M. et al. (2019). *New J. Chem.* 43: 1626–1631.
51 Misaki, Y., Noda, S., Kato, M. et al. (2020). *ChemSusChem* 13: 2312–2320.
52 Shadike, Z., Lee, H.S., Tian, C. et al. (2019). *Adv. Energy Mater.* 9: 1900705.
53 Geng, F., Yang, Q., Li, C. et al. (2021). *Chem. Mater.* 33: 8223–8234.
54 Ling, H., Shen, L., Huang, Y. et al. (2020). *ACS Appl. Mater. Interfaces* 12: 56995–57002.
55 Ma, J., Zhong, G., Shi, P. et al. (2022). *Energy Environ. Sci.* 15: 1503–1511.

5

Radical-Based Organic Cathodes

5.1 Introduction

Organic radicals, a class of organic molecules or polymers with unpaired electrons, are highly reactive and easy to be quenched by environmental factors, and thus the life of most free radicals is short [1]. As the lower bandgap of frontier orbital of free radicals compared with other organic molecules, radicals have excellent optical and electrochemical properties and are a kind of promising optoelectronic device materials [2]. Therefore, the design and synthesis of stable radicals, especially air-stable radicals, are attractive to chemists. The ideal radical should have good environmental stability and reaction inertia, which should have three key factors: (i) The air-stable radical must bear high oxidation potential to prevent electrons from being taken; (ii) A highly conjugated structure of radicals enables the electrons to be delocalized into a larger π system to relieve dimerization or polymerization; (iii) There is a large steric hindrance near the position of the spin of unpaired electron [1]. Most air-stable radicals such as nitroxyl, phenoxyl, verdazyl, or galvinoxyls are extensively explored [3].

As attractive electrode materials, radicals are a typical subclass of high-rate and stable organic electrodes due to their fast redox kinetics, in which the reaction involves a singly occupied molecular orbital enabled by favorable electron transfer and thus a low voltage polarization [4, 5]. The radicals usually exhibit bipolar properties, which can be either reduced to negatively charged states or oxidized to positively charged states through a one-electron reaction, as shown in Figure 5.1, accompanied by the intercalation of cations or anions [6]. Generally, organic radicals are used as cathode through the p-type reaction because of the high redox voltage above 3 V (versus Li/Li^+), while as an anode with the n-type reaction, the redox potential of which is about 2 V versus Li/Li^+. Meanwhile, n-type reactions of radicals are more susceptible to side reactions than the p-type process, thus leading to lowered reversibility [7].

Despite the rapid reaction kinetics, radical molecule faces the problem of high solubility, resulting in poor cycling stability and serious self-discharge. To mitigate the dissolution, the common strategy is attaching radical units to a polymer backbone, which is nonconductive and thus compromises the reaction kinetics of

Rechargeable Organic Batteries: Materials, Mechanisms, and Prospects, First Edition.
Yongzhu Fu, Xiang Li, Shuai Tang, and Wei Guo.

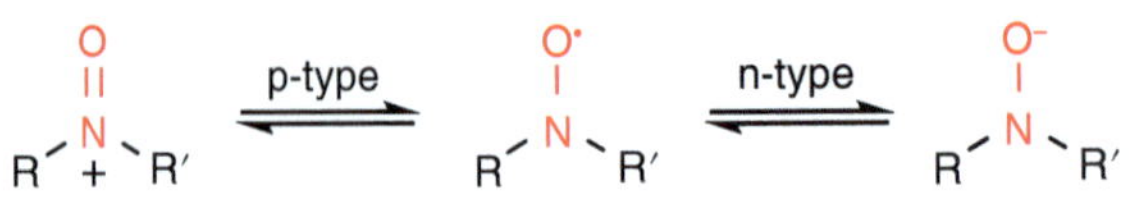

Figure 5.1 Redox mechanisms of radicals.

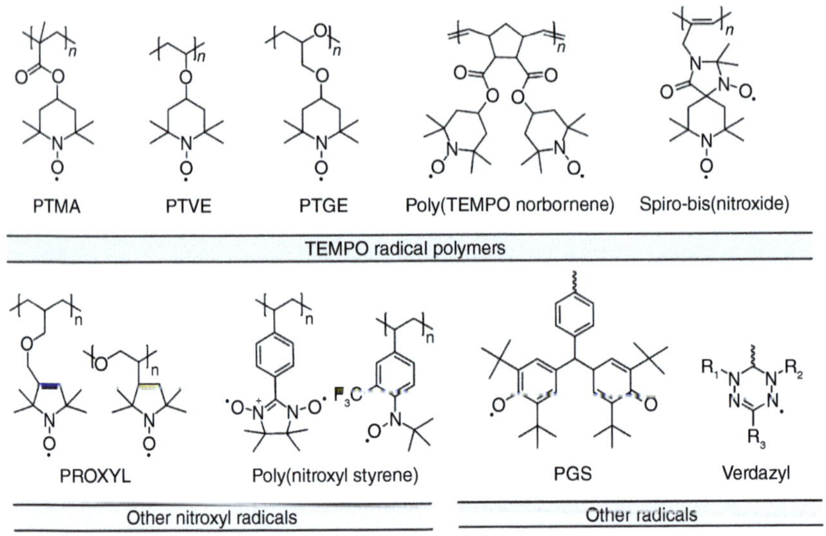

Figure 5.2 TEMPO radical polymers, other nitroxyl radicals, and other radicals.

the electrodes. The first battery based on radicals that utilized this concept was reported by Nakahara et al. in 2002 [8]. After that, various radical active materials with multifarious polymer chains have sprung up. Particularly for nitroxyl radical-bearing structures, like 2,2,6,6-tetramethylpiperidinyl-Noxyl (TEMPO) and relative derivatives, they have been successfully applied to different battery systems such as metal ion batteries and redox flow batteries. Given that most radical electrode materials are polymers, this chapter mainly reviews some breakthroughs and the latest research about active radical materials reported based on the type of polymer skeletons [9]. The structures of the described active radical materials are listed in Figure 5.2.

5.2 Radical for Metal-Ion Battery

Since the TEMPO-based electrochemically active material was first proposed in 2002, it has become the most prominent radical compound in organic battery material research so far, in which most radical polymers contain TEMPO moieties as the electrochemically active sites in the side chain. The reason that TEMPO has attracted much attention is also attributed to its excellent stability in electrolyte and electrochemical properties, where the molecule structure nearly has no obvious change during a fast one-electron reaction. Figure 5.3 shows the structure of TEMPO and the corresponding redox mechanism.

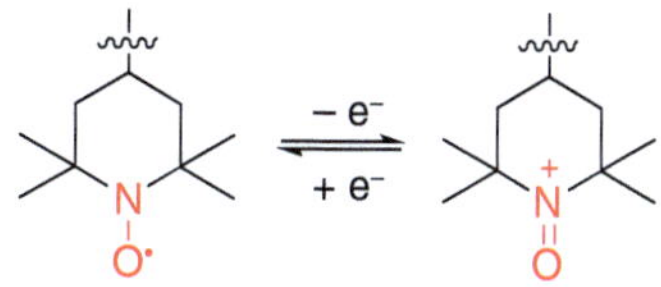

Figure 5.3 Structure redox mechanism of TEMPO.

5.2.1 PTVE Radical

The most common organic radical polymer with a TEMPO unit is poly(2,2,6,6-tetramethylpiperidinyloxy-4-ylmethacrylate) (PTMA), as shown in Figure 5.4a. The first reported PTMA/Li battery has a discharge voltage of 3.5 V and an initial discharge capacity of 77 mAh g^{-1}. Excellent cycle performance proves the prospect of this electrode material [8]. However, PTMA still suffers from low discharge capacity and fast capacity decay ascribed to the dissolution and lack of good electronic conductivity.

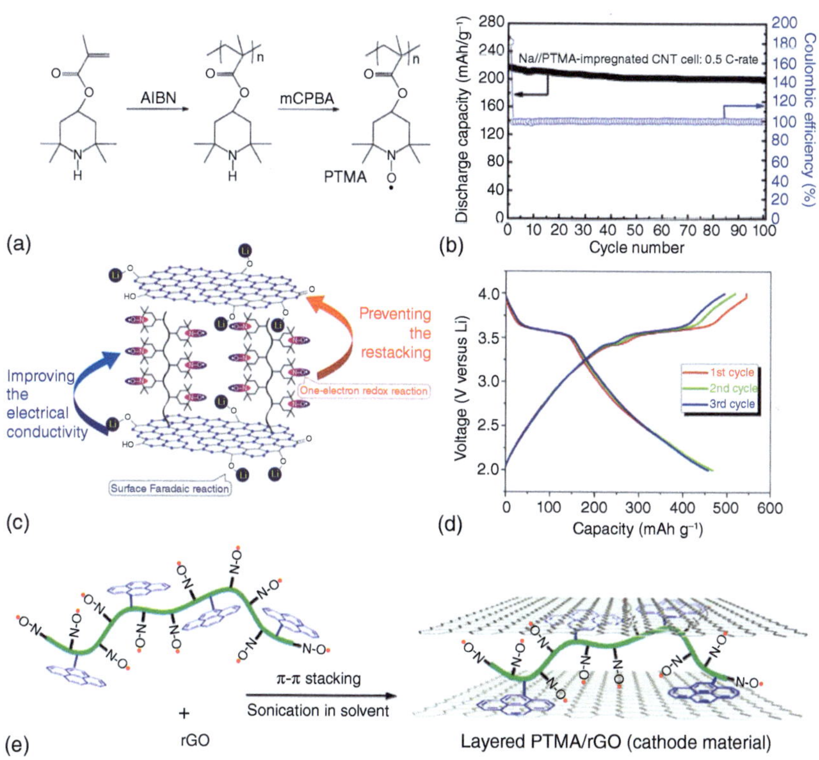

Figure 5.4 (a) Synthesis procedure of PTMA. Source: Adapted with permission from Lu et al. [7], Elsevier. (b) The cycle performance of PTMA-impregnated CNT sodium battery. Source: Reproduced with permission Kim [10], The Royal Society of Chemistry.
(c) Illustration of the energy storage and interaction mechanism of G-g-PTMA/rGO cathode. (d) Charge/discharge profile of G-g-PTMA/rGO at a current of 0.2 A g^{-1}. Source: Reproduced with permission Kim et al. [11], American Chemical Society. (e) Formation of a layered PTMA/rGO composite through noncovalent π–π stacking between pyrene groups and rGO sheets. Source: Reproduced with permission Li et al. [12], American Chemical Society.

Many insoluble and highly conductive inorganic hosts, such as carbon nanotubes (CNTs), reduced graphene oxide (rGO), carbon black, and graphene, were proposed to improve the conductivity of PTMA electrodes. Kim et al. reported a micro-fibrous organic radical electrode film consisting of PTMA, poly(vinylidene fluoride-*co*-hexafluoropropylene) (PVDF), and carbon black power, displaying a discharge capacity of 109 mAh g^{-1} and excellent rate performance [10]. Then, they prepared a PTMA-impregnated CNT composite electrode [11]. The fast ion and electron diffusion and the trap capacity of active polymers are achieved by interlaced network and narrow vacant spaces of CNT. The fabricated Na/PTMA cell impregnated with CNT presents a reversible two-electron redox reaction corresponding to two redox plateaus at 3.3 and 2.1 V, exhibiting a high initial capacity of 222 mAh g^{-1} at 0.1 C and a capacity retention of 93% after 100 cycles (Figure 5.4b).

In addition to directly mixing with inorganic conductive materials, radical polymers can also be stabilized through chemical bonding interactions with inorganic frameworks. Li et al. used functionalized graphene sheets covalently linked with PTMA, graphene-grafted-PTMA (G-g-PTMA), as an active material complex with rGO to be adopted as a composite cathode (Figure 5.4c). As displayed in Figure 5.4d, the battery delivers a capacity of up to 466 mAh g^{-1} based on the mass of PTMA [12]. The excess capacity higher than the theoretical value of PTMA is attributed to the fast electron transport redox reaction of composite cathode and surface Faradaic reaction of rGO. Meanwhile, Zhang et al. synthesize a functionalized PTMA with pyrene groups via a rapid nitroxide radical coupling reaction. The resulting copolymers have amounts of pyrene units beside the backbone chain, producing greater π–π stacking interaction between rGO, and thus the polymer can be efficiently stabilized through molecular-level bonding by rGO (Figure 5.4e). The prepared PTMA/rGO composite achieved rapid charge storage and long-cycle performance in Li-ion batteries [13].

5.2.2 Other TEMPO-Based Nitroxyl Radicals

The second kind of active nitroxyl radical polymer with TEMPO units is poly(TEMPO-substituted norbornene), in which the respective norbornene monomers contain two TEMPO structure units. Suga et al. assembled an all-organic battery with poly(TEMPO-substituted norbornene) as a cathode, which could maintain almost identical discharge capacity at 5–360 C rate [14]. Then, Dai et al. also utilized poly(TEMPO-substituted norbornene) as a cathode and applied it in NIBs [15]. The maximum initial discharge capacity reaches 75 mAh g^{-1} at a current of 50 mA g^{-1}.

Another common polymer backbone based on TEMPO units is poly(2,2,6,6-tetramethylpiperidinyl-*N*-oxyl vinyl ether) (PTVE), as shown in Figure 5.5a, in which PTVE is a representative polymer applied in protic electrolytes. The hydrophilic properties of PTVE facilitate electrochemical processes in aqueous electrolytes, which possess high ionic conductivity and lower cost than most radical polymers. However, the pristine PTVE material exhibited a low activity because of its low electronic conductivity, which is the common challenge of nonconductive polymer backbone.

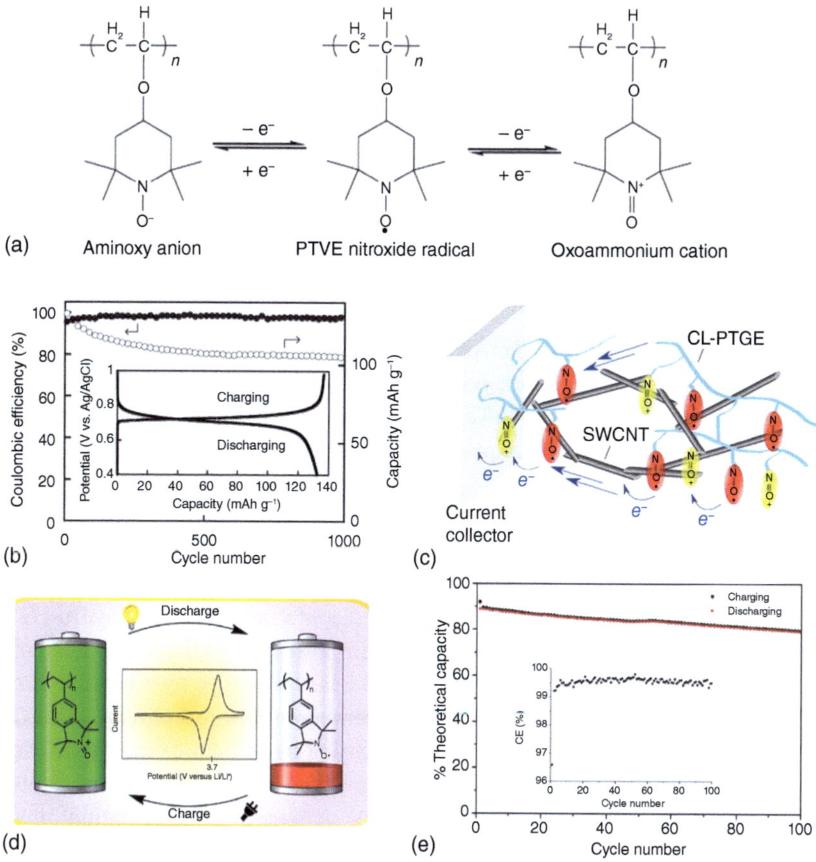

Figure 5.5 (a) Redox reaction of PTVE. Source: Reproduced with permission from Dai et al. [15], Elsevier. (b) Cycle performance and charging/discharging of the PTVE at the rate of 60 C. Source: Reproduced with permission from Guo et al. [16] John Wiley & Sons. (c) Schematic illustration of charge transportation in the CL-PTGE/SWCNT. Source: Reproduced with permission Koshika et al. [17], The Royal Society of Chemistry. (d) Schematic diagram of the reaction of PVTMIO. (e) Cycle performance of PVTMIO/Li battery. Source: Reproduced with permission Suga et al. [18], American Chemical Society.

Benefiting from the high electronic conductivity and structural stability provided by carbon materials like graphene, Guo et al. synthesized a nanocomposite cathode of PTVE and graphene via co-deposition method [16]. The composite electrode shows a reversible two-electron redox process, in which a high specific capacity of 261 mAh g^{-1} in LIBs is achieved. An excellent rate performance of up to 200 C and a long cycle life over 20 000 times suggest that the electrochemical performance and structural stability of PTVE are tremendously enhanced by 3D networks of graphene.

Since PTVE undergoes a reversible redox reaction in neutral and acidic electrolytes, rechargeable aqueous zinc batteries with PTVE as cathode were developed. The homogeneous solvation of radicals and adequate supply of counterion by the hydrophilicity of PTVE render a superior electrochemical performance. As shown in Figure 5.5b, the as-prepared zinc battery displays a capacity of 131 mAh g^{-1}

at 60 C and a 65% capacity retention after 500 cycles. Even the applied charging rates of half-cell was enhanced up to 1200 C, the thin PTVE-film electrode also displayed excellent cycling stability, indicating the extraordinary rate performance of PTVE [17].

Other nitroxide radicals based on the TEMPO structure have also been studied. The ion conductor backbone PEO was introduced in the cross-linked radical polymer poly(2,2,6,6-tetramethylpiperidinyl-*N*-oxyl glycidyl ether) (PL-PTGE). The polymer has better swelling ability and ionic conductivity, benefiting from the PEO chain, which allows for thicker cathode materials and a lower proportion of conductive additive (Figure 5.5c). The composite cathode of SWCNT and PTGE achieves a discharge capacity of 93 mAh g^{-1} at 10 C [19]. Oyaizu et al. proposed that with high electronegative backbone polymer chains, the PTGE cathode presents a high plateau voltage of about 3.63 V. The capacity has no significant decay after 100 cycles, and the capacity and voltage plateau are still kept unchanged even at a large rate of 10 C [20].

5.2.3 Other Nitroxyl Radicals

One typical organic radical is derived from 2,2,5,5-tetramethyl-2,5-dihydro-1*H*-pyrrol-1-oxyl-3-yl (PROXYL), which possesses ultrafast redox kinetics because of the rapid electron transfer. The PROXYL radical based on a five-membered cyclic nitroxide has less molecular mass than the six-membered ring structure of TEMPO and thus gets a higher theoretical specific capacity of more than 85 mAh g^{-1}. The discharge voltage of typical PROXYL is about 3.6 V in a Li metal battery, which is slightly higher than that of common PTMA [20]. A 90% capacity retention after 100 cycles is obtained even at a current of 10 C rate with a discharge capacity of 80 mAh g^{-1}. Oyaizu et al. synthesized a PROXYL-contained molecule based on the PEO chain, the theoretical capacity of which reaches 147 mAh g^{-1}, close to that of traditional intercalation electrodes like $LiCoO_2$. Although the practical discharge capacity is lower than 100 mAh g^{-1}, a reduction potential of 3.7 V and 80% capacity retention after 1000 cycles are obtained, which still has the advantages of energy density and cycling stability [21].

In addition to reducing the number of atoms on the radical ring, the introduction of a second nitroxide radical group per polymer unit is another way to improve the capacity, such as in the radical polymer of spiro-bis(nitroxide)s and poly(nitroxyl styrene). However, the lithium battery tests revealed that only radicals based on six-membered rings enable a reversible redox process. For the latter, the two NO-radicals are stabilized by a strengthening π conjugate structure from the phenyl ring, by which the charge can be effectively delocalized. Meanwhile, *tert*-butyl groups nearby also bring strong steric hindrance to radical sites [22]. Suga et al. further proposed that electron-withdrawing groups (e.g. $-CF_3$) on benzene rings can enhance the reversibility of the n-type reaction, but such radical materials have not been applied to batteries [18]. In 2018, Hansen et al. synthesized a novel styrenic nitroxide polymer, poly(5-vinyl-1,1,3,3-tetramethylisoindolin-2-yloxyl) (PVTMIO), and applied it in Li metal battery (Figure 5.5d). Compared to styrene-type TMIO monomer, the polymer state PVTMIO exhibited a high redox potential of 3.69 V (versus Li/Li^+), which is the best kind of this class of radical materials. Moreover, Figure 5.5e indicates that the battery combined with Li metal anode achieved a capacity of 104.7 mAh g^{-1} and a 99% CE over 100 cycles [23].

5.2.4 Other Radical Electrode Materials

Another type of organic radical is galvinoxyl radical with different structures and redox mechanisms from nitroxide radicals, of which the electron transfer reaction is located on benzyloxy. Galvinoxyl can be reversibly reduced to phenolate ion through an n-type reaction, which acts as an organic anode material and is applied in all-organic battery systems, but the prepared battery delivers a low capacity of only 51 mAh g^{-1} due to the high molar mass and single electron transfer [14].

Jähnert et al. fabricated an aqueous all-organic battery utilizing a poly(ethynyl phenyl galvinoxyl) anode and PTMA cathode, but the battery failed to achieve a desirable performance because PTMA is more suitable for neutral or weak acid electrolytes, while galvinoxyl is more compatible with alkaline conditions [24]. They also reported a battery based on a poly(galvinoxyl styrene) (PGS) anode and a TEMPO-based poly(norbornene) cathode, and the battery only exhibited an output voltage of 0.6 V and a discharge capacity of 32 mAh g^{-1}, while a gratifying cycling stability and rate performance of 10 C were obtained. It is worth noting that, besides being a king of anode materials, PGS can work as a traditional oxide cathode but also as an organic cation intercalation cathode. This allows the battery to operate in a "rocking chair" mode, in which only Li^+ participates in charge transport. Therefore, a small amount of supporting electrolyte is enough to maintain the redox process of the electrode. Suga et al. prepared a battery with poly(TEMOO norbornene) as anode and PGS as cathode, which illustrates a reversible capacity of 29 mAh g^{-1} and an output voltage of 0.6 V. When utilizing lithium metal as anode, the Li/PGS battery obtained a capacity of 42 mAh g^{-1} (based on the mass of cathode) and a working voltage of 3.1 V, and there is no obvious capacity decay during 500 cycles [14].

The last kind of stable radical to mention is verdazyls, the only family of neutral radicals that has outstanding stability compared to nitroxides [25]. In addition to the properties of magnetic and coordination chemistry, verdazyls possess electrochemical redox properties. Gilroy et al. found that the redox potential and reversibility of verdazyls are related to substituents in the molecular structure, in which the methylene-bridged verdazyls show a low oxidation potential, so that can be a great anode material, and the capacity of the cell based on verdazyls is higher than that for other neutral radicals. There are also some promising stable radical materials such as phenoxyl radical [26] and Blatter's radical [27]. However, many of them have not been applied in batteries yet, and it is prospective to exhibit promising electrochemical behaviors in the future.

5.2.5 Other Effect of TEMPO

As a redox-active species, TEMPO can be used not only as an electrode material but also as a redox mediator in other battery systems. For lithium-oxygen batteries, the insulating nature of deposited Li_2O_2 will passivate the surface of the cathode and result in unwanted side products, which can be efficiently addressed by the addition of redox mediators. In 2016, Zhang et al. adopted PTMA as an organic catalyst in Li-O_2 batteries [28]. They found that the n-type reaction of PTMA accelerated the formation of Li_2O_2 during the discharge process, while the oxidation of L_2O_2 is also catalyzed by the p-type conversion of PTMA. Taking advantage of reversible charge/discharge reaction and applicable redox potential, PTMA facilitated the capacity release and stable cycling of the Li-O_2 battery. As shown in Figure 5.6a and

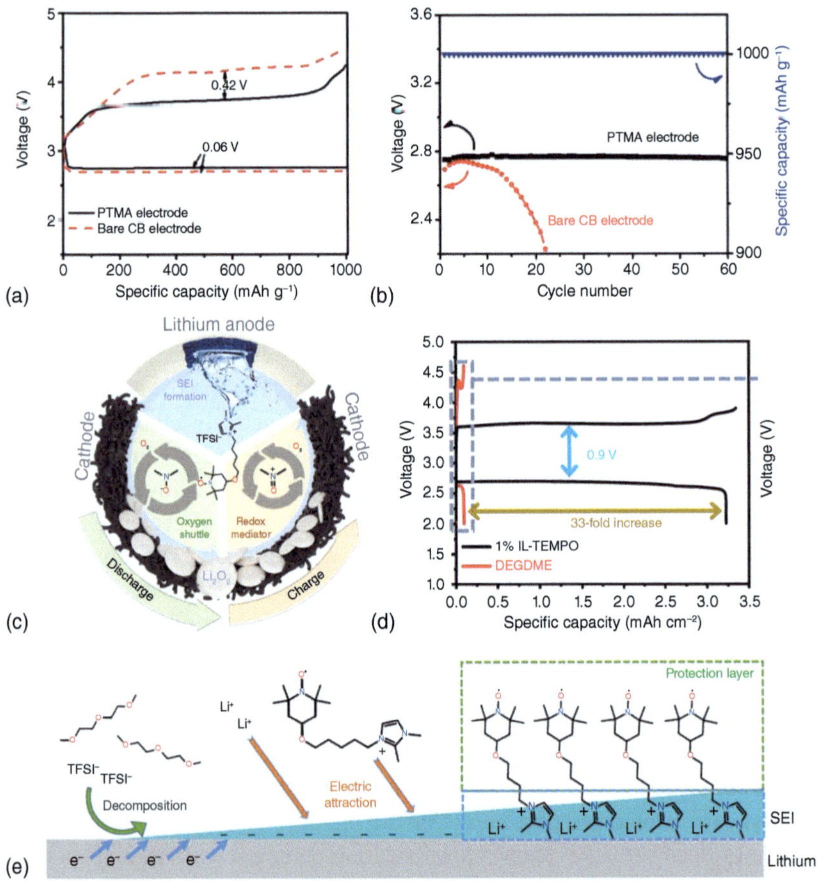

Figure 5.6 (a) Discharge/charge and corresponding (b) profiles of Li-O_2 batteries with and without PTMA. Source: Berezin et al. [27]/John Wiley & Sons/CC BY 4.0. (c) Schematic illustration of IL-TEMPO catalyzing the Li-O_2 battery. (d) The discharge/charge profiles of a Li-O_2 battery with the 1% IL-TEMPO. (e) The schematic illustration of the SEI formation on lithium metal anode within the IL-TEMPO electrolyte. Source: Zhang et al. [28]/Springer Nature/CC BY 4.0.

b, the battery achieves a capacity of 1000 mAh g^{-1} over 50 cycles and the charge overpotential dwindles from 4 to 3.58 V.

However, soluble mediators will corrode the lithium anode. Based on this challenge, Zhang et al. reported a multifunctional TEMPO-grafted ionic liquid, incorporating the function of the redox mediators and lithium metal protection as well as the design of electrolyte solvent in one strategy [29]. The as-synthesized ionic liquid 1,2-dimethyl-3-(4-(2,2,6,6-tetramethyl-1-oxyl-4-piperidoxyl)-pentyl) imidazolium bis(trifluoro methane) sulfonimide (IL-TEMPO) has highly redox reversibility with its n- and p-type reactions, acting as the redox mediator and accelerating the discharge and charge process of Li-O_2 battery (Figure 5.6c). Furthermore, highly reactive imidazolium groups tend to react with Li metal and yield a superior SEI, as exhibited in Figure 5.6e, facilitating a uniform deposition and dissolution of Li. Compared to the Li-O_2 battery with a commonly used ether electrolyte, the IL-TEMPO achieved a 33-fold increase in capacity and a low charge voltage of 3.6 V (Figure 5.6d). Robust SEI formed on Li anode guarantees a more stable cycle performance over 200 cycles. Moreover, IL-TEMPO acting as an electrolyte solvent enables a high work temperature of 70 °C for batteries with excellent electrochemical performances. Similar to Li-O_2 batteries, TEMPO can also be used in electrochemical double-layer capacitors and improve the capacitance of electrodes. Hu et al. introduced TEMPO in an aqueous electrolyte to energize extra capacitance of activated carbon through the large electron-transfer rate constant and rate performance of TEMPO [30].

5.3 Radicals for Redox Flow Batteries

Redox flow battery (RFB) is a new type of large-scale and high-capacity electrochemical energy storage device in which the active material is dissolved in the flowing liquid electrolyte [31]. At present, modern society has an increasing demand for energy storage technology and calls for extensive use of renewable energy. In this context, RFB with low cost and wide applicability has great potential applications in the field of large-scale energy storage. Redox-active materials as carrier of energy conversion, play a vital role in RFB [32]. Compared with inorganic electrode materials with high cost and limited resources, organic molecules have the advantages of accessibility, low cost, and richness in elements. Moreover, the organic molecular structure is easy to design and control, providing a variety of electrochemically active materials [32].

In recent years, a series of organic active molecules have been developed. The redox radical species have also been successfully used in static organic batteries, and they are expected to be explored in RFB fields. The high activity and fast kinetics of radicals indeed afford unique power density; however, more efforts must be made to study how to improve the stability and solubility of radical materials [33]. The introduction of different functional groups or polymer backbones through molecular design is proposed to optimize the compatibility of radical materials. Many studies have confirmed that nitroxyl radicals like TEMPO, as representative stable radicals, have a high redox potential and excellent reversibility. TEMPO is also the most widely studied type of radical in RFB, in which the charge/discharge mechanism, like that in traditional organic batteries, mainly involves the p-type conversion of TEMPO [34].

5.3.1 Functionalization for Radicals

One of the key factors to improve the performance of RFB is to enhance the solubility of active materials to obtain a higher energy density. Generally, the solubility of organic molecules mainly depends on intermolecular interaction and solvent polarity, and molecules with weak intermolecular interaction usually have higher solubility, such as TEMPO. Wei et al. found that the maximum concentration of TEMPO in a mixed carbonate solvent of EC/PC/EMC reached 5.2 M. Even in the presence of supporting lithium salt of $LiPF_6$, the solubility still exceeded 2 M. With the lithium metal as counter electrode and 15 wt% FEC additive, an average capacity fade of 0.2% per cycle was obtained [35]. The battery voltage is 3.5 V, and the energy density exceeds 120 Wh l^{-1}. However, the electrolyte shows increased viscosity due to its high concentration, causing adverse effects for pumps to drive the electrolytes. In addition, the limitation of power density also restricts its application in actual scenarios. Then, Milshtein et al. enhanced the battery voltage by introducing 4-acetamido functional group in TEMPO; the cycle stability and redox kinetics were also improved. However, the solubility of AcNH-TEMPO decreases to 0.5 M, decreasing the energy density of the battery [36].

It is reported that the interaction between redox molecules is related to their melting points, and low melting points are expected to obtain higher solubility. In addition, when the melting point drops below room temperature, the liquid redox species can even be used as an auxiliary solvent. Based on this, Takechi et al. introduce a high-concentration catholyte of solvated ionic liquid called redox supercooled liquid, which is formed by a mixture of stable radicals and lithium salts with a melting point lower than room temperature [37]. They chose TEMPO (and MeO-TEMPO) as redox-active materials and LiTFSI as supporting salt. As exhibited in Figure 5.7a, the formed Li^+-TEMPO complex coupling with $TFSI^-$ anion shows the property of a deep eutectic electrolyte. It is noted that the prepared catholyte showed the best energy density when the ratio of LiTFSI and MeO-TEMPO was 1 : 1, but the viscosity also reached 10 Pas because of the high concentration of MeO-TEMPO. Interestingly, the viscosity will reduce by introducing a small amount of water, accompanied by an increasing ionic conductivity of 3.1 mS cm^{-1}. In the flow cell tests, benefiting from the high concentration of active materials, the complex catholyte with 17 wt% water addition combined with lithium metal anode delivers a high CE of 99% and an energy density of 200 Wh l^{-1}.

Although organic electrolytes have the advantages of high solubility and wide work potential, their high cost and probable pollution of the environment make them unsuitable for large-scale application. In this case, aqueous electrolytes show great superiority. To adapt to the aqueous solvation environment and narrow voltage window (1.23 V), it is necessary to modify the redox-active molecules. Winsberg et al.

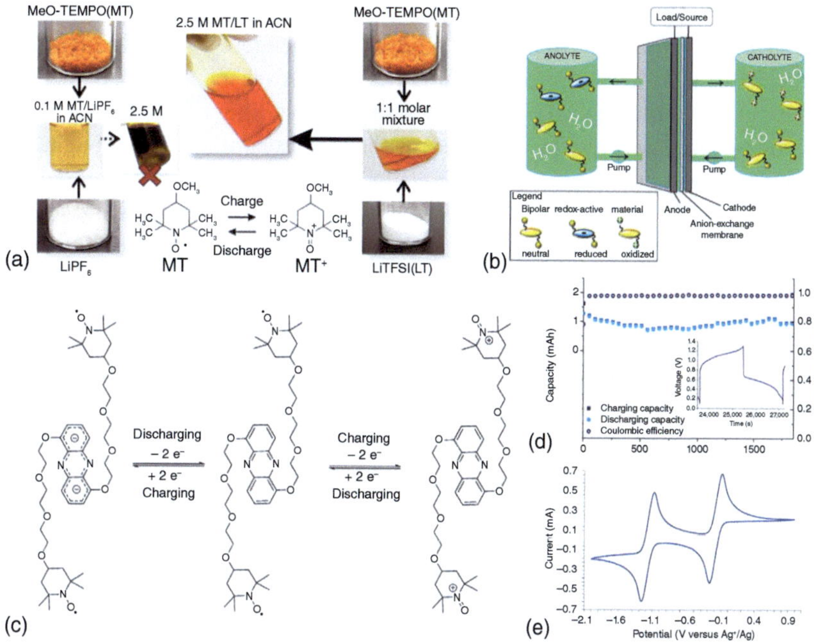

Figure 5.7 (a) Optical photo of the materials for both routes in each step and the reaction of TEMPO. Source: Adapted with permission from Milshtein et al. [36]/Elsevier. (b) Schematic graph of a symmetric aqueous RFB with bipolar combi-molecule. (c) The redox mechanism of TEG-linked TEMPO and VM. (d) Long-cycle performance of RFB in (c). Source: Reproduced with permission Takechi et al. [37], American Chemical Society. (e) CV test of verdazyl radical in 0.5 M $TBAPF_6/CH_3CN$. Source: Reproduced with permission Duan et al. [38], Elsevier.

used a triethylene glycol (TEG) unit to connect two independent active molecules of TEMPO and phenazine through a covalent bond (Figure 5.7b). This strategy promotes the solubility of the synthesized electrode materials in aqueous solution [39]. In addition, different redox-active sites in the molecular structure can release capacity in isolated potential intervals, called ambipolar redox molecules, which can be used as both anode and cathode (Figure 5.7c). It has been proved that both TEMPO and phenazine have excellent redox reversibility. Meanwhile, the same electrode materials eliminate the adverse effects of the shuttle effect between the cathode and anode. Therefore, the assembled RFB showed good electrochemical stability with a voltage of 1.2 V. As shown in Figure 5.7d, there is no obvious capacity decay more than 1800 cycles with a CE of 98.3%, however; the voltage drops of charge and discharge limit the energy efficiency.

As an ambipolar active molecular, only the p-type reaction of TEMPO is commonly considered stable in aqueous and nonaqueous electrolytes. The irreversibility of the n-type reaction of TEMPO originates from the easy attack of aminoxy anion by proton donor solvent. To address the block of reduction process of TEMPO, Wei et al. reported a novel nitronyl nitroxide radical, 2-phenyl-4,4,5,5-tetramethylimidazoline-1-oxyl 3-oxide (PTIO), as a stable ambipolar charge storage material for RFB [38]. The N–O$^-$ anion group produced in an n-type reaction was stabilized by π-conjugation and steric hindrance engendered in molecule. The introduction of TBA$^+$ cation instead of Li$^+$ further improves the redox reversibility of PTIO$^-$ anion. Although PTIO exhibits a sizable solubility of 2.6 M, the high electrolyte viscosity and resistance as well as the serious shuttle effect brought by high PTIO concentration lead to low CE. Thus, the amount of active material in the flow cell was limited to 0.5 M. Nevertheless, two redox processes afford a cell voltage of 1.73 V as well as an energy density of 9 Wh l^{-1} during charge.

Besides PTIO, verdazyl radicals are another kind of bipolar electrochemical active material and exhibit good air- and water-stability. As mentioned above, the redox character relates to its structure, which has a 1,2,4,5-tetrazine hexatomic circular central. The unpaired electrons that resonated among the N and C atoms in the ring provide the stability of Verdazyl. Dyker et al. found that the carbonyl-substituted oxo-verdazyl generally behaved more stably because the conjugated structure of C=O bond could afford extra resonance area [40]. They also reported a modified verdazyl radical, 3-phenyl-1,5-di-*p*-tolylverdazyl (PDTV), as both catholyte and anolyte active materials in symmetric flow batteries. As shown in Figure 5.7e, the electron accepting and losing processes occur at −0.17 and −1.15 V (versus Ag/Ag$^+$), respectively. The flow batteries with $TBAPF_6$/MeCN electrolyte show a cell voltage of 0.97 V and high utilization of active species for the initial charge and discharge. The CE was still among 94–97% throughout the experiment. However, the very low verdazyl concentration of 0.04 M and short cycle life span of 50 cycles restrict the application, in which the instability is mainly caused by the decomposition of PDTV molecule. After that, Korshunov et al. studied a series of verdazyl radicals to improve molecular stability, among which 1,5-diphenyl-3-isopropyl-6-oxoverdazyl (isoV) shows better solubility and CE in nonaqueous electrolyte [41]. Unfortunately, the bad compatibility between isoV and separator leads to serious resistance increases and performance deterioration.

Benefiting from its nonflammable properties and high ionic conductivity, aqueous electrolyte is designed as a low-cost substitution for redox flow batteries. PH-natural aqueous electrolytes with noncorrosive and moderate nature offer a safe option for large-scale energy storage. Liu et al. reported an all-organic aqueous RFB with sustainable methyl viologen (MV) as anolyte and 4-hydroxy-2,2,6,6-tetramethylpiperidin-1-oxyl (HO-TEMPO) as catholyte (Figure 5.8a). It has been demonstrated that MV possesses a high solubility of 3 M in water and is excellently redox-reversible [42]. Moreover, compared to TEMPO, hydrophilic –OH group also has greater solubility for the whole molecule. Combined with a NaCl-supporting electrolyte and Cl$^-$-exchange membrane, the batteries show a high voltage of 1.25 V and excellent stability for 100 cycles with nearly 100% CE (Figure 5.8b). However, the shuttle effect of soluble small molecules of HO-TEMPO results in significant capacity decay.

Besides OH-TEMPO, Zhou et al. investigated the effect of different substitutive TEMPO radicals, such as carbonyl, amino, nitrile, and carboxyl, through studying the redox potential, kinetics process, stability, and solubility of these derivatives in aqueous electrolytes [43] (Figure 5.8c). The result demonstrates that the key

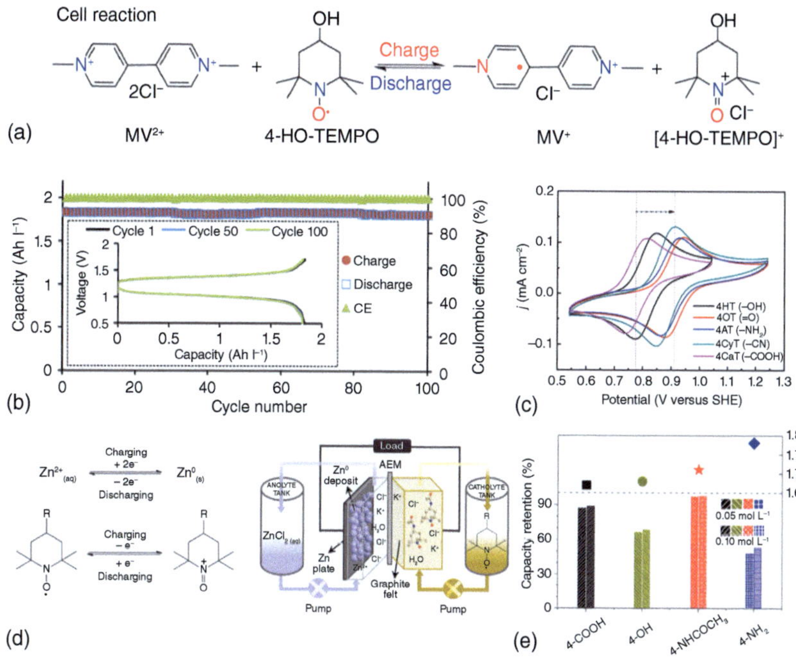

Figure 5.8 (a) Discharged and charged reaction of 4-OH-TEMPO and MV^{2+}. (b) Cycle performance and corresponding charge/discharge profile of 4-OH-TEMPO/MV^{2+} RFB. Source: Reproduced with permission from Korshunov et al. [41], Wiley-VCH. (c) CV test of five TEMPO derivatives. Source: Reproduced with permission Liu et al. [42], The Royal Society of Chemistry. (d) Redox mechanisms and schematic illustration of R-TEMPO/Zn hybrid AFB. (e) Capacity retention of R-TEMPO/Zn AFBs with different derivatives. Source: Reproduced with permission Zhou et al. [43], American Chemical Society.

challenge is the crossover of the TEMPO compounds and the resultant considerable performance degradation. Then, Fan et al. further figured out the capacity fading mechanism of different 4-position substituent R groups on TEMPO in zinc hybrid RFB [44]. The so-called hybrid flow cell refers to the fact that at least one of the oxidation and reduction states of active materials is insoluble. A typical example is the zinc hybrid flow cell, in which the conversion between zinc and Zn^{2+} occurs at the anode during the cycle, while fully soluble organic or inorganic materials are generally used in the cathode (Figure 5.8d). They found that the redox potential and stability of R-TEMPO depend on radical charge amount and energy levels, in which the N-acetyl on 4-position brings lower radical energy and larger steric hindrance, effectively suppressing the side reactions (Figure 5.8e). Therefore, the 4-$NHCOCH_3$–TEMPO/Zn flow battery with $ZnCl_2$ supporting electrolyte delivers a high capacity retention of 99.65% per day at a current density of 10 mA cm^{-2}. The stable cathode material enables a long-term cycle for more than 800 hours and a continuous energy density of 11.4 Wh l^{-1}.

Abominable cross-contamination issues during cycling existing in symmetric redox flow batteries are always an annoyance. In 2021, Nambafu et al. reported a low-cost Riboflavin-TEMPO (RIBOTEMPO) molecule and successfully used it in symmetric flow batteries. Riboflavin-TEMPO combines with the advantage of both Riboflavin and TEMPO units, for which PH-neutral supporting electrolyte and inexpensive anion exchange membrane can meet the requirement [45]. CV tests demonstrate that two reversible redox couples at −0.72 and 0.51 V (versus Hg/Hg_2SO_4) correspond to Riboflavin and TEMPO respectively, exhibiting a voltage of 1.23 V. The RIBOTEMPO batteries could stably work for more than 100 cycles with a CE of 80% at a current density of 2.5 mA cm^{-2}. In spite of its low cost and environmentally friendly features, the decomposition of RIBOTEMPO in an aqueous solution leads to rapid capacity decay.

One way to increase the energy density of a flow battery is to increase the operating potential of electrode materials. Hu et al. proposed a new class of radical molecules with five-membered ring pyrrolidine and pyrroline moieties [46] (Figure 5.9a). Through introducing a C=C double bond in a radical ring structure, the synthesized 3-carbamoyl-2,2,5,5-tetramethylpyrroline-1-oxyl (CPL) shows a higher redox potential of 0.76 V (versus SSCE) than the origin TEMPO-type radical (0.60 V). They found that the ring structures have a great impact on Gibbs free energy for the electrochemical reaction of radical and the charge amount population on N–O radical head, and the lower free energy in reduction reaction and charge population sum of N–O radical head significantly improve the electrode potential. Combined with a stable BTMAP-Vi ionic liquid anolyte, the CPL/BTMAP-viologen aqueous RFB exhibits a voltage of 1.31 V (Figure 5.9b) and high capacity retention of 99.96% for more than 500 cycles.

To further address the narrow potential gap and separator fouling, Gao et al. recently reported a high-performance PH-neutral aqueous flow battery with functionalized viologen derivative 1,1′-bis(3-phosphonopropyl)-4,4′-(1,4-phenylene) bipyridinium dibromide ($(PPBPy)Br_2$) as anode and longer-side chain 4-(propoxy-3-sodium sulfonate)-TEMPO (PSS-TEMPO) as cathode [48] (Figure 5.9c). The phenylene-bridged viologen structure displays a redox potential of −0.801 V (versus SHE) and low-level shuttle issue through cation-exchange membranes; thereby, the half-cell based on $(PPBPy)Br_2$ achieves a stable long-cycle performance over 1000 times. On the cathode side, the longer sulfonate chain effectively relieves the effect of the electron-withdrawing group on the skeleton stability and suppresses the side reaction of TEMPO radicals, meanwhile, providing a lower permeability through the separator. Hydrophilic sulfonate unit also enables a high solubility of PSS-TEMPO reach to 2.55 M in a PH-neutral aqueous solution. Therefore, the PSS-TEMPO in catholyte presents a slow capacity degradation of 0.012% per cycle. The prepared $(PPBPy)Br_2$/PSS-TEMPO all-organic flow battery displays a cell voltage of 1.61 V and a high power density of 509 mW cm^{-2}.

5.3.2 Ionization for Radicals

To address the crossover issue in RFB, a series of ionized radical molecules were developed to reduce the permeability in ion-selective membranes by introducing a positive charge on the TEMPO structure. Schubert et al. functionalized

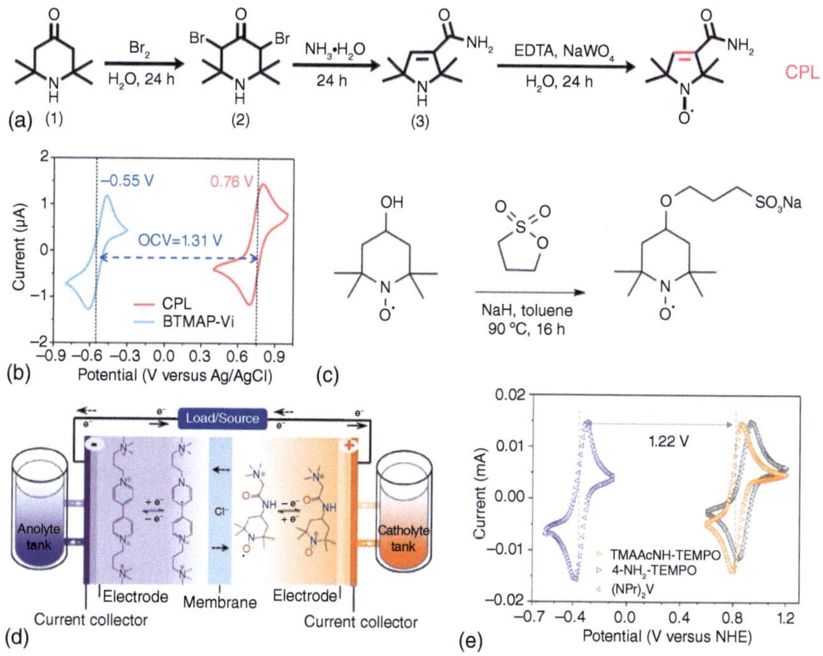

Figure 5.9 (a) Synthetic routes for CPL. (b) CV curves of CPL (red trace) and BTMAP-Vi (blue trace). Source: Reproduced with permission from Nambafu et al. [45], Wiley-VCH. (c) Synthetic routes for PSS-TEMPO. Source: Reproduced with permission from Hu et al. [46], Wiley-VCH. (d) Schematic of RFB with TMAAcNH-TEMPO catholyte and $(NPr)_2V$ anolyte with an anion exchange membrane. (e) CV curves of TMAAcNH-TEMPO, 4-NH_2-TEMPO, and $(NPr)_2V$. Source: Reproduced with permission from Chang et al. [47], Wiley-VCH.

TEMPO with a cationic trimethylammonium group to obtain a more robust *N*,*N*,*N*-2,2,6,6-heptamethylpiperidinyl oxy-4-ammonium chloride (TEMPTMA). The introduction of trimethylammonium chloride substituent ($-N(CH_3)_3{}^+Cl^-$) enhances the voltage window of TEMPO via its strong electron-withdrawing effect. In addition, the hydrophilic ionic group improves the solubility in common NaCl solution [49]. It is noted that the approach of ionization alleviates the permeability in the anion exchange separator and endows an improved cycle performance. After that, Chang et al. synthesized a cationic grafted 4-OH-TEMPO (g^+-TEMPO) through a one-step ring-opening reaction between glycidyl trimethylammonium cation ($GTMA^+$) and 4-OH-TEMPO [47]. The positively charged g+-TEMPO decreases the possible crossover of active species in a cross-linked methylated polybenzimidazole (PBI) membrane. Meanwhile, the lower charging voltage of g^+-TEMPO relieves the side reactions and achieves an enhanced kinetic process. The g^+-TEMPO/Zn hybrid RFB shows improved energy efficiency and cycling stability, with a capacity fade of 0.046% per cycle for more than 140 cycles.

Besides OH-TEMPO, Fan et al. also reported a quaternary ammonium cationic $-NH_2$ substitutional TEMPO molecular, 4-[3-(trimethylammonium)acetylamino]-TEMPO chloride (TMAAcNH-TEMPO), via the strategy of grafting the acetylamino and trimethylammonium units on 4-position [50] (Figure 5.9d). The solubility of the as-prepared compound in water reaches 4.3 M, and the redox potential is located at 0.84 V, as shown in Figure 5.9e, enabling a higher energy density. Compared to 4-NH-TEMPO, theoretical arithmetic results show that the possibility of nucleophilic and electrophilic attack for TMAAcNH-TEMPO decreases so that the side reaction and crossover are effectively relieved. In pair with 1,1′-bis[3-(trimethylammonium)propyl]-4,4′-bipyridinium tetrachloride anolyte and 1 M KCl electrolyte, the RFB delivers an open-circuit voltage of 1.22 V and a long-term lifetime over 40 days (1500 cycles) with a low capacity fade of 0.0144% per hour.

In current RFB system, the choice of catholyte is limited, and aqueous-soluble organic molecules commonly suffer from a short lifetime. Liu et al. proposed an extremely stable TEMPO derivative, 4-[3-(trimethylammonio)propoxy]-2,2,6,6-tetramethylpiperidine-1-oxyl (TMAP-TEMPO) chloride, and used it in all organic RFB (Figure 5.10a). Intermolecular strong coulombic repulsion restrains the bimolecular side reaction and suppresses the molecular decomposition [52]. Functionalized TMAP-TEMPO chloride shows a high solubility of 4.62 M in pH-neutral aqueous solution and a reversible single-electron transfer process at 0.81 V (versus SHE). Coupled with an anolyte material of bis-(trimethylammonio) propyl viologen (BTMAP-Vi) with ultra-long cycle life, as shown in Figure 5.10b, the TMAP-TEMPO/BTMAP-Vi battery exhibits a voltage of 1.1 V and a CE of 99.73%. The capacity retention even reaches 93% over 1000 cycles, corresponding to a capacity fade rate of 0.026% per hour.

To date, quaternary ammonium ionic unit for redox organic compound have been broadly explored, and Chang et al. further studied the effect of imidazolium-based substituents on TEMPO [51]. They synthesized an imidazolium-grafted TEMPO via a simple two-step reaction and applied it in zinc hybrid aqueous RFB. In combination with 1 M NaCl aqueous solution and a Cl^--conducting membrane, the battery achieved a cell voltage of 1.64 V and a high CE of 99.8%, which is superior to the original 4-OH-TEMPO. Moreover, with 0.1 M catholyte concentration at a current density of 50 mA cm^{-2}, stable cycle performance was demonstrated over 400 cycles with a capacity fade of 0.15% per cycle. Similar molecular engineering strategies have recently been reported. Fan et al. replaced the hydroxyl group with the amino group and synthesized a "p–π" conjugated acetylamino functionalized TEMPO [53] (MIAcNH-TEMPO). Together with the "π–π" conjugated effect imidazolium, as exhibited in Figure 5.10c, the electron density on the N–O head was effectively delocalized and thereby improved the stability of radical. Figure 5.10d shows that the MIAcNH-TEMPO/Zn battery with 1.5 M active materials achieves an excellent capacity retention of 99.95% per day over 16.7 days (400 cycles). The cell voltage of the system and the theoretical energy density for catholyte are up to 1.71 V and 68.7 Wh l^{-1} respectively. More importantly, simple synthesis reactions and high yield of raw materials are conducive to industrial-scale production.

It is reported that pyrrolidinium group could act as cationic part of ionic liquids, in which chemical stability and strong polarity usually result in broaden electrochemical window and great solubility in aqueous solution. Based on this,

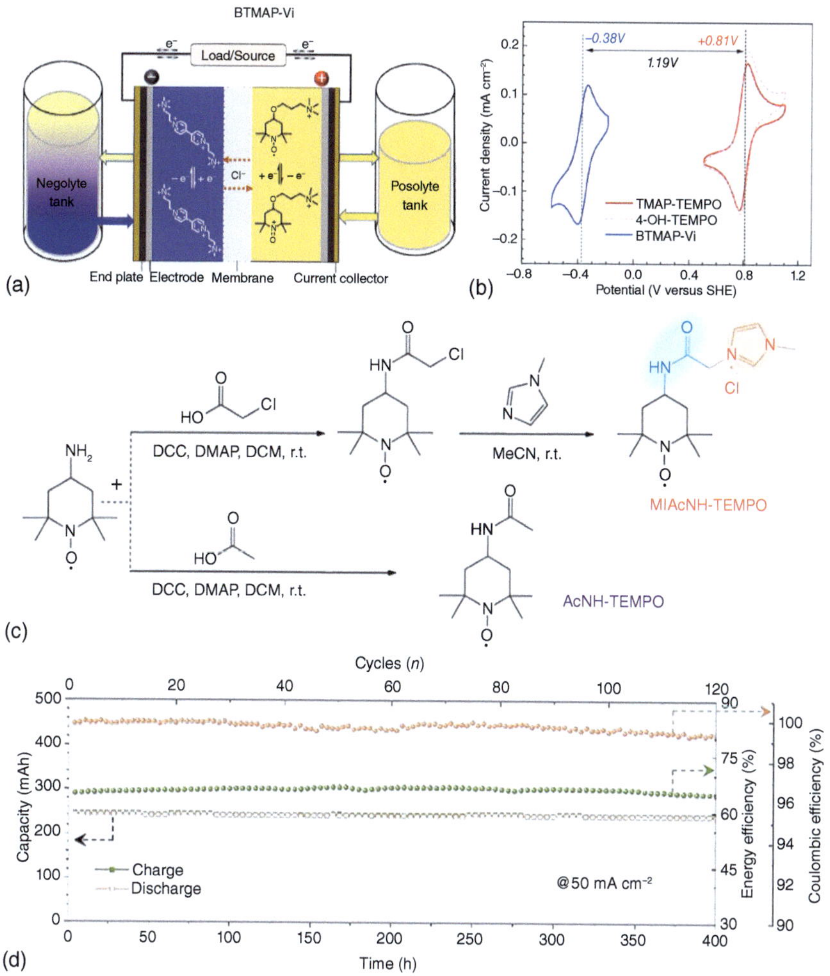

Figure 5.10 (a) Schematic of RFB assembled with TMAP-TEMPO and BTMAP-Vi an anion-selective membrane. (b) CV test of TMAP-TEMPO, 4-OH-TEMPO, and BTMAP-Vi. Source: Reproduced with permission Fan et al. [50], Elsevier. (c) Different synthetic routes. Trapezoid: p–π conjugate group. Pentagon: π–π conjugate group. (d) Cycle performance of MIAcNH-TEMPO/Zn RFB. Source: Reproduced with permission from Chang et al. [51], Wiley-VCH.

Pan et al. proposed pyrrolidinium cation functionalized TEMPO (Pyr-TEMPO) and extended viologen ((PyrPV)Cl_4), respectively, acting as cathode and anode in RFB [54] (Figure 5.11a). Taking advantage of the hydrophilic pyrrolidinium unit, two active materials both achieve outstanding water-solubility of 3.35 and 1.13 M respectively. The as-prepared RFB obtained a high energy density of 16.8 Wh l^{-1} under a cell voltage of 1.57 V (Figure 5.11b), which is extremely attractive in current PH-neutral aqueous RFB. Benefited from the superb stability and reversibility of electrode materials, the battery shows a long cycling life of over 1000 cycles with a low capacity decay of 0.05% per cycle, of which the CE is almost 100%.

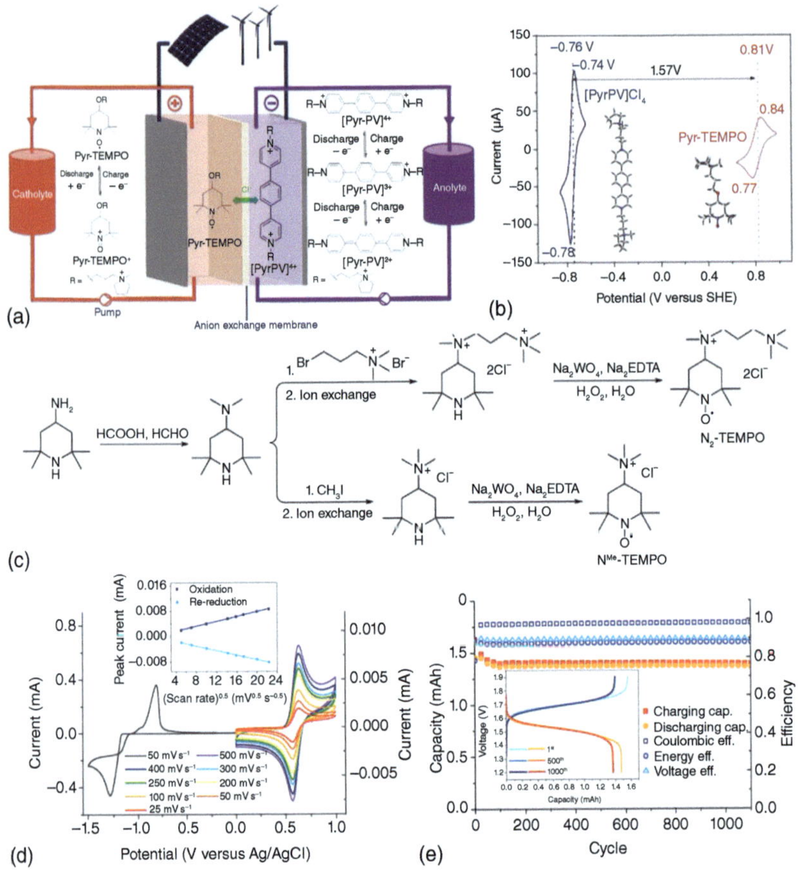

Figure 5.11 (a) Structural configuration and redox processes of Pyr-TEMPO/[PyrPV]Cl_4 RFB. (b) CV curves of[PyrPV]Cl_4 and Pyr-TEMPO dissolved in 0.5 m NaCl solutions. Source: Reproduced with permission from Fan et al. [53], Wiley-VCH. (c) Synthesis of N_2-TEMPO and NMe-TEMPO. Source: Reproduced with permission from Pan et al. [54], Wiley-VCH. (d) CV measurements of 1 mM TEMPO-4-sulfate potassium in aqueous electrolyte. (e) Long-term performance and charge–discharge profile of TEMPO-4-sulfate potassium. Source: Reproduced with permission Hu et al. [55], American Chemical Society.

The introduction of cation groups in organic molecules has been identified as a promising approach to improve the performance of radical materials. However, whether increasing the number of cation sites is beneficial is unclear. In 2021, Hu et al. reported a new TEMPO derivative with a dual-ammonium di-cationic group grafted on 4-position of TEMPO (Figure 5.11c), namely N_1, N_1, N_1, N_3, N_3, 2, 2, 6, 6-nonamethyl-N_3-(piperidinyloxy)propane-1,3-bis(ammonium) dichloride (N_2-TEMPO) [55]. Profited from the stronger electrostatic repulsion and steric effect of two cationic centers, the bimolecular interactions and serious crossover of organic molecules are immensely inhibited. Moreover, the strong electron-withdrawing group enhances the potential of TEMPO-based materials. Large molecular size could further suppress the permeation of catholyte through the membrane. Therefore, with the incorporation of a viologen ($(NPr)_2V$) anolyte, the battery achieves a high energy density of 18.1 Wh l^{-1} with a voltage of 1.35 V and 1 M active materials. With an electrolyte concentration of 0.5 M, an excellent power density of 114 mW cm^{-2} at 60 mA cm^{-2} was obtained, accompanying nearly 100% capacity retention for 400 cycles.

In addition to the cationic group, anions can also act as a feasible substituent group on TEMPO to obtain similar results. Winsberg et al. synthesized TEMPO-4-sulfate potassium salt by introducing sulfonic acid anion on 4-HO-TEMPO (Figure 5.11d). The high solubility of TEMPO-4-sulfate potassium in $ZnCl_2$ solution enables an enhanced capacity and energy density of 26.7 Ah l^{-1} and 20.4 Wh l^{-1}. As shown in Figure 5.11e, a long-cycle performance over 1000 cycles with a capacity retention of 94% confirms the stability of this TEMPO-derivative. However, TEMPO-OSO_3^- may form large aggregates of zwitterionic salt and precipitate in a charged state, which can be mitigated by an adequate amount of supporting electrolyte [56].

5.3.3 Radicals Polymer

In addition to using TEMPO monomers as active materials, as mentioned above, TEMPO groups can also be grafted onto polymer backbone to enhance the functionality of radical molecules. Furthermore, the crossover of active materials is a key challenge of RFB, which leads to the loss of capacity and energy efficiency; therefore, a permselective separator for RFB is usually necessary. Compared with the development of expensive and complex ion-selective separators, polymers with larger sizes can effectively alleviate the crossover of active materials. However, polymers with long chain and high molecular weight bring other problems, such as low solubility, high viscosity, and low conductivity. Practical requirements can be fulfilled through reasonable molecular design, including the modification of polymer chains and side chain groups.

Sukegawa et al. introduced a radical group-riched bottlebrush-type polymer, poly (norbornene)-g-poly(4-methacryloyloxy2,2,6,6-tetramethylpiperidin-1-oxyl) (PNB-g-PTMA), which is synthesized through the anionic polymerization of norbornene-substituted diphenyl hexyl lithium and TEMPO monomer as well as the subsequent ring-opening polymerization of norbornene unit [57]. The polymer obtained by grafting-through has better structural adjustability, and the size can be tuned by the lengths of both side chain and main chain. Importantly, bottlebrush polymer shows a lower viscosity compared with single linear polymer and thereby enabling an increased solubility in organic solvent. The half-cell containing 0.1 M PNB-g-PTMA through a flow channel shows a 95% charge capacity of theoretical value, and less than 1% of active materials penetrate the porous separator after working for 24 hours. These results suggested that the size effect of polymer molecules is also the key to improve the charge/discharge performance of RFB.

To address the challenge of the high viscosity of electrolytes caused by polymers, polymer molecules with special shapes have been developed. Winsberg et al. introduced a micellar structure containing radical groups to relieve the solubility problem in carbonates. The micellar structures were obtained by block copolymerization of polar PTMA and unpolarized poly(styrene) (PS) block [58]. Then, the as-prepared PTMA-PS diblock copolymer catholyte was successfully applied in polymer/zinc hybrid RFB. The assembled PTMA-PS/Zn batteries show stable cycle performance between 0.5 and 2.0 V with a high CE of 99.8% as well as a capacity retention of 95% over 1000 cycles, displaying an excellent redox reversibility of the cathode active material. The special structure of micellar aggregates enables good solubility and a decreasing viscosity of electrolyte (Figure 5.12a), and furthermore, the active molecule is easy to be confined by a porous separator.

The excellent performance of TEMPO in organic electrolytes is obvious, but from the perspective of ecology and economy, the extensive use of organic solvents is not a long-term solution. Generally, aqueous electrolyte has higher conductivity and lower cost; therefore, for broad applicability and safety, similar design concepts above should also be applied in aqueous electrolyte. In 2015, Janosch et al. reported a safe, scalable, low-cost all-organic battery system, which uses polymer as electrode material and NaCl aqueous solution as electrolyte [59]. Combined with a cheap dialysis membrane, the RFB, completely based on aqueous solution and green polymer, obtained an energy density of 10 Wh l^{-1}, and the current density reached 100 mA cm^{-2}. Specifically, the constructed polymer exhibits enhanced water solubility and enough to remain dissolved during the charge/discharge process. The active polymers in the cathode and anode are derived from TEMPO and viologen, respectively, and their water solubility is improved by quaternary ammonium cation on the polymer chain (Figure 5.12b). This RFB system not only obtains a high capacity of 10 Ah l^{-1} but also maintains a low viscosity of 17 and 5 mPa s for catholyte and anolyte, facilitating the efficient transport of the solutions through the flow channel. Moreover, as exhibited in Figure 5.12c, an unpumped cell based on the electrode materials also delivers a long-term performance of 10 000 cycles with a high capacity retention of 80%. This study further indicates that the introduction of branched polymers instead of linear polymers can generally reduce the viscosity of electrolytes at a higher concentration, thus increasing the energy density.

The application of halogens in aqueous electrolyte RFB has been studied in the early stage, but the inherent corrosivity of halogens brings a series of safety risks. For this problem, Winsberg et al. proposed a green TEMPO-based/zinc hybrid aqueous flow battery. Compared with halogens, the TEMPO molecules improved by the Winsberg approach made great progress in the reduction of toxicity and environmental pollution. They synthesized different TEMPO-containing copolymer molecules as cathode, and the supporting electrolytes were mainly composed of nontoxic $ZnCl_2$, NH_4Cl, and NaCl [60]. These active polymers show a cell voltage of 2 V in aqueous solution. Among these, p(TEMPO-co-METAC) (METAC is[2-(methacryloyloxy) ethyl] trimethylammonium chloride) shows a capacity retention rate of 79% after 1000 cycles at a current of 12 mA cm^{-2}. The energy density is 1.9 Wh l^{-1} and the energy efficiency is up to 80%. In addition, these materials can also obtain 99.7% CE and 500 stable cycles in conventional carbonate solvent. More importantly, this strategy requires no inert atmosphere (e.g. O_2, H_2O) and is more suitable for large-scale energy storage.

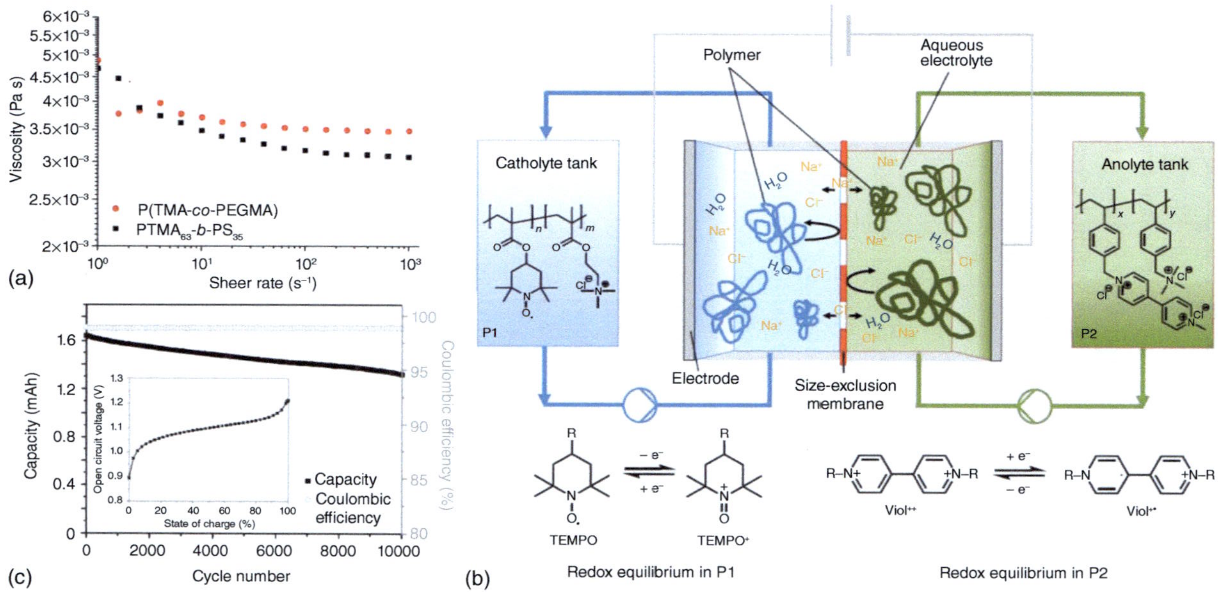

Figure 5.12 (a) Rheological measurements of 13 mg ml^{-1} PTMA$_{63}$-b-PS$_{35}$ and P(TMA-*co*-PEGMA)$_{24}$ in carbonate electrolyte. Source: Reproduced with permission Sukegawa et al. [57], The Royal Society of Chemistry. (b) Schematic representation of a polymer-based RFB with a semipermeable size-exclusion membrane and corresponding electrode reactions of two polymers. (c) The long-term stability of the polymer-based battery in an unpumped test cell. Source: Reproduced with permission from Winsberg et al. [58], Springer Nature.

The cation-suspended radical polymer has achieved better performance; the same effect may occur for the anion group. Hagemann et al. selected zwitterionic[(2-(methacryloxy)ethyl) dimethyl-(3-sulfopropyl)]ammonium hydroxide as copolymerization monomer to synthesize a new TEMPO-containing polymer [61]. Due to low viscosity and high solubility of zwitterionic group, the polymer obtains a high charge storage capacity of $20\,\mathrm{Ah\,l^{-1}}$ and a better kinetic process, of which the diffusion coefficient and electron transfer rate constant are up to 3.77×10^{-7} and $9.72\times10^{-4}\,\mathrm{cm\,s^{-1}}$, respectively. Incorporation with a size exclusion and anion-exchange membrane, the TEMPO polymer symmetric cell achieves a stable 1000-cycle performance with a capacity decay of 0.08% per cycle. The all-organic RFB was further investigated with MV as an anolyte, which delivers a stable performance of 125 cycles with a low capacity loss of 0.29% per cycle. Moreover, such RFB exhibits a high CE and energy efficiency of 99.01% and 93% during the whole cycle process.

The development of organic redox-active materials has certainly injected new vitality into the field of energy storage, but the question now is whether these materials can be degraded in nature and environmentally friendly or can be recycled. Based on these, Nguyen1 et al. designed and synthesized degradable polypeptide-based cathode and anode materials and successfully applied them in all organic RFB [62]. Starting from L-glutamic acid and esterification reactions and then undergoing two divergent multi-step synthetic pathways, viologen and bi-TEMPO polypeptide anodes and cathodes materials can be obtained respectively. The viologen chloride polypeptide showed two reversible redox potentials at 2.23 and 2.69 V and the viologen-chloride (viol-Cl) polypeptide exhibited one redox peak at 3.36 V (versus Li^+/Li), providing a voltage of 1.1–1.6 V for a full battery. The battery obtained an initial charge capacity of 37.8 and $7.5\,\mathrm{mAh\,g^{-1}}$ after 250 cycles, which is mainly due to the dissolution of active materials in the electrolyte. Furthermore, the designed polypeptides chain contains amide and ester units, which are easy to decompose in harmless material by enzymatic, basic, and acidic conditions. This design concept is essential for the sustainable development of energy materials.

5.4 Summary and Prospect

Recently, electrochemically activated organic radical materials have been widely used in the field of energy storage, and reports on the design and synthesis of stable radical molecules have increased dramatically. Thanks to the excellent reaction kinetics and high working potential of radical materials, it provides a possible choice for the development of a new electrode system. This chapter reviews the main research progress of organic radical electrode materials up to now and summarizes the molecular structure, solubility, stability, and corresponding electrochemical features of each type of radical, including redox potential, kinetic process, cycling performance, etc. In addition, we also discuss the application of different redox radical materials in various battery systems. The relevant contents are summarized in Figure 5.13.

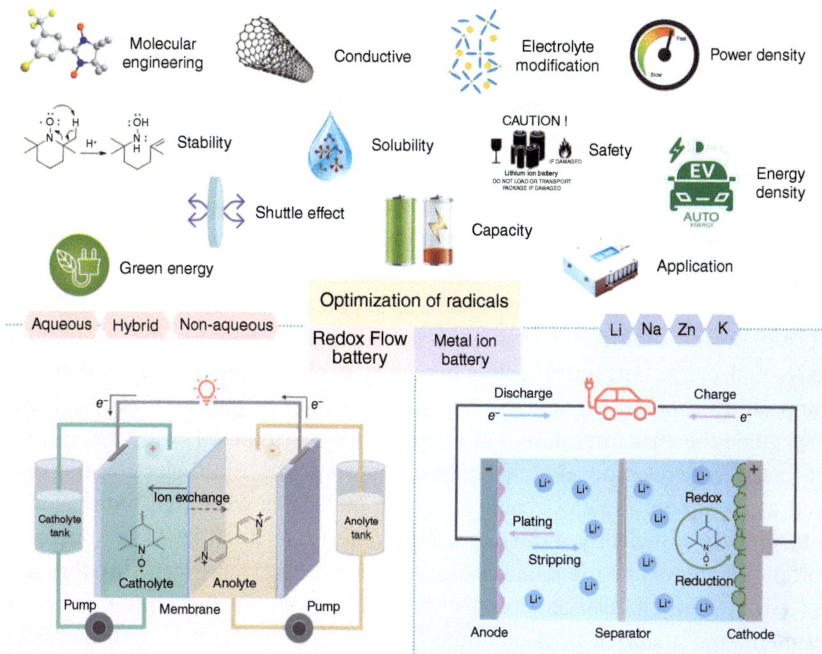

Figure 5.13 Summary of Design and Application for the active radical materials in energy storage.

Redox radical materials inherit the structural diversity and molecularly tunable characteristics of organic molecules. By introducing flexible functional groups or polymer chains, a variety of functionally specific radical molecules can be obtained for different application situations and requirements, such as aqueous or nonaqueous electrolyte systems. Based on the property of redox potential and chemical stability of the designed molecules, it can also be used to match various counter electrode materials like alkali metal, zinc metal viologen, etc.

However, despite remarkable progress, the development of electrochemically active radical compounds still faces several challenges. For general static batteries, the main obstacle is the solubility of organic molecules, of which the challenge can be eased to some extent by combining with polymer skeleton. Nevertheless, there have been few fresh reports recently, and there are no obvious advantages of electrochemical performance over other organic molecules. Therefore, this chapter mainly focuses on flow cells.

Firstly, although it has been found that many organic molecules possess high solubility (>2 M), most reported practical battery performance is based on low concentration (<0.1 M), especially for the radical polymers with large molecular weight, which seriously affects the energy density of the battery. Secondly, it is still worth noting the stability of the highly active free radical molecule. The interaction between the active molecules, or the decomposition caused by the reaction with solvent or air, is unfavorable to the long-cycle performance of the battery. Another problem to be addressed is the nature of the low capacity of radical molecules, which is mainly limited by their one-electron electrochemical reactions and the poor reversibility of n-type reactions. Moreover, most applications of organic electrode materials lack consideration for cost and environmental impact, which are necessary for large-scale application and production.

Fortunately, it is possible to effectively address these issues with suitable strategies. There are two ways to improve the solubility of the materials. One is to optimize the active molecule by introducing solvophilic functional groups (for hydrophilic groups, such as quaternary ammonium cation and sulfonate anion). The second is the utility of compatible electrolytes such as support salts, solvents, or additives, and generally, the additional solvation interaction can provide sufficient solubility of active molecules. Furthermore, it is necessary to clarify the pathway of material decomposition through the analysis and identification of parasitic products and then to guide the structural design of highly stable molecules. Thirdly, there is an urgent need to develop novel active radical materials with multi-electron transfer, as well as to explore the electroactive organic molecules with high redox potentials to provide sufficiently attractive energy density. In addition, it is extremely important to understand the relationships between property and structure through advanced in situ characterization techniques as well as the combination of experiment and theoretical calculations. Most importantly, it is indispensable to transfer from the field of basic study to practical application, establish a new technological-economic-ecological model, and promote the development of radical organic electrode materials.

In summary, remarkable progress has been made in the development of radical electrode materials. An understanding of the basic structure and properties of such molecules provides valuable guidance for further optimization and design. However, it should also be clear that any breakthrough in scientific research requires long-term exploration, and the practical application of organic redox radicals in the field of energy storage is still not clear, which requires more investigation and efforts to achieve scale production and wide application.

Acknowledgments

The work was supported by the National Natural Science Foundation of China (Grant Nos. U2004214 and 21975225) and the National Key Research and Development Program of China (2022YFB2402200).

References

1 Ji, L., Shi, J., Wei, J. et al. (2020). *Adv. Mater.* 32: 1908015.
2 Janoschka, T., Hager, M.D., and Schubert, U.S. (2012). *Adv. Mater.* 24: 6397–6409.
3 Yu, Q., Xue, Z., Li, M. et al. (2021). *Adv. Energy Mater.* 11: 2002523.
4 Xu, D., Liang, M., Qi, S. et al. (2021). *ACS Nano* 15: 47–80.

5 Lu, Y. and Chen, J. (2020). *Nat. Rev. Chem.* 4: 127–142.
6 Shea, J.J. and Luo, C. (2020). *ACS Appl. Mater. Interfaces* 12: 5361–5380.
7 Lu, Y., Zhang, Q., Li, L. et al. (2018). *Chemistry* 4: 2786–2813.
8 Nakahara, K., Iwasa, S., Satoh, M. et al. (2002). *Chem. Phys. Lett.* 359: 351–354.
9 Muench, S., Wild, A., Friebe, C. et al. (2016). *Chem. Rev.* 116: 9438–9484.
10 Kim, J.-K. (2013). *J. Power Sources* 242: 683–686.
11 Kim, J.-K., Kim, Y., Park, S. et al. (2016). *Energy Environ. Sci.* 9: 1264–1269.
12 Li, Y., Jian, Z., Lang, M. et al. (2016). *ACS Appl. Mater. Interfaces* 8: 17352–17359.
13 Zhang, K., Hu, Y., Wang, L. et al. (2017). *ACS Appl. Mater. Interfaces* 9: 34900–34908.
14 Suga, T., Ohshiro, H., Sugita, S. et al. (2009). *Adv. Mater.* 21: 1627–1630.
15 Dai, Y., Zhang, Y., Gao, L. et al. (2010). *Electrochem. Solid-State Lett.* 13: A22.
16 Guo, W., Su, J., Li, Y.-H. et al. (2012). *Electrochim. Acta* 72: 81–86.
17 Koshika, K., Sano, N., Oyaizu, K., and Nishide, H. (2009). *Macromol. Chem. Phys.* 210: 1989–1995.
18 Suga, T., Pu, Y.-J., Kasatori, S., and Nishide, H. (2007). *Macromolecules* 40: 3167–3173.
19 Sukegawa, T., Sato, K., Oyaizu, K., and Nishide, H. (2015). *RSC Adv.* 5: 15448–15452.
20 Oyaizu, K., Suga, T., Yoshimura, K., and Nishide, H. (2008). *Macromolecules* 41: 6646–6652.
21 Oyaizu, K., Kawamoto, T., Suga, T., and Nishide, H. (2010). *Macromolecules* 43: 10382–10389.
22 Suga, T., Konishi, H., and Nishide, H. (2007). *Chem. Commun.* 1730–1732.
23 Hansen, K.-A., Nerkar, J., Thomas, K. et al. (2018). *ACS Appl. Mater. Interfaces* 10: 7982–7988.
24 Jähnert, T., Häupler, B., Janoschka, T. et al. (2013). *Macromol. Chem. Phys.* 214: 2616–2623.
25 Jähnert, T., Hager, M.D., and Schubert, U.S. (2016). *Macromol. Rapid Commun.* 37: 725–730.
26 Price, J.T., Paquette, J.A., Harrison, C.S. et al. (2014). *Polym. Chem.* 5: 5223–5226.
27 Berezin, A.A., Constantinides, C.P., Drouza, C. et al. (2012). *Org. Lett.* 14: 5586–5589.
28 Zhang, J., Sun, B., Xie, X. et al. (2016). *Adv. Sci.* 3: 1500285.
29 Zhang, J., Sun, B., Zhao, Y. et al. (2019). *Nat. Commun.* 10: 602.
30 Hu, L., Shi, C., Guo, K. et al. (2018). *Angew. Chem. Int. Ed.* 57: 8214–8218.
31 Luo, J., Hu, B., Hu, M. et al. (2019). *ACS Energy Lett.* 4: 2220–2240.
32 Fang, X., Li, Z., Zhao, Y. et al. (2022). *ACS Mater. Lett.* 4: 277–306.
33 Wei, X., Pan, W., Duan, W. et al. (2017). *ACS Energy Lett.* 2: 2187–2204.
34 Zhang, L., Feng, R., Wang, W., and Yu, G. (2022). *Nat. Rev. Chem.* 6: 524–543.
35 Wei, X., Xu, W., Vijayakumar, M. et al. (2014). *Adv. Mater.* 26: 7649–7653.
36 Milshtein, J.D., Barton, J.L., Darling, R.M., and Brushett, F.R. (2016). *J. Power Sources* 327: 151–159.
37 Takechi, K., Kato, Y., and Hase, Y. (2015). *Adv. Mater.* 27: 2501–2506.

38 Duan, W., Vemuri, R.S., Milshtein, J.D. et al. (2016). *J. Mater. Chem. A* 4: 5448–5456.

39 Winsberg, J., Stolze, C., Muench, S. et al. (2016). *ACS Energy Lett.* 1: 976–980.

40 Charlton, G.D., Barbon, S.M., Gilroy, J.B., and Dyker, C.A. (2019). *J. Energy Chem.* 34: 52–56.

41 Korshunov, A., Milner, M.J., Grünebaum, M. et al. (2020). *J. Mater. Chem. A* 8: 22280–22291.

42 Liu, T., Wei, X., Nie, Z. et al. (2016). *Adv. Energy Mater.* 6: 1501449.

43 Zhou, W., Liu, W., Qin, M. et al. (2020). *RSC Adv.* 10: 21839–21844.

44 Fan, H., Zhang, J., Ravivarma, M. et al. (2020). *ACS Appl. Mater. Interfaces* 12: 43568–43575.

45 Nambafu, G.S., Siddharth, K., Zhang, C. et al. (2021). *Nano Energy* 89: 106422.

46 Hu, B., Fan, H., Li, H. et al. (2021). *Adv. Funct. Mater.* 31: 2102734.

47 Chang, Z., Henkensmeier, D., and Chen, R. (2017). *ChemSusChem* 10: 3193–3197.

48 Gao, M., Salla, M., Song, Y., and Wang, Q. (2022). *Angew. Chem. Int. Ed.* 61: e202208223.

49 Janoschka, T., Martin, N., Hager, M.D., and Schubert, U.S. (2016). *Angew. Chem. Int. Ed.* 55: 14427–14430.

50 Fan, H., Wu, W., Ravivarma, M. et al. (2022). *Adv. Funct. Mater.* 32: 2203032.

51 Chang, Z., Henkensmeier, D., and Chen, R. (2019). *J. Power Sources* 418: 11–16.

52 Liu, Y., Goulet, M.-A., Tong, L. et al. (2019). *Chem* 5: 1861–1870.

53 Fan, H., Hu, B., Li, H. et al. (2022). *Angew. Chem. Int. Ed.* 61: e202115908.

54 Pan, M., Gao, L., Liang, J. et al. (2022). *Adv. Energy Mater.* 12: 2103478.

55 Hu, B., Hu, M., Luo, J., and Liu, T.L. (2022). *Adv. Energy Mater.* 12: 2102577.

56 Winsberg, J., Stolze, C., Schwenke, A. et al. (2017). *ACS Energy Lett.* 2: 411–416.

57 Sukegawa, T., Masuko, I., Oyaizu, K., and Nishide, H. (2014). *Macromolecules* 47: 8611–8617.

58 Winsberg, J., Muench, S., Hagemann, T. et al. (2016). *Polym. Chem.* 7: 1711–1718.

59 Janoschka, T., Martin, N., Martin, U. et al. (2015). *Nature* 527: 78–81.

60 Winsberg, J., Janoschka, T., Morgenstern, S. et al. (2016). *Adv. Mater.* 28: 2238–2243.

61 Hagemann, T., Strumpf, M., Schröter, E. et al. (2019). *Chem. Mater.* 31: 7987–7999.

62 Nguyen, T.P., Easley, A.D., Kang, N. et al. (2021). *Nature* 593: 61–66.

6

Organometallic Complexes-Based Electrodes

6.1 Introduction

Organic electrode materials have the advantages of tunable structures, high capacity, low costs, etc., which make them promising candidates for next-generation energy storage [1–3]. However, the dissolution of the organic electrode materials induces a severe loss of the active species, and thus the capacity of the battery fades fast. In addition, the dissolved organic electrode materials may shuttle between the cathode and anode during the charge and discharge process, lowering the Coulombic efficiency. There are several strategies to overcome the issues of organic electrode materials, including polymerization, salification, and other methods like separator modification, electrolyte engineering, etc. The polymerization of the organic species may decrease the kinetics, and the salification needs to introduce extra inactive groups such as sulfonate or hydroxyl groups, abating the specific capacity of the resultant electrode materials.

Besides, the organometallic complexes (OMCs) supply opportunities to alter the electrochemical behaviors of the organic materials [4]. Small molecules like metal porphyrin and metal phthalocyanine coordination complexes can bring new charge and discharge mechanisms and decrease the dissolution. In addition, the metallocenes, especially the ferrocenes, with their high electrochemical reversibility, also represent a kind of OMC to circumvent the challenge of organic materials. The ferrocenes are also extensively explored in the redox flow batteries, which will be discussed in detail in the following chapter. Highly ordered crystals with topological structure, for example, metal-organic frameworks (MOFs), emerged in recent years. The large pores and the metal nodes in the materials make them widely studied in the catalytic reaction, separation, sensing, and so on. The Prussian blue is one kind of MOF with active metal nodes to store the electrons by the valence change. Other MOFs are often investigated as the framework to adsorb and catalyze the electrode materials, like high-capacity sulfur or oxygen electrodes.

It is generally considered that the low electric conductivity of MOFs makes them difficult to be directly adopted as electrode materials except for the Prussian blues. However, the new one- or two-dimensional MOFs with ameliorated electric conductivity are found to be highly active as the electrode materials with the active centers of both the metal nodes and the organic ligands [5, 6]. The chances are

Rechargeable Organic Batteries: Materials, Mechanisms, and Prospects, First Edition.
Yongzhu Fu, Xiang Li, Shuai Tang, and Wei Guo.

significantly expanded to improve the performance of organic electrode materials by designing MOFs with various structures.

In this chapter, it is focused on the fundamentals and advances of OMC-based electrode materials, including small molecules and crystalline materials with topological structures (MOFs). It is hoped that the chapter can become a fast guide in the area of OMC-based electrode materials.

6.2 Small Molecules

6.2.1 Porphyrin Complex

The small OMCs can be divided into porphyrin, phthalocyanine, ferrocene, etc. The porphyrins generally have a low HOMO–LUMO gap; therefore, the electrode has fast kinetics and may be used in supercapacitors. As illustrated in Figure 6.1, the dimethyl-substituted norcorrole nickel(II) complex (NiNC) was reported as the electrochemically active material in 2014, illustrating the 4-electron transfer process [7]. Although the antiaromatic NiNC was thought to be unstable, because of the 4n π electrons, the NiNC cathode is capable of cycling for 100 cycles. The authors also demonstrate that the bipolar NiNC can be integrated into a NiNC-NiNC full cell.

Gao et al. designed a [5,15-bis(ethynyl)-10,20-diphenylporphinato]copper(II) (CuDEPP) as the cathode for Li batteries, which shares the fast charge–discharge rate and long cycling lifespan because the porphyrins generally possess low LUMO–HOMO energy gap [8]. Over 8000 stable cycles are achieved for the Li-CuDEEP batteries. The cathode also illustrates the four-electron reaction, and the upper-potential two-electron process corresponds to the anion insertion. Therefore, the working voltage ranges from 1.8 to 4.5 V, while the slope-based voltage profile is obtained. Later, Zhang et al. polymerized the small porphyrin molecules to avoid the dissolution of the organic electrode materials. It is reported that Zn(II) 5,10,15,20-tetrakis[(carbazol-9-yl)phenyl]porphyrin (Zn-mTCPP) can be electrochemically polymerized using the N-substituted carbazole as the linker. Lin et al. further confirm the redox mechanism of CuDEPP in the ionic liquid electrolyte through operando Raman spectroscopy. For the lower-potential region of the CuDEPP electrode, the electrochemically active group is the porphyrin ring and ethynyl substituent. Therefore, it can be inferred that the valence change of the Cu node is the origin of the electrochemical activity in the upper potential region [9]. The CuDEPP cathode is also successfully tried in the potassium-ion batteries (KIBs) to address the lack of the cathode for KIBs. The four-electron process is similarly observed in the KIBs. The polymerized CuDEPP cathode exhibits stable cycling for 300 times with a reversible capacity of over 120 mAh g^{-1} [10]. Yuan et al. designed a porphyrin-based polymer cathode for Li storage through the linkage of Br and the carbon–carbon triple bonds [11]. Thus, the cathode shows over 2000 cycles with a reversible capacity of c. 100 mAh g^{-1}. Differently, the four-electron reaction completely occurs on the nitrogen atoms in the porphyrin structure, and the Zn

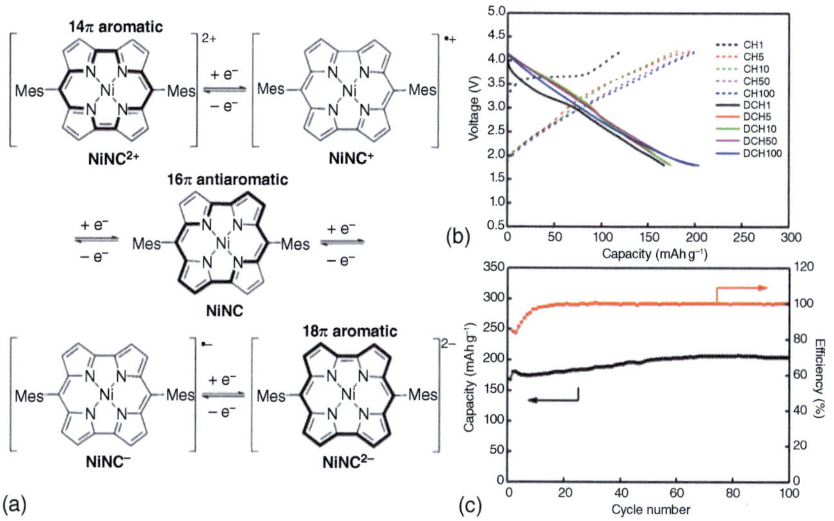

Figure 6.1 (a) The charge and discharge mechanism of the NiNC in Li batteries. (b) The voltage profiles and (c) capacity retention of the Li|NiNC batteries. Source: Reproduced with permission Shin et al. [7]/John Wiley & Sons.

node does not share the electrochemical activity. Therefore, the electrochemical activities of porphyrin-based cathodes are not identical, and more spectral and theoretical work needs to be carried out to understand the differences.

6.2.2 Phthalocyanine Complex

With a similar structure to porphyrin, phthalocyanine complexes are also investigated in energy storage systems. Early in 1982, Yamaki et al. reported the high discharge capacity of 1440 mAh g^{-1} for Fe phthalocyanine, and the activity is mainly attributed to the intercalation between the phthalocyanine compounds [12]. Crowther et al. demonstrated the advantage of the Fe-phthalocyanine cathode in the primary Li batteries, and the discharge capacity reached more than 2000 mAh g^{-1} with the nanographene platelets, which corresponds to 43 Li^+ per Fe-phthalocyanine [13]. More recently, Sanchez et al. elucidated the intercalation mechanism of Li ions in the Fe-phthalocyanine compounds by the DFT calculations, and the intercalation was mainly attributed to the electrostatic interaction between the Li cations and the negatively charged Fe-phthalocyanine [14].

Furthermore, Wang et al. studied the effect of the metal nodes and the substitutional groups on the electrochemical performance of the resultant electrodes [15]. They unravel that Cu tetraaminephthalocyanine (CuTAPc) is the most promising candidate because of the high binding energy between CuTAPc and PF_6^- anions and the electropolymerization of CuTAPc monomers (Figure 6.2). The four-electron electrochemical process and the wide potential region render CuTAPc a kind of

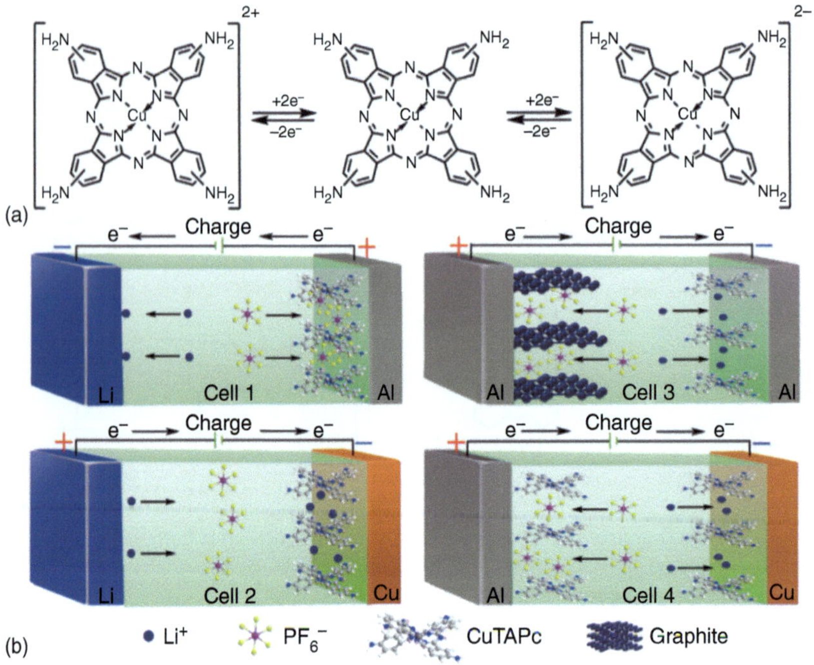

Figure 6.2 (a) The bipolar mechanism of the CuTAPc electrode in Li batteries and (b) the four kinds of batteries constructed with the CuTAPc electrodes. Source: Reproduced with permission Wang et al. [15]/John Wiley & Sons.

bipolar electrode material. The resultant all-organic symmetric cell can achieve a high energy density of 239 Wh kg^{-1}.

6.2.3 Ferrocene

Metallocenes are another main kind of OMC, among which ferrocene is a commonly used reversible electrode in classical aqueous three-electrode electrochemistry. The Fe(III) and Fe(II) can be reversibly transferred, and the theoretical capacity reaches 144 mAh g^{-1} when adopted as electrode materials. However, ferrocene faces dissolution and shuttle issues like other organic electrodes. Therefore, the polymerized ferrocenes are synthesized by linkage with various methods. Not only is stability enhanced, but the discharge voltage, as high as 3.5 V, is also maintained. Poly(vinylferrocene), poly(ethylferrocene), and poly(ferrocene) are synthesized to mitigate the dissolution (Figure 6.3), among which poly(vinylferrocene) illustrates the highest capacity of about 100 mAh g^{-1} over 300 cycles [16]. Meanwhile, a discharge voltage of 3.5 V is achieved for the Li-ferrocene batteries.

In addition, triphenylamine was introduced as the substituent of ferrocene, and the polymerized ferrocene can show a specific capacity of 129.1 mAh g^{-1} [18]. Furthermore, Su et al. synthesized the 4-(1*H*-pyrrol-1-yl) phenyl ferrocenecarboxylate (FcPy) as the monomer and polymerized the monomers into the polypyrrole

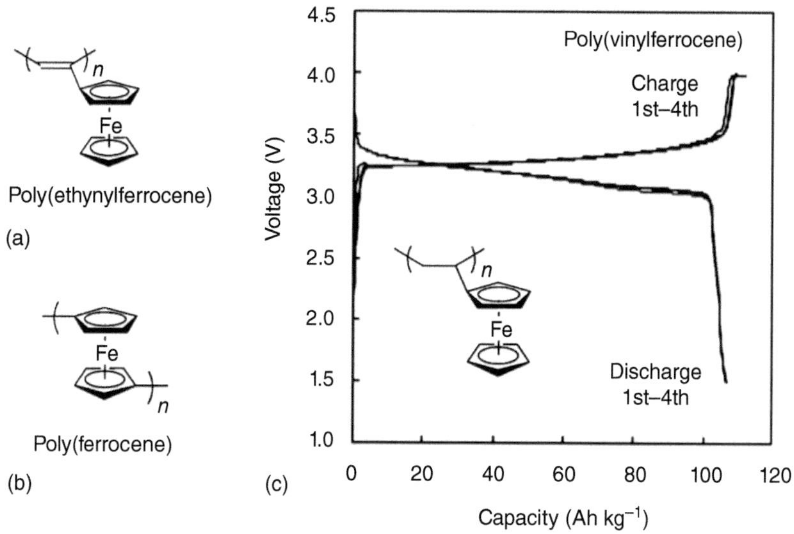

Figure 6.3 The molecular structure of (a) poly(ethynylferrocene) and (b) poly(ferrocene). (c) The voltage profile of the Li|poly(vinylferrocene) batteries, and the inset is the molecular structure of the poly(vinylferrocene). Source: Reproduced with permission Tamura et al. [16]/John Wiley & Sons.

skeleton [19]. Similarly, the discharge plateau at 3.5 V is observed. However, the reversible capacity is only about 50 mAh g^{-1}. Up to now, the research on ferrocene as the cathode material for Li batteries has not gained great breakthroughs, due to its high solubility and low specific capacities. Beladi-Mousavi et al. designed an all-metallocene battery with the poly-cobaltocenium as the anode and poly-ferrocene/reduced graphene oxide as the cathode [17]. The resultant battery displays a discharge voltage of 1.3 V and fast electron transfer with the $LiClO_4$ salt. Different from the classical lithium-ion batteries, the charge carrier is an anion rather than Li^+. The reciprocal volume variations of the two electrodes can be seen due to the directional migration of the anions during the charge/discharge process, which is claimed to be "ion breathing," unlike the "rocking chair" mechanism of Li-ion batteries (Figure 6.4a). However, the volume change did not destroy the cycling stability, and a capacity decay rate below 0.5% was obtained for 500 cycles (Figure 6.4b).

On the other hand, the ferrocenes were investigated as the cathode materials in the redox flow batteries, taking advantage of the high solubility in the organic electrolyte. As shown in Figure 6.5, Yu et al. designed the Li-ferrocene flow batteries with the solid-state electrolyte as the separator [20]. The kinetics of ferrocene in various solvent systems are studied with rotation disk electrodes. The aprotic electrolyte enables ferrocene to have a highly reversible capacity of 130 mAh g^{-1} and a discharge voltage of 3.5 V (Figure 6.5c). After 250 cycles, 90% of the reversible capacity is retained (Figure 6.5d). Aqueous redox flow batteries are a kind of promising and safe system in redox flow batteries. However, the ferrocene has a very low solubility in the aqueous solution, and thus a low energy density is obtained.

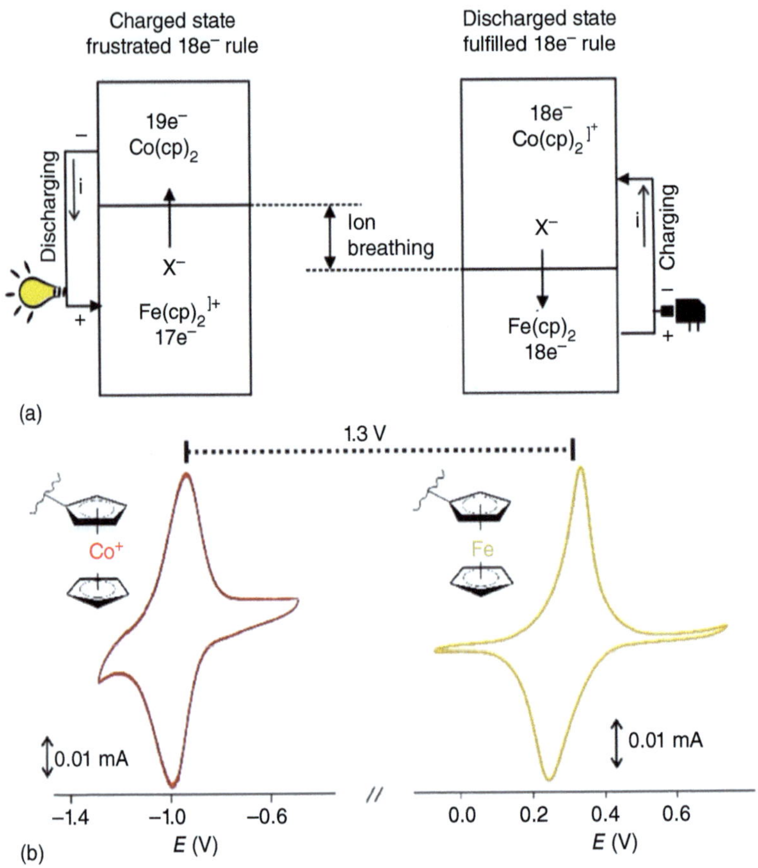

Figure 6.4 (a) The scheme of the ion breathing mode of the all-organic battery with the $Fe(CP)_2$ and $Co(CP)_2$ electrodes. (b) The voltammograms of the $Fe(CP)_2$ and $Co(CP)_2$ electrodes. Source: Reproduced with permission Beladi-Mousavi et al. [17]/John Wiley & Sons.

To address the issue, (ferrocenylmethyl)trimethylammonium chloride (FcNCl) and N1-ferrocenylmethyl-N1, N1, N2, N2, N2-pentamethylpropane-1,2-diaminium dibromide (FcN_2Br_2) are introduced in the aqueous redox flow batteries because of their high solubilities [21]. With the anolyte of methyl viologen, the theoretical energy density of the resultant redox batteries reaches 44.5 Wh l^{-1}, and 93% of the active capacity is maintained after 700 cycles.

Besides, Li et al. synthesized a series of 1,2-benzenedithiol (BDT)-based small-molecule OMCs with Sn and Cu metals [22]. The ionic bonds are formed between Sn or Cu and organothiol anions. The valence state of Sn changes from +4 to 0, and that of organothiols keeps no change during the charge and discharge processes (Figure 6.6). The Sn-$(BDT)_2$ complex shows a high reversible capacity of 221 mAh g^{-1} for 200 cycles. In addition, the Cu-BDT illustrates better cycling stability, although only the two-electron reaction is involved. As shown in Figure 6.6d,

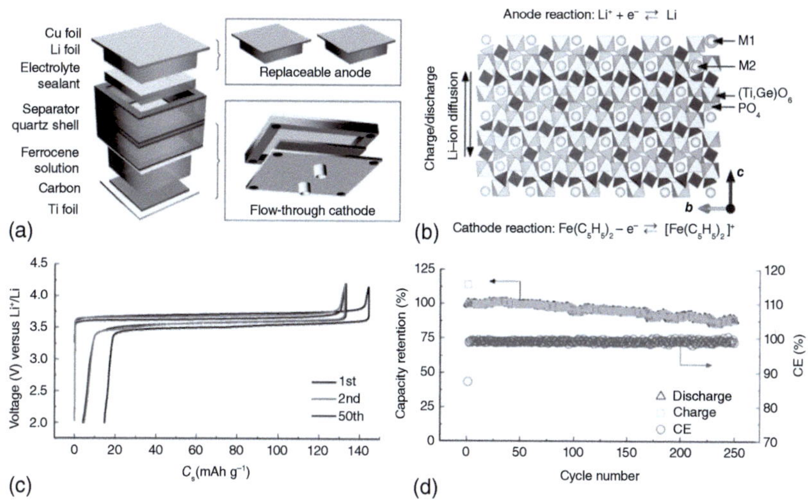

Figure 6.5 (a) The structure and (b) the electrode reactions of the redox flow batteries with the ferrocene cathode. (c) The voltage profiles and (d) the cycling performance of the Li|ferreocene flow battery. Source: Reproduced with permission Zhao et al. [20]/John Wiley & Sons.

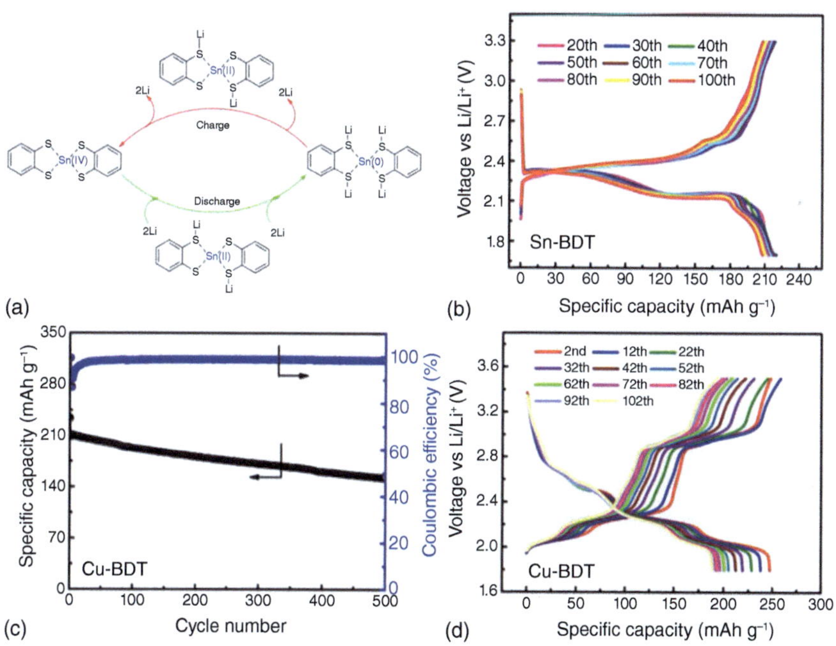

Figure 6.6 (a) The scheme of the charge and discharge mechanism of the Sn-BDT cathode in Li batteries. (b) The voltage profiles of the Li|Sn-BDT batteries. (c) The capacity retention and (d) voltage profiles of the Li|Cu-BDT batteries. Source: Reproduced from Li et al. [22]/Elsevier.

500 stable cycles are achieved for Cu-BDT with a reversible capacity of more than 150 mAh g^{-1}.

6.3 1D MOF

MOFs are a kind of topological OMC materials formed through the coordination between the metal nodes and organic ligands [23]. The MOF materials supply sufficient channels for the transportation of ions and molecules, and meanwhile, the metal nodes illustrate the capability of catalysis. Since the end of the twentieth century, MOFs have been extensively investigated in the area of catalysis [24], separation [25], sensing [26], and so on. For energy storage, MOFs can be used as the host for the Li sulfur and Li oxygen batteries to accommodate and catalyze the conversion of polysulfides and lithium oxides. Meanwhile, the MOF-based separators can confine the soluble electrode materials because of the small and tunable channels [4, 27]. In addition, MOFs can also be directly adopted as electrochemical active materials, among which the Prussian blue (white) materials are almost going to be commercialized in sodium-ion batteries. Besides, one- (1D) or two- (2D) dimensional MOFs have also been investigated as electrode materials in recent years. In this chapter, we will talk about the research progress of 1, 2, and 3D MOF-based electrode materials in detail, respectively.

For the 1D MOFs, the tetradentate ligands are generally adopted, and the metal nodes are usually Ni, Cu, Co, and so on. Kapaev et al. used 1,2,4,5-tetraaminobenzene as the ligand and synthesized Cu- and Ni-centered 1D MOFs [29]. The Ni-based materials show a reversible capacity of 80 mAh g^{-1} at 20 A g^{-1} and can be stably cycled for 20 000 times in lithium-ion batteries. Besides, the Cu-based materials illustrate the high capacity of 262 mAh g^{-1} with inferior capacity retention than that of Ni-based ones. Later, Chen et al. prepared the Ni-based 1D MOFs as the electrode materials in Na-ion batteries using the same ligands [28]. The authors found that the slower reaction rate can promote better crystalline, and the vapor-pressure method can achieve the best crystalline (Figure 6.7a). The conversion between C—N single bond and double bond contributes to two reversible electron transfers, and the conversion between Ni(I) and Ni(II) contributes to the third reversible electron transfer. The Ni-based materials exhibit discharge voltage plateaus at 1.5 and 0.8 V and a slope from 0.7 to 0 V (Figure 6.7b). The reversible capacity reaches 400 mAh g^{-1} for 400 cycles with a weak decay rate (Figure 6.7c). Wu et al. further adopted ethenetetrathiolate (ETT) as the ligand and Ni as the metal nodes to synthesize the 1D MOFs [30]. They unravel that the Ni-ETT MOF can have two reversible electron transfers with the chemical state of Ni unchanged when used as the electrode material in Na-ion batteries. Meanwhile, the resonance structure is also revealed for the reduced product, Ni-TTO. The resultant battery displays the discharge voltage plateaus at 2.6 and 1.7 V with a reversible capacity of 140 mAh g^{-1}. Xie et al. emphasized the effect of ligands on the cycling performance of the resultant electrode materials [31]. They prepared the two kinds of Ni-based MOF with pure amino-group or amino- and thiol-group mixed ligands,

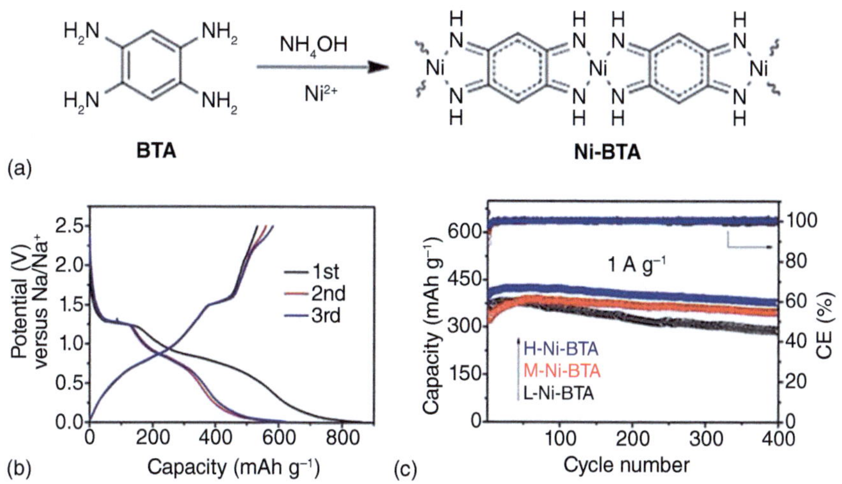

Figure 6.7 (a) The synthesis route of Ni-BTA. The (b) voltage profile and the cycling performance (c) of the Ni-BTA electrodes. Source: Reproduced with permission Chen et al. [28]/John Wiley & Sons.

$Ni[C_6H_2(NH)_4]_n$ (NiNH) and $Ni[C_6H_2(NH)_2S_2]_n$ (Ni—S). The Ni can be reduced to a valence of 0, and thus total four electron transfer is achieved for each unit of both MOFs. Impressively, the Ni—S cathode shows greatly enhanced capacity retention than NiNH, with a reversible capacity of 700 mAh g^{-1} after 180 cycles. The authors ascribe the superior capacity retention of Ni—S to the capacitive contribution. Wu et al. observed similar superior performances of the MOFs with the mixed chelating atoms to the single type of chelating atoms [32]. However, they attributed it to its superior electronic conductivity and structural stability.

Besides, the effects of different metal centers on the electrochemical performance of MOFs were also explored. The isostructural 1D MOFs with the 5-hydroxyisophthalic acid (H_3IPA) ligand and different metal nodes (Co, Ni, Fe, Zn, Mn) were synthesized and denoted as M-HIPA ($M(HIPA)\text{-}(H_2O)_3]_n$), in which M represents the metal centers and HIPA represents the ligand. [33] The resultant 1D MOFs show distinct electrochemical behaviors (Figure 6.8). The Co...- and Ni-HIPA show good stability in the electrolyte due to the high Gibbs free energy, while the other three became amorphous after being soaked in the electrolyte for 10 days. Co-HIPA obtains a superior capacity to Ni-HIPA, and the reversible capacity reaches ~1000 mAh g^{-1} after 200 cycles. The 5 Li^+ coordinate to the Co center, and the other five-electron transfer corresponds to the carbonyl group in the ligands for each unit of Co-HIPA (Figure 6.8d). The outstanding performance was attributed to the higher coordination ability of Co to Li^+ and the capacitive contribution, which confirms that the electronic state of the metal centers can directly affect the electrochemical behaviors of MOFs. Likewise, the 1D M-DHBQ was prepared, in which M represents Ni, Zn, and Mn, and DHBQ represents 2,5-dihydroxy-1,4-benzoquinone [34]. The Ni-DHBQ can illustrate the three-electron transfer, including the C—O/C=O bonds and Ni(II)/(I), while the other two ones can only show the charge transfer between

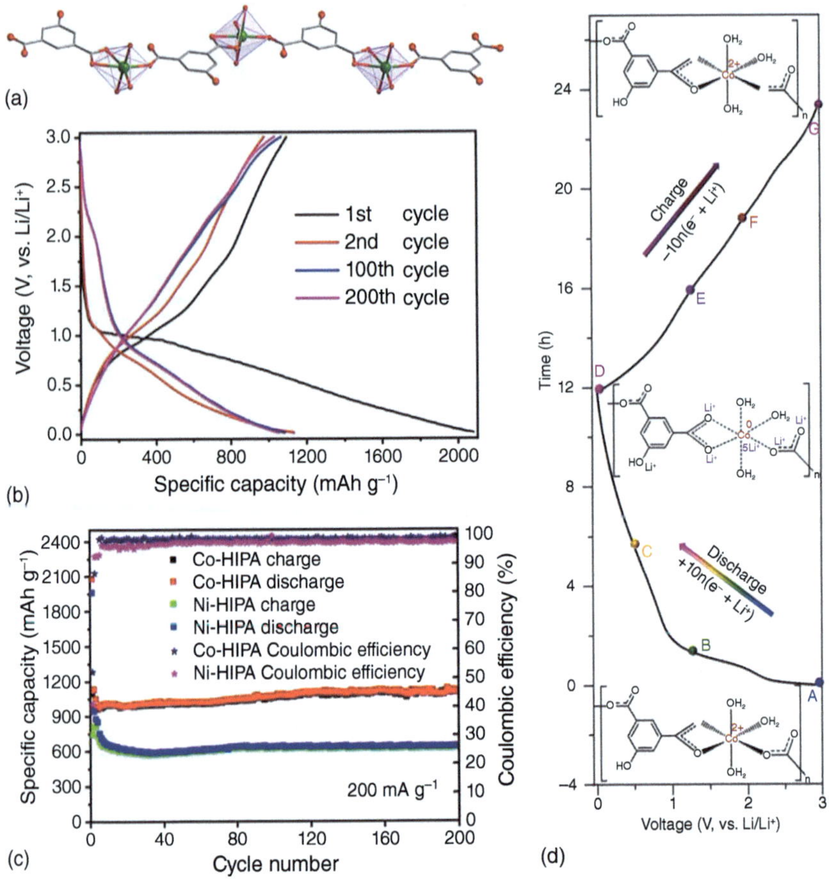

Figure 6.8 (a) The structure of the M-HIPA MOF. (b) The voltage profile and (c) capacity retention of the M-HIPA. (d) The charge/discharge mechanism of the Co-HIPA. Source: Reproduced with permission Du et al. [33]/John Wiley & Sons.

C—O and C=O bonds. It also emphasized that the metal center plays a key role in the electrochemical behaviors.

The 1D MOFs also illustrate the electrochemical activity in the KIBs. The same MOF with the Ni nodes and the amino-based ligands also displays highly reversible electrochemical behavior, supplying a promising choice for the scarce cathodes of potassium batteries [35]. There are two electron transfers in the potential range from 0.5 to 2.0 V versus K/K^+ and the chemical state of Ni nodes does not change. The reversible capacity of the Ni-based MOF reaches 200 mAh g^{-1} during 200 cycles, which indicates the high compatibility of MOFs with the electrochemical storage of different ions, i.e. Li^+, Na^+, K^+, etc.

The redox-active sites of the MOFs above lie at the chelating atom-related C—N, C—O, and C—S bonds or the metal centers. Li et al. substituted the ligands based on the benzene ring with the benzoquinones, and thus the resultant Ni-TABQ (TABQ represents tetraaminobenzoquinone) can have the extra active C=O bonds to improve the specific capacity [36]. The Ni-TABQ displays a reversible capacity of 318.7 mAh g^{-1} as the cathode. and c. 170 mAh g^{-1} as the anode. The C—Li bonds were observed in the potential range near 0 V, illustrating the ability of Ni-TABQ as the anode material for Li-ion batteries. Therefore, the all-organic symmetric batteries were constructed as proof of concept and show promising performance as the flexible batteries.

According to the summary above, we can see that the storage mechanism of the alkali metal ions in the 1D MOFs is still not clear and is even controversial in different studies. Kapaev et al. adopted in-situ Raman and XRD to understand the mechanism of the 1D MOF, combined with the theoretical simulation [37]. They unravel that only the ligands can show electrochemical activity and, thus a total of two-electron reactions occur for each repeating unit of the Ni-BTA (BTA represents benzenetetramine). Meanwhile, not only the Li^+ but also the Na^+ and K^+ show the same storage behaviors in Ni-BTA. The Ni(II) cannot be reduced even when the potential reaches 0 V versus Li/Li^+, Na/Na^+, and K/K^+, respectively. It indicates that it still needs more work to comprehensively understand the behaviors of the emerging electrochemically activc 1D MOFs.

6.4 2D MOF

The planar multidentate ligands with coordination atoms higher than 4 can facilitate the formation of the two-dimensional MOFs with the π-d conjugation structure, which could have good electrical conductivity and stability. The 2D MOFs can be used in various areas, like energy storage materials, catalysts, and optical-electronic devices, due to their tunable structures, unlike the general inorganic 2D materials like graphene, boron nitrides, metal sulfides, etc. Huang et al. found that the 2D MOFs with the Cu nodes and the benzenehexathiol (BHT) ligands can have an electronic conductivity of 1580 S cm^{-1} at room temperature [5]. Cu(II) with hexahydroxybenzene (HHB) can enable the semi-conductive MOF with a bandgap of 1 eV [38].

The hexaaminobenzene (HAB) ligands were first introduced in the redox-active 2D MOFs. Wada et al. synthesized the metallically conductive bis(diimino)nickel framework (NiDI) and found the multielectron transfer reaction (Figure 6.9a) [39]. In the charge state, the PF_6^- anions were observed in the cathode, indicating the insertion of the anion. Meanwhile, the XRD patterns of the cathodes in oxidation and reduction states almost show no change, illustrating that the insertion of anions and Li ions may occur in the cavity of the MOFs rather than the space in the layered structure of graphite in traditional Li-ion batteries. The Li|NiDI batteries show good cycling performance in the carbonate electrolyte, and a high capacity of 155 mAh g^{-1} is achieved. However, the charge and discharge mechanisms are not fully understood. At the same time, the Co-HAB MOFs were synthesized by Park et al. [40]. The ratio of the NH_4OH to HAB directly determines the crystallinity of the resultant MOFs, and the alkaline medium is beneficial to the high yield and crystallinity. The high electric conductivity of 1.57 S cm^{-1} facilitates the high rate performance of the MOFs for storage of Na^+. The reversible capacity reaches 230 and 160 mAh g^{-1} at

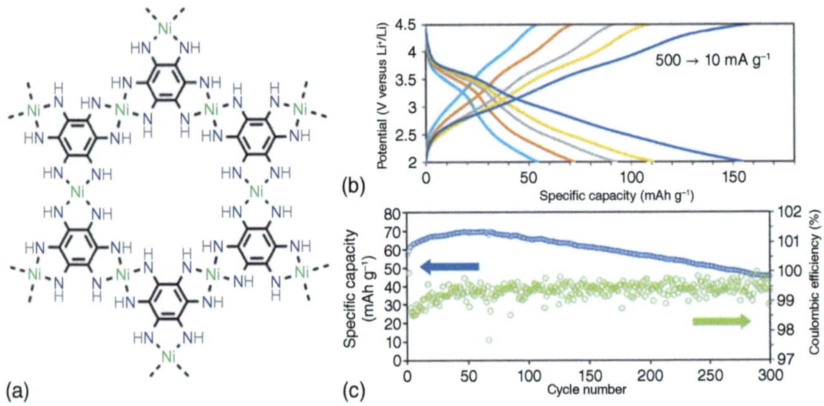

Figure 6.9 (a) The molecular structure of the Ni-HAB. (b) The voltage profile and (c) capacity retention of the Ni-HAB electrode. Source: Reproduced with permission Wada et al. [39]/John Wiley & Sons.

1 and 10 A g^{-1} respectively. The redox activity mainly originates from the ligands, and the Co center almost shows no change in the redox process, according to the XPS results. Meanwhile, due to the porous structure of the Co-HAB cathode, the capacitive contribution reaches 47%, which is also the origin of the high rate performance.

Besides the HAB ligands, the 2,3,6,7,10,11-hexahydroxytriphenylene (HHTP) ligands are also introduced in the 2D MOFs. The Cu-HHTP cathode in the Li-ion batteries shows a capacity of ~95 mAh g^{-1} in the voltage range from 1.7 to 3.5 V, which is close to the theoretical value based on one-electron transfer between Cu(II) and Cu(I) [41]. The XPS results confirmed the valence change of the Cu center during the redox process. The ligands keep inert in the voltage range, which enables a low specific capacity, and the wider voltage range should be investigated to obtain a higher capacity. Later, the Cu-HHTP was explored in the aqueous Zn batteries. The high capacity of 228 mAh g^{-1} and the stable cycling over 500 times at 18C were achieved [42]. Different from the Cu-HHTP in Li batteries, the Cu center in the Zn batteries is partially involved in the redox process, and the ligands also show electrochemical activity. Chen et al. prepared the Cu-HHTP and Zn-HHTP 2D MOF cathodes in Na batteries [43]. They unravel that the Cu-HHTP can enable the higher specific capacity because both Cu and the ligands can participate in the redox reaction while only the ligands of Zn-HHTP are redox active, which facilitates an inferior capacity (Figure 6.10). However, the Zn-HHTP cathode illustrates a much longer cycling life than that of Cu-HHTP because of the structural stability caused by the inertness of Zn nodes. It should be noted that the three ligands of Zn-HHTP are redox active in the repeat unit, while only two ligands are active in other MOFs. Therefore, both the ligands and the metal nodes can affect the redox behaviors, which should be more extensively studied to obtain a deeper understanding.

In addition, the benzenehexathiolate (BHT) enables a highly conductive 2D Cu-BHT MOF, in which four electrons are transferred in one repeated unit for Li storage [44]. The theoretical capacity reaches 236 mAh g^{-1} and the reversible

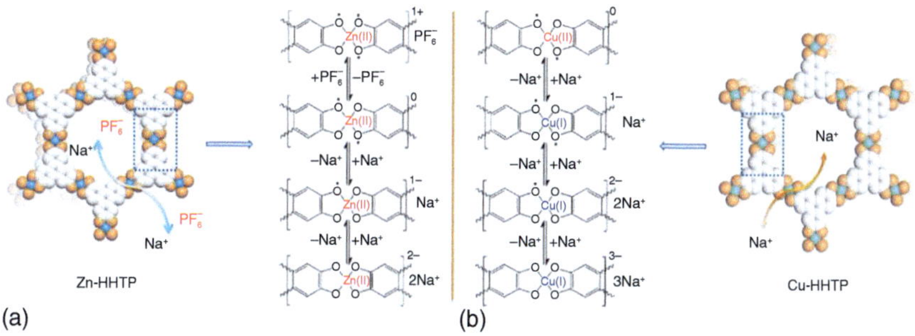

Figure 6.10 The charge/discharge schematic of the (a) Zn-HHTP and (b) Cu-HHTP. Source: Reproduced with permission Chen et al. [43]/John Wiley & Sons.

capacity retains ~140 mAh g^{-1} at 300 mA g^{-1} after 500 cycles. More importantly, the obvious potential platforms are observed for the Cu-BHT cathode, unlike the other conductive MOFs, which generally have slope-shaped voltage profiles. More complicated ligands were introduced in the redox-active 2D MOFs. The nitrogen-rich TQ was combined with the carbonyl groups to form the 2,3,7,8,12,13-hexahydroxytricycloquinazoline (HHTQ) ligands, and thus the resultant 2D Cu-HHTQ MOFs can have redox centers from the carbonyl and C=N groups, significantly improving the specific capacity [45]. Therefore, the high capacity of 658 mAh g^{-1} at 600 mA g^{-1} is achieved. Bipolar ligands are designed in the MOFs, aiming at storing both the anions and cations to ameliorate the specific capacity. Guo et al. synthesized the Cu-based MOFs with N-containing ligands linked by cupric acetate, and the MOFs successfully worked in the Al batteries [46]. In the discharge process, the $AlCl_4^-$ is extracted from the MOF, then $AlCl_2^+$ is inserted, and at last, $AlCl_2^+$ is adsorbed, in which two electrons are transferred in each step. The transition of the organic ligands between 16π and 20π facilitates the 4-electron reaction. The valence changes from Cu(II) to Cu(I) promote the third-step 2-electron transfer. The stable cycling of Cu-based MOFs with a capacity of 150 mAh g^{-1} over 1000 times indicates the great promise of MOF-based materials to enrich the scarce electrode candidates of Al, K batteries, etc.

From the examples above, we can see that the 2D conductive MOFs supply new chances to design high-capacity and high-rate performance electrode materials. However, the voltage profiles are evidently distinct, illustrating the different mechanisms of charge storage. More mechanistic studies should be carried out to understand the difference in designing higher-performance electrode materials.

6.5 3D MOF

3D MOFs are the most extensively studied ones since the appearance of MOFs, involving applications in catalysis, gas absorption, separation, energy storage, etc. The Prussian blue or white electrodes are highly attractive materials that are declared to be adopted in Na-ion batteries [47, 48]. However, we will not discuss the

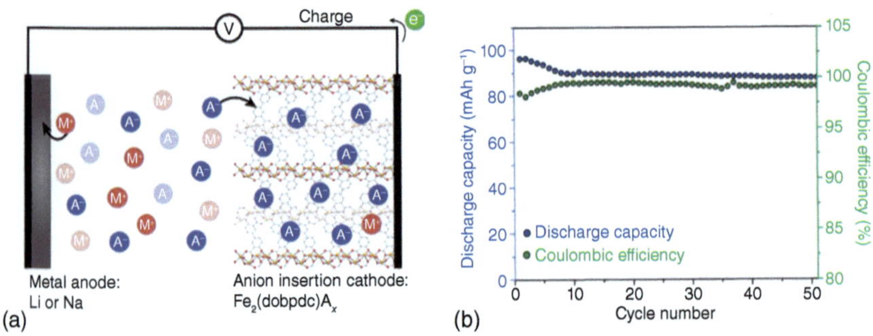

Figure 6.11 Schematic of energy storage (a) and the cycling performance (b) of the $Fe_2(dobpdc)A_x$ cathode. Source: Reproduced with permission Aubrey and Long [49]/American Chemical Society.

development of the Prussian blue or white materials in the chapter because only the metal nodes are redox-active and the ligands are inert. In this part, we focus on the research advances of 3D MOF-based electrode materials with redox-active ligands.

Aubrey et al. found that the 4,4′-dioxidobiphenyl-3,3′-dicarboxylate (dobpdc) ligands can form Fe_2(dobpdc) MOFs with the Fe centers [49]. Generally, lithium or sodium ions can intercalate into most of the Prussian blue materials with the reduction of the metal nodes such as Fe and Mn. However, the BF_4^- or PF_6^- anions can be reversibly inserted into the Fe_2(dobpdc) with the oxidation of the Fe from the valence state of +2 to +3 (Figure 6.11a). The quinoidal ligand oxidation can also be observed but with poor reversibility. A proof-of-concept Na–Fe_2(dobpdc) dual-ion battery is prepared, and it shows 50 stable cycles with the highest energy density of 316 Wh kg^{-1} (Figure 6.11b). It is also reported that the anions can be intercalated in the Mn-based MOFs with the valence state transition of Mn from (II) to (III). Synergistic with the redox reaction of the carbonyl groups, the Mn-based MOFs exhibited a reversible capacity of about 160 mAh g^{-1}.

Similar carboxyl-based ligands, 1,4-benzenedicarboxylic, can coordinate with Mn and Ni metal centers, forming the ultrathin MOF nanosheets, of which both metal centers and the ligands illustrate the redox activity [50]. Specifically, the Mn-based MOFs show the highest capacity of 1187 mAh g^{-1} when used as the anode of Li batteries. Due to the high diffusion coefficient (2.48×10^{-9} cm^2 s^{-1}), the anode shows a reversible capacity of 701 mAh g^{-1} at the high rate of 2 A g^{-1}. In addition, zinc perylenetetracarboxylates (Zn-PTCA) show the ability to store the Na ions in the aromatic rings, and the carbonyls also keep electrochemically active (Figure 6.12) [51]. Therefore, the reversible capacity of Zn-PTCA reaches 357 mAh g^{-1} at 50 mA g^{-1} for storage of Na^+. The aromatic ring in the Zn-PTCA is wavy-layered rather than pillared-layered, and thus the Na^+ is accessible to the aromatic rings. The method of increasing the space to store Na^+ in the sp^2 hybridized carbons uncovers a new way to activate the aromatic organic materials for Na^+ storage.

The amino-group-ended ligands with the Cu centers enable a kind of 3D MOF with both the redox-active metal centers and the nitrogen atoms [52]. Because the oxidation reaction of the nitrogen atoms occurs, the redox potential reaches 4.3 V

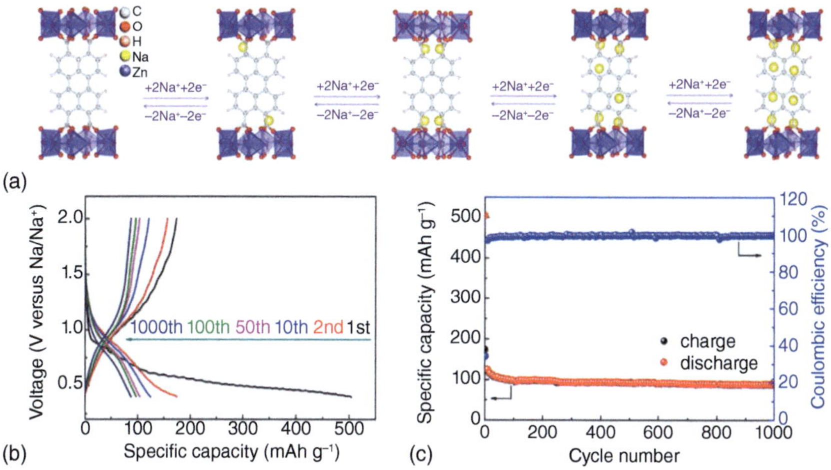

Figure 6.12 (a) The schematic of the energy storage mechanism of the Zn-PTCA electrode. (b) The voltage profile and (c) the cycling performance of the Na/Zn-PTCA battery. Source: Reproduced from Liu et al. [51]/Elsevier.

accompanied by the insertion of the anions. The valence state transition of Cu from +2 to +1 emerges at about 2 V. However, the redox reversibility of the metal nodes is inferior to that of ligands, and the cathode can be cycled 200 times for Li storage.

6.6 Summary and Outlook

In summary, the OMC-based electrodes in the secondary batteries have achieved obvious progresses in recent years, from small compounds to ordered topological structures with different dimensions. The OMCs generally have excellent kinetics due to their highly porous structures capable of fast ionic transport. Some of the OMC electrodes have illustrated stable cycling with more than thousands of cycles, indicating the great promises of the OMC electrodes in secondary batteries. However, it also should be noteworthy that challenges also exist, like the low energy density, slope-shaped voltage profiles, and dependency on expensive transition metals like nickel and copper. Several outlooks are supplied, as follows:

The distinctive cycling stabilities are obtained due to the different metal nodes or ligands. Therefore, it is important to figure out the relation between the structure and the corresponding functions of the OMCs as electrodes to prepare high-performance OMC-based electrodes more efficiently. Advanced characterization tools like in-situ TEM and in-situ synchrotron XRD should be introduced to understand the ionic transport and charge transfer of OMCs in the batteries.

The OMC electrodes usually show voltage slopes without obvious plateaus, like the commercial $LiFePO_4$ or transition metal oxides. It means that OMCs cannot supply stable output voltage during work, which is not beneficial to the applications on many occasions. Thus, understanding the mechanism of energy storage in the

OMC electrodes is important to design more structures with plateau-shaped voltage profiles.

It also should be noteworthy that the specific capacity of OMC electrodes is not high, which induces the low energy density of the resultant batteries. The activity of the ligands and metal nodes should be improved to enhance the capacity, or else the OMCs cannot have obvious advantages over commercial electrodes like $LiFePO_4$, etc. Moreover, the OMCs should avoid the adoption of expensive transition metals like Ni, Mn, Co, etc., as the resources of which are not abundant on the Earth. The adoption of them will obviate the advantage of organic electrodes. More low-cost OMCs based on metal nodes like Fe, Zn, etc., should be developed.

Acknowledgments

The work was supported by the National Natural Science Foundation of China (Grant Nos. 22209152, U2004214, and 21975225), the National Key Research and Development Program of China (2022YFB2402200), and China Postdoctoral Science Foundation (Grant Nos. 2022T150588, 2020M682330).

References

1 Chen, Y. and Wang, C. (2020). *Acc. Chem. Res.* 53: 2636–2647.
2 Poizot, P., Gaubicher, J., Renault, S. et al. (2020). *Chem. Rev.* 120: 6490–6557.
3 Lu, Y. and Chen, J. (2020). *Nat. Rev. Chem.* 4: 127–142.
4 Wang, D.Y., Liu, R.L., Guo, W. et al. (2021). *Coord. Chem. Rev.* 429: 213650.
5 Huang, X., Sheng, P., Tu, Z. et al. (2015). *Nat. Commun.* 6: 7408.
6 Maiti, S., Pramanik, A., Manju, U., and Mahanty, S. (2016). *Microporous Mesoporous Mater.* 226: 353–359.
7 Shin, J.Y., Yamada, T., Yoshikawa, H. et al. (2014). *Angew. Chem. Int. Ed.* 53: 3096–3101.
8 Gao, P., Chen, Z., Zhao-Karger, Z. et al. (2017). *Angew. Chem. Int. Ed.* 56: 10341–10346.
9 Lin, X.-M., Wu, D.-Y., Gao, P. et al. (2019). *Chem. Mater.* 31: 3239–3247.
10 Lv, S., Yuan, J., Chen, Z. et al. (2020). *ChemSusChem* 13: 2286–2294.
11 Yuan, J., Ren, B., Feng, X. et al. (2020). *Chem. Commun.* 56: 5437–5440.
12 Yamaki, J.i. and Yamaji, A. (2019). *J. Electrochem. Soc.* 129: 5–9.
13 Crowther, O., Du, L.-S., Moureau, D.M. et al. (2012). *J. Power Sources* 217: 92–97.
14 Ramos-Sanchez, G., Callejas-Tovar, A., Scanlon, L.G., and Balbuena, P.B. (2014). *Phys. Chem. Chem. Phys.* 16: 743–752.
15 Wang, H.G., Wang, H., Si, Z. et al. (2019). *Angew. Chem. Int. Ed.* 58: 10204–10208.
16 Tamura, K., Akutagawa, N., Satoh, M. et al. (2008). *Macromol. Rapid Commun.* 29: 1944–1949.
17 Beladi-Mousavi, S.M., Sadaf, S., Hennecke, A.K. et al. (2021). *Angew. Chem. Int. Ed.* 60: 13554–13558.

18 Su, C., Yang, F., Ji, L. et al. (2014). *J. Mater. Chem. A* 2: 20083–20088.

19 Su, C., Wang, L., Xu, L., and Zhang, C. (2013). *Electrochim. Acta* 104: 302–307.

20 Zhao, Y., Ding, Y., Song, J. et al. (2014). *Angew. Chem. Int. Ed.* 53: 11036–11040.

21 Hu, B., DeBruler, C., Rhodes, Z., and Liu, T.L. (2017). *J. Am. Chem. Soc.* 139: 1207–1214.

22 Li, F., Guo, W., Si, Y. et al. (2021). *Electrochim. Acta* 370: 137757.

23 Furukawa, H., Cordova, K.E., O'Keeffe, M., and Yaghi, O.M. (2013). *Science* 341: 974.

24 Farrusseng, D., Aguado, S., and Pinel, C. (2009). *Angew. Chem. Int. Ed.* 48: 7502–7513.

25 Li, J.-R., Kuppler, R.J., and Zhou, H.-C. (2009). *Chem. Soc. Rev.* 38: 1477–1504.

26 Lustig, W.P., Mukherjee, S., Rudd, N.D. et al. (2017). *Chem. Soc. Rev.* 46: 3242–3285.

27 Zhao, R., Liang, Z., Zou, R., and Xu, Q. (2018). *Joule* 2: 2235–2259.

28 Chen, Y., Tang, M., Wu, Y. et al. (2019). *Angew. Chem. Int. Ed.* 58: 14731–14739.

29 Kapaev, R.R., Olthof, S., Zhidkov, I.S. et al. (2019). *Chem. Mater.* 31: 5197–5205.

30 Wu, Y., Chen, Y., Tang, M. et al. (2019). *Chem. Commun.* 55: 10856–10859.

31 Xie, J., Cheng, X.F., Cao, X. et al. (2019). *Small* 15: e1903188.

32 Wu, Y., Zhang, Y., Chen, Y. et al. (2021). *Energy Environ. Sci.* 14: 6514–6525.

33 Du, J., Ren, J., Shu, M. et al. (2021). *Angew. Chem. Int. Ed.* 60: 4142–4149.

34 Fan, K., Zhang, C., Chen, Y. et al. (2022). *J. Mater. Chem. C* 10: 2592–2599.

35 Kapaev, R.R., Zhidkov, I.S., Kurmaev, E.Z. et al. (2020). *Chem. Commun.* 56: 1541–1544.

36 Li, K., Yu, J., Si, Z. et al. (2022). *Chem. Eng. J.* 450.

37 Kapaev, R.R., Zhugayevych, A., Ryazantsev, S.V. et al. (2022). *Chem. Sci.* 13: 8161–8170.

38 Park, J., Hinckley, A.C., Huang, Z. et al. (2018). *J. Am. Chem. Soc.* 140: 14533–14537.

39 Wada, K., Sakaushi, K., Sasaki, S., and Nishihara, H. (2018). *Angew. Chem. Int. Ed.* 57: 8886–8890.

40 Park, J., Lee, M., Feng, D. et al. (2018). *J. Am. Chem. Soc.* 140: 10315–10323.

41 Gu, S., Bai, Z., Majumder, S. et al. (2019). *J. Power Sources* 429: 22–29.

42 Nam, K.W., Park, S.S., Dos Reis, R. et al. (2019). *Nat. Commun.* 10: 4948.

43 Chen, Y., Zhu, Q., Fan, K. et al. (2021). *Angew. Chem. Int. Ed.* 60: 18769–18776.

44 Wu, Z., Adekoya, D., Huang, X. et al. (2020). *ACS Nano* 14: 12016–12026.

45 Yan, J., Cui, Y., Xie, M. et al. (2021). *Angew. Chem. Int. Ed.* 60: 24467–24472.

46 Guo, Y., Wang, W., Lei, H. et al. (2022). *Adv. Mater.* 34: e2110109.

47 Kundu, D., Talaie, E., Duffort, V., and Nazar, L.F. (2015). *Angew. Chem. Int. Ed.* 54: 3431–3448.

48 Lu, Y., Wang, L., Cheng, J., and Goodenough, J.B. (2012). *Chem. Commun.* 48: 6544–6546.

49 Aubrey, M.L. and Long, J.R. (2015). *J. Am. Chem. Soc.* 137: 13594–13602.

50 Li, C., Hu, X., Tong, W. et al. (2017). *ACS Appl. Mater. Interfaces* 9: 29829–29838.

51 Liu, Y., Zhao, X., Fang, C. et al. (2018). *Chemistry* 4: 2463–2478.

52 Peng, Z., Yi, X., Liu, Z. et al. (2016). *ACS Appl. Mater. Interfaces* 8: 14578–14585.

7

Polymer-Based Organic Cathodes

7.1 Introduction

With the rapid promotion of carbon neutrality and the vigorous development of new energy industry, the energy storage system, as an important carrier of energy storage and transformation, has become a key link to support the energy revolution. Since 1991, lithium-ion batteries (LIBs) have become one of the most mature and widely used energy storage systems. However, there are inherent weaknesses in energy density, power density, and cost, which make it difficult to meet the new challenges of energy structure reform [1–5]. Polymers are considered as promising energy storage materials due to their rich resources, designable structure, high specific capacity, mechanical flexibility, and good electrochemical performance, and have been continuously applied to the field of energy storage by researchers in recent years. However, there are still some shortcomings in the structure stability, conductivity, and solubility in electrolytes, which are urgent problems to be solved in future development [6–11].

Here, we systematically summarize the latest research progress of polymer cathode materials in the field of energy storage and analyze their synthesis methods, redox mechanisms, and electrochemical properties from the perspective of internal structural differences. As shown in Figure 7.1, in view of the function of polymer materials in batteries, we summarize the different types of polymers, including organosulfur, carbonyl-derived, covalent organic framework-derived, organic radical-derived, and other polymers, which have different structures and redox sites, thus exhibiting unique electrochemical properties. Among them, the first part introduces the application of organosulfur polymers in rechargeable metal batteries, mainly including unsaturated bond-derived polymers (UB-DPs), thiol-derived polymers (SH-DPs), sulfurized polyacrylonitrile (SPAN) and covalent organosulfur polymers (COSPs). Their reaction mechanism in charge–discharge cycle is mainly based on the reversible breaking and formation of S—S bonds in molecules, in which each broken S—S bond would combine two lithium ions (Li^+) and two electrons (e^-). The second part is carbonyl-derived polymers, which can be divided into polyquinones (PQs), polyimides (PIs), and polyanhydrides (PAs). They take carbonyl groups as active sites, which show the combination of oxygen

Rechargeable Organic Batteries: Materials, Mechanisms, and Prospects, First Edition.
Yongzhu Fu, Xiang Li, Shuai Tang, and Wei Guo.

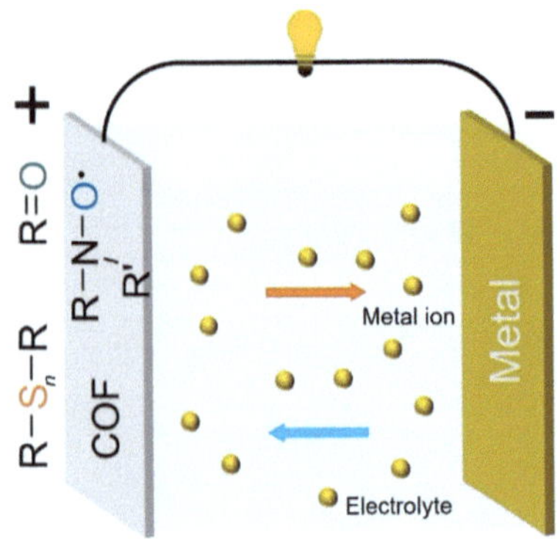

Figure 7.1 The different polymer cathodes in metal batteries.

atoms in carbonyl groups and metal ions during discharge. In the third part, the polymers derived from covalent organic frameworks are introduced, which have attracted wide attention because of their low density, large specific surface area, adjustable structure, easy functionalization, and good physical and chemical stability. In the fourth part, several kinds of 2,2,6,6-tetramethyl-1-piperidinyloxy-based nitroxide radical polymers are introduced. In the fifth part, some other polymers are briefly introduced, mainly including some compounds with active sites on nitrogen atoms, e.g. triphenylamine and hexaazatrinaphthalene-based polymers. Finally, the challenges faced by polymer-based organic cathodes for batteries are discussed, and the future development trends and prospects are prospected. This chapter has important guiding significance for solving the problems that hinder the development of polymer electrode materials at present and also provides certain reference values for the development of polymer electrode materials.

7.2 Organosulfur Polymers

7.2.1 Unsaturated Bond-Derived Organosulfur Polymers

The elemental sulfur (S_8) mainly exists in the form of an eight-membered ring; it melts at 120–124 °C to form a clear yellow liquid phase, and further heating above 159 °C results in a ring-opening polymerization (ROP) reaction to form a linear polysulfane with diradical chain ends, which then would be polymerized into polymeric sulfur with a high molecular weight. The stabilization of this double-radical polymerized sulfur can be achieved by copolymerization with unsaturated bonds such as alkenes and alkynes. Due to the intensive interaction between sulfur and carbon skeleton, the dissolution and diffusion of polysulfide can be effectively inhibited by chemical bonding [12–14].

Pyun et al. synthesized poly(sulfur-random-1,3-diisopropenylbenzene) (poly(S-r-DIB)) with high sulfur content by copolymerization between sulfur and 1,3-diisopropenylbenzene (DIB) through inverse vulcanization at 185 °C (Figure 7.2a), in which the sulfur content can be controlled by adjusting the mass ratio of

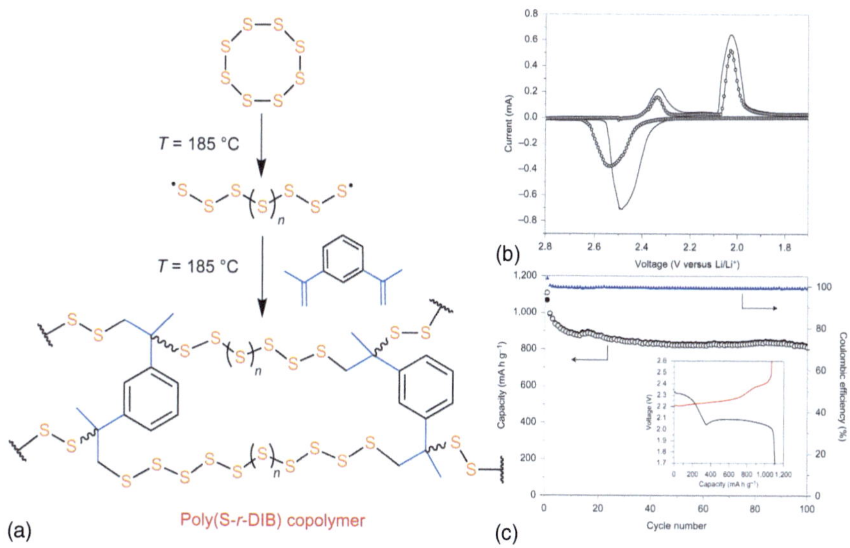

Figure 7.2 (a) The copolymerization of S_8 with DIB. (b) The CV curves of Li–S (solid black line) and Li/poly(S-r-DIB) (10 wt% DIB) (open circles) batteries. (c) Cycling performance of poly(S-r-DIB) at 0.1 C. Source: Reproduced with permission from Chung et al. [15]/Springer Nature.

precursor sulfur to DIB monomer [15]. The cyclic voltammetry (CV) curves of lithium–sulfur (Li–S) and Li/poly(S-r-DIB) batteries are very similar (Figure 7.2b), in which the peak at 2.3–2.4 V represents the reduction of sulfur to high-order linear polysulfide, while the peak at 2.0–2.1 V is attributed to the formation of low-order sulfides down to Li_2S, which indicates that the poly(S-r-DIB) copolymer shows very similar electrochemical behavior to S_8. The poly(S-r-DIB) copolymer exhibits an initial discharge-specific capacity of 1100 mAh g^{-1} with a high capacity of 823 mAh g^{-1} after 100 cycles (Figure 7.2c). Although the electrochemical performance of poly(S-r-DIB) is not particularly excellent, the proposed inverse vulcanization method provides an idea for the subsequent synthesis of organosulfur polymers. Since then, a large number of organosulfur polymer materials synthesized by inverse vulcanization have been widely studied and applied to the energy storage field [16–19].

Graphene sheets are usually added to polymer matrixes as conductive nano-fillers because of their high conductivity (7200 S m^{-1}). In order to improve the conductivity of the electrode, Chang et al. prepared vinyl group-functionalized reduced graphene oxide (V-rGO) sheets, which can be used for in-situ diradical copolymerization with sulfur and DIB by simple inverse vulcanization at a temperature higher than 185 °C (Figure 7.3a) [18]. The prepared electroactive polymer/graphene nanocomposite (a covalently grafted polysulfur-graphene nanocomposite (polySGN)) shows excellent electrochemical properties. The conductive V-rGO sheets are uniformly incorporated into the polyS matrix, effectively increasing the discharge capacity from 632 to 978 mAh g^{-1} at 0.2 C. In addition, it can be noted that the reduced polarization has

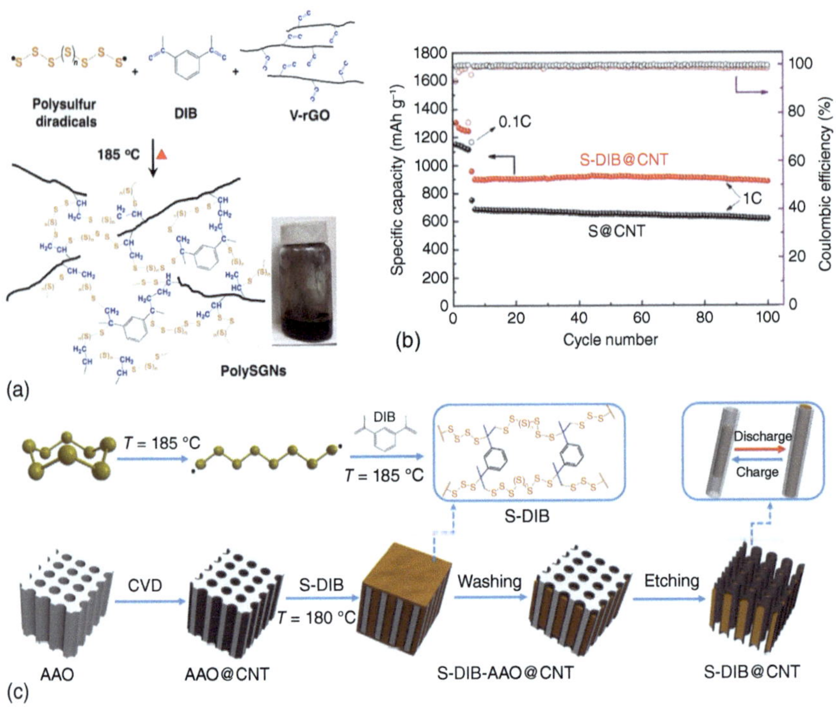

Figure 7.3 (a) Synthesis scheme of polySGNs. Source: Adapted with permission from Chang and Manthiram [18]/American Chemical Society. (b) The cycling performance of the S@CNT and S-DIB@CNT hybrids at 1 C. (c) Synthesis scheme of the S-DIB@CNT copolymer. Source: Reproduced with permission from Hu et al. [19]/John Wiley & Sons.

decreased from 180 to 138 mV. The capacity retention rate of polySGN cathode after 100 cycles is 67%, while that of the cell with polyS cathode is 46%. In the long-cycle performance test of polySGN with 300 cycles, the capacity attenuation rate per cycle can be as low as 0.18%. Meanwhile, the cell with a high sulfur loading of 10.5 mg cm^{-2} shows a high initial discharge capacity of 1135, 992, and 827 mAh g^{-1} at 0.05, 0.1, and 0.2 C, respectively.

Hu et al. infiltrated the S-DIB polymer pre-synthesized by the inverse vulcanization method into anodic alumina oxide@carbon nanotube (AAO@CNT) by melt diffusion and then removed the AAO template to prepare S-DIB@CNT composite electrode rich in sulfur atoms, which was used in high-performance rechargeable lithium batteries (Figure 7.3c) [19]. The designed structure effectively inhibits the dissolution and diffusion of polysulfide through the physical limitation of hollow CNT wall and the strengthening interaction of the C—S bond in copolymer. The S-DIB@CNT electrode shows better electrochemical performance than S@CNT, such as better rate performance and smaller voltage polarization. After 100 cycles, it still has a discharge capacity of 880 mAh g^{-1} with a capacity retention rate of 98% at 1 C (Figure 7.3b).

In addition to DIB monomer, there are some other organic compounds containing C=C double bonds as monomers for synthesizing organosulfur copolymers. Yu

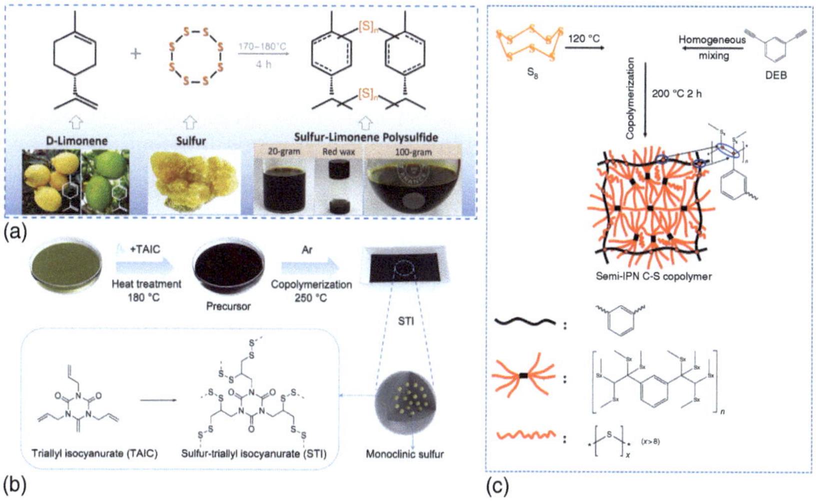

Figure 7.4 (a) One-pot synthesis of SLP from D-limonene and sulfur at heat treatment. Source: Adapted with permission from Wu et al. [20]/John Wiley & Sons, Inc. (b) The preparation route of STI. Source: Adapted with permission from Li et al. [21]/Elsevier. (c) Schematic illustration of the synthesis of semi-IPN C–S copolymer. Source: Adapted with permission from Sun et al. [22]/Royal Society of Chemistry.

et al. prepared sulfur-limonene polysulfide (SLP) by the reaction of cheap limonene (the extract of lemon, grapefruit, and orange peel) with S_8 (Figure 7.4a) [20]. This material is easy to synthesize and can be compounded with various carbon materials. The SLP electrode has achieved excellent electrochemical performance, with a discharge capacity of 1000 mAh g^{-1} at 0.5 C, and the capacity attenuation per cycle is only 0.008% after 300 cycles. S-triallyl isocyanurate (STI) organosulfur polymer composite was synthesized by copolymerization of S_8 and triallyl isocyanurate (TAIC) by a simple two-step heating method (Figure 7.4b) [21], in which the ultrafine monoclinic S particles are embedded in organosulfur polymer matrix. The C—S bonds contribute to the uniform distribution of S and improve the electrochemical activity of S on the molecular level. The synthesized STI polymer complex has a unique "watermelon seed" structure, in which the "seed" is the monoclinic S nanocrystals and the "wrapper" is the organosulfur polymer. STI cathode with S content up to 90 wt% can present an initial capacity of 904 mAh g^{-1} at 0.5 C and a low capacity attenuation rate of 0.017% over 350 cycles at 1 C. In addition, aliphatic compounds containing unsaturated bonds can also be used as linkers. Fu et al. synthesized the vulcanized polyisoprene (SPIP) nanowires, which are composed of sulfur chain cross-linked polyisoprene (PIP) main chain [23]. Electrochemical analysis shows that the bridged sulfur chains (S_n, $n \geq 2$) have unique electrochemical activity due to their close combination with the PIP matrix. The products of 250 °C (SPIP-250), 300 °C (SPIP-300), and 350 °C (SPIP-350) correspond to the sulfur contents of 49.1, 48.2, and 25.4 wt%. The SPIP-300 exhibits an initial capacity of nearly 1400 mAh g^{-1}, and the capacity decays to 1000 mAh g^{-1} after 20 cycles, which is unstable compared with the above C–S polymers. The capacity decay of SPIP compounds may be due to the formation of irreversible C=S groups and possible mercaptan fragments. Besides, SPIP compounds also exhibit limited rate performance due to their low conductivity.

Similar to alkenes, alkynes can also be used as linkers for copolymerization with elemental sulfur. The sulfur-rich copolymer with a cage-like semi-interpenetrating network (semi-IPN) structure was prepared utilizing the sulfur and 1,3-diethynylbenzene (DEB) by Sun et al. [22] (Figure 7.4c). The strong chemical interaction between sulfur and carbon skeleton and the unique cage structure in the C–S copolymer can effectively inhibit the dissolution and diffusion of polysulfides from the cathode by chemical and physical methods. The copolymer delivers an initial capacity of 1143 mAh g^{-1} at 0.1 C, and the capacity retention rate is 70% after 500 cycles with a Coulombic efficiency (CE) of over 99%.

Organosulfur polymers can also be used as electrode materials for low-temperature batteries [24]. In a low-temperature environment, the initial discharge process of sulfur is easier because of its higher reaction potential and faster kinetics. With the progress of the reaction, the low-order intermediate (Li_2S_2) tends to aggregate in the electrolyte to form particles with high crystallinity and prevent the electrode reaction. Wang et al. synthesized a new highly crosslinked organosulfur polymer cathode material with a sulfur content of 75 wt%, sulfur-rich vulcanized clay rubber (SRVCR), by reaction between S_8 and cis-1,4-polyisoprene (RUB) molecular chains (Figure 7.5a), which made Li–S battery have excellent performance at room temperature and low temperature. The SRVCR electrode shows a discharge capacity of 780 mAh g^{-1} at 1 C with a low capacity fade rate of 0.022% over 300 cycles

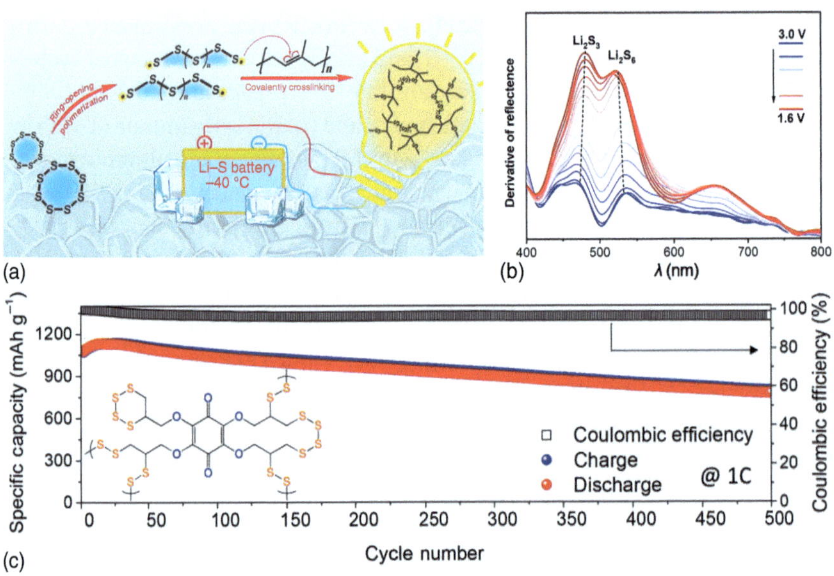

Figure 7.5 (a) Schematic diagram of the synthesis of SRVCR. (b) In-situ UV/Vis spectra of SRVCR cathode during the discharging process. Source: Reproduced with permission from Wang et al. [24]/Elsevier. (c) The long-cycle performance of the Li/poly(S-TABQ) cell at 1 C for 500 cycles. Source: Reproduced with permission from Kang et al. [25]/John Wiley & Sons.

at room temperature. It also has a discharge capacity of 587 mAh g^{-1} even after 50 cycles at −40 °C. The organosulfur cathode material experiences a completely different redox mechanism from the traditional Li–S battery during discharge. Li_2S_6 and Li_2S_3 are the main discharge intermediates (Figure 7.5b), and all the discharge intermediates exist in an amorphous state without long-range order, which promotes a large number of transformations of the discharge intermediates and greatly improves the low-temperature performance of the Li–S battery.

The studies have shown that it is a good strategy to introduce heteroatoms into organosulfur polymers with adjustable molecular structures to regulate their electrochemical behavior. Oxygen-containing unsaturated bonds can also be used to synthesize sulfur-rich polymers [25]. The role of four different linking molecules based on quinone and triazine in the synthesis of sulfur-rich polymers was studied by Kang et al. Among them, (tetra(allyloxy)-1,4-benzoquinone)$_2$-S_8 ((TABQ)$_2$-S_8) shows the lowest bandgap energy of 2.27 eV, which is beneficial to improve sulfur utilization and redox kinetics. Compared with the elemental sulfur cathode, the use of a tetra(allyloxy)-1,4-benzoquinone (TABQ) linker increase the conductivity by a factor of 450, and the lithium diffusion kinetics increases by two orders of magnitude during the redox reaction. Meanwhile, the interaction between TABQ and LiPSs effectively suppresses the shuttle effect. The poly(S-TABQ) cathode with a sulfur content of 75% retains 74% of the initial capacity after 500 cycles at 1 C (Figure 7.5c) and a record high rate performance of 833 mAh g^{-1} at 10 C.

As a homologous element of sulfur, selenium (Se) has higher electronic conductivity (1×10^{-3} S m^{-1}) and comparable theoretical volume capacity density (3253 mAh cm^{-3}). Therefore, selenium-doped sulfur molecules are attractive candidates for batteries. At present, many inorganic selenium@carbon (Se@C) or sulfur-selenium solid solution@carbon (S_xSe_y@C) complexes have been widely studied, but there are few reports on selenium-doped organosulfur polymers [26–28]. The chain-like Se-doped poly(diallyl tetrasulfide) (PDATtSSe) was prepared utilizing the diallyl sulfide (DADS) as a precursor and selenium disulfide (SeS_2) powders as a doping source (Figure 7.6a) [29]. Selenium doping can improve the conductivity, promote lithium ion transport, and completely inhibit the formation of long-chain LiPSs. The CV curve shows two cathodic peaks at about 2.25 and 2.12 V in the negative scan, followed by the cathode peak with a starting potential of 2.07 V and another peak of 1.75 V, which corresponds to the lithiation process. In the subsequent anodic scan, the peak value of 2.3 V corresponds to the delithiation. The PDATtSSe as cathode can provide a specific capacity of 700 mAh g^{-1} with a high retention of 92% after 400 cycles and a CE close to 100%. Meanwhile, the PDATtSSe also exhibits a high volumetric capacity of 2457 mAh cm^{-3} and a high areal capacity of 5.0 mAh cm^{-2} with an active material loading of 7.07 mg cm^{-2}.

Dong et al. reported the preparation of selenium-containing sulfur-rich copolymer cathode material by inverse vulcanization utilizing the S_8, SeS_2, and DIB [30]. The divinyl functional groups of DIB are chemically cross-linked with S/SeS_2 chain radicals by ROP (Figure 7.6b). The newly formed copolymers with C–S, C–Se, and S—Se bonds can effectively alleviate the shuttle effect of polyselenosulfide. The existence of selenium in sulfur-rich copolymer not only improves the electrochemical

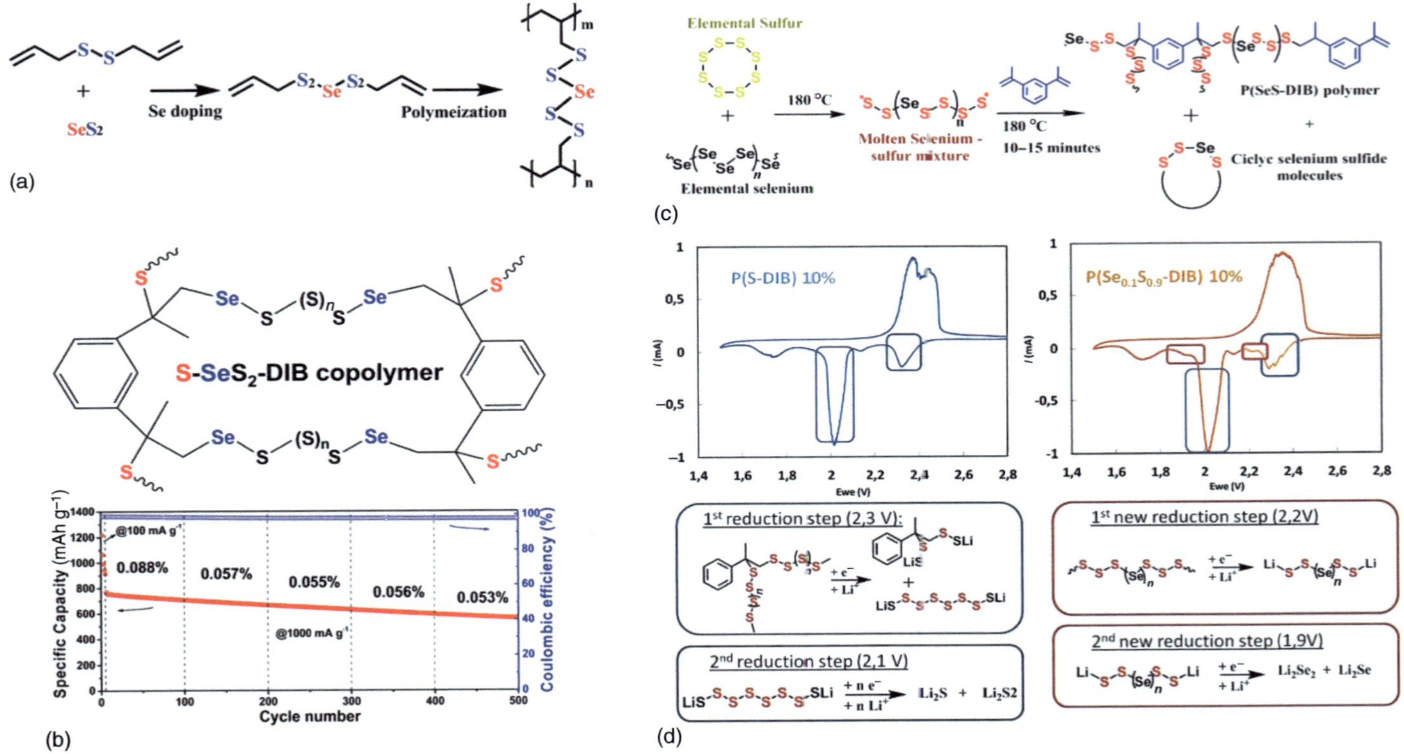

Figure 7.6 (a) The synthetic mechanism of selenium-doped PDATtSSe polymer. Source: Adapted with permission from Zhou et al. [29]/Wiley-VCH. (b) Synthesis route of polymeric selenium sulfides via inverse vulcanization method. Source: Reproduced with permission from Dong et al. [30]/American Chemical Society. (c) Synthesis processes of poly($Se_{0.1}S_{0.9}$-DIB). (d) The CV curves of poly(S-DIB) 10% and poly($Se_{0.1}S_{0.9}$-DIB) 10% correspond to the redox process. Source: Reproduced with permission from Gomez et al. [31]/John Wiley & Sons.

utilization and CE but also enhances the electrode reaction kinetics and forms a more stable solid electrolyte interface (SEI) layer with low charge-transfer resistance. The sulfur-rich polymeric selenium sulfides@Ketjenblack EC600JD (S-SeS_2-DIB@KB600) cathode with an optimum S/SeS_2 mass ratio of 9 : 1 (wt%/wt%) shows an initial specific capacity of 1218 mAh g^{-1} at 100 mA g^{-1}. At the same time, more than 98.2% of high overall CE is obtained for 500 cycles with a low attenuation rate of 0.0549% at 1000 mA g^{-1}.

In 2018, Gomez et al. synthesized sulfur-selenium hybrid copolymers poly ($Se_{0.1}S_{0.9}$-DIB) 10% and poly(S-DIB) 10% with DIB, S_8, and Se_8 as precursors at 180 °C (Figure 7.6c) [31]. It can be seen from the CV curves in Figure 7.6d that there are two similar reduction peaks at 2.30–2.35 V and 2.0 V, which correspond to the transformation of sulfur into high-order linear polysulfide and high-order polysulfide into low-order polysulfide, respectively. In addition, two new peaks at 2.2 and 1.9 V appeared during the reduction of poly($Se_{0.1}S_{0.9}$-DIB), which are caused by the formation of polyselenosulfide and Li_2Se. In the subsequent charging process, the oxidation peak of the copolymer containing selenium shifted to a lower potential, indicating that the addition of selenium improved the reaction kinetics. The cathode containing 5 to 7.5 mol% of Se within S and 10 wt% of DIB shows an initial capacity of 860 and 880 mAh g^{-1} at 0.2 C, respectively. Compared with the polymer without selenium, the capacity loss per cycle is only 0.14% and 0.4%, respectively.

Sodium–sulfur (Na–S) battery based on conversion chemistry has attracted much attention due to its high energy density and low-cost sulfur cathode. However, the dissolution and shuttle effect of high-order polysulfide (Na_2S_n, $4 \leq n \leq 8$), the irreversible deposition of low-order polysulfide (Na_2S_n, $1 \leq n \leq 2$) on the electrode, and the loss of a large amount of cathode active materials have seriously hindered the development of room-temperature Na–S batteries [32]. Ghosh et al. reported that cardanol-based benzoxazine (Ca), a naturally occurring phenolic compound present in agro-waste, can be covalently linked with sulfur through the solvent-free inverse vulcanization method (Figure 7.7a) [33]. Ca with unsaturated bonds and sulfur copolymer form a polymer CS90 with a sulfur content of 90 wt%, and then compounded with rGO to further improve the conductivity of the electrode. The CS90 cathode can exhibit an initial capacity of 542 mAh g^{-1} and retain 335 mAh g^{-1} after 50 cycles, with an average fading rate of 0.76% per cycle. Zhou et al. reported a poly(S-pentaerythritol tetraacrylate) (poly(S-PETEA))-based cathode, combining with a PETEA-tris[2-(acryloyloxy)ethyl] isocyanurate (PETEA-THEICTA)-based gel polymer electrolyte (GPE) for quasi-solid-state Na–S batteries (Figure 7.7b) [34]. The strong chemical combination between sulfur and polymer skeleton in sulfur-rich copolymer electrode effectively inhibits the dissolution and shuttle of sodium polysulfide. Functional GPE with high ionic conductivity and enhanced safety formed in situ can not only help to form a stable Na/GPE interface during cycling but also strongly fix sodium polysulfide to prevent it from diffusing into electrolyte solvents. The Na/GPE/poly(S-PETEA)@C cell can deliver a specific capacity of 877 mAh g^{-1} at 0.1 C and retain 736 mAh g^{-1} after 100 cycles.

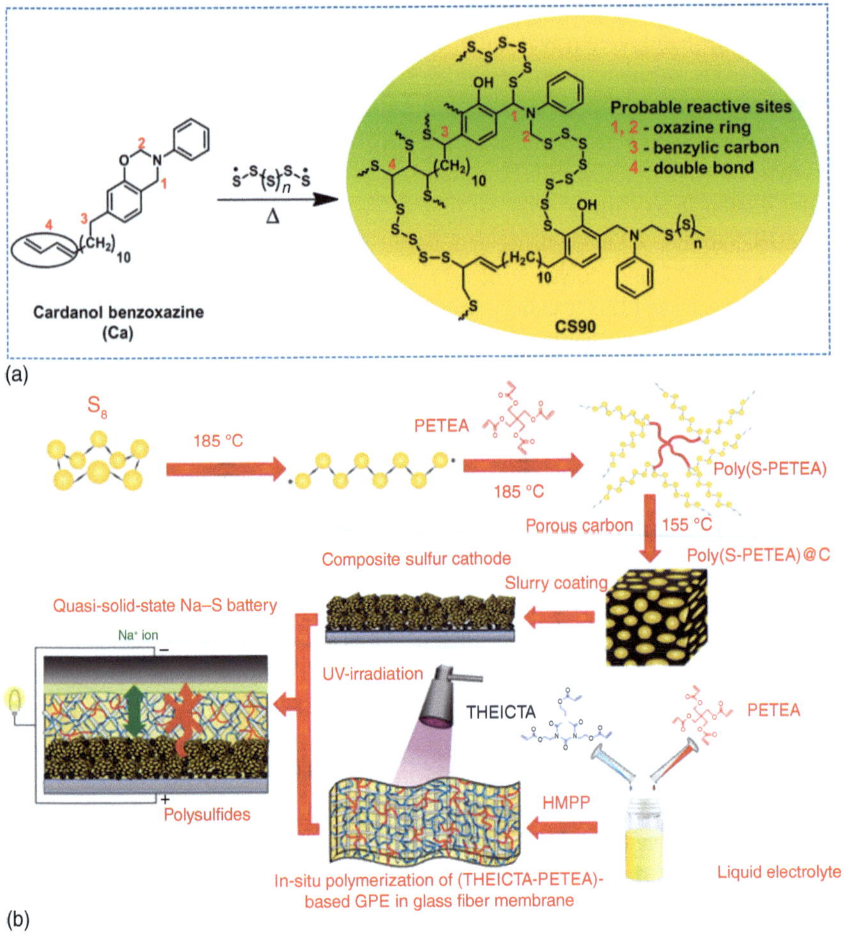

Figure 7.7 (a) The chemical structure of copolymer by reaction of Ca monomer with S_8. Source: Reproduced with permission from Ghosh et al. [33]/Springer Nature. (b) Schematic illustration of the preparation of a quasi-solid state Na–S battery. Source: Reproduced with permission from Zhou et al. [34]/John Wiley & Sons.

7.2.2 -SH-Derived Organosulfur Polymers

Due to the high reactivity between thiols and sulfur, thiol-containing organics are another linker for sulfur-rich copolymers in addition to the unsaturated hydrocarbons mentioned above, allowing the formation of linear sulfur chains by polymerization, thus allowing a homogeneous distribution of sulfur in the polymer matrix [13].

Naoi et al. believed that in order to obtain good electrochemical performance, organosulfur compounds must meet three conditions [35]. Firstly, the electrode structure should allow the rapid breakage and formation of S—S bonds at room temperature. Secondly, it contains polysulfide chains ($-S_n-$, $n > 2$), the larger the n value, the higher the specific capacity. Thirdly, the recombination rate of all

S—S bonds during recharging should be high. Based on this, they designed and synthesized side chain polyorganosulfide (poly(dithiodianiline), PDTDA). In its structure, there are two polyaniline (PAn) backbones, each with an -SH attached to the benzene ring, and the corresponding two -SHs are combined to form an S—S bond. The S—S bonds of PDTDA are broken at discharge state and recombined during the charging process. Li–PDTDA cell can release a specific capacity of 270 mAh g^{-1} and an energy density of 675 Wh kg^{-1}.

Later, Park and his colleagues used porous trithiocyanuric acid (TTCA) crystals as soft templates to synthesize three-dimensional (3D) interconnected organosulfur polymers (S-TTCA-I) [36], which deliver a content of 63 wt% sulfur (Figure 7.8a). The amine groups of TTCA are found to be helpful for the rapid migration of lithium ions in the cathode framework during cycling, and it is closely related to the improvement of the capacity of these Li–S batteries. The S-TTCA-I shows a discharge capacity of 1210 mAh g^{-1} at 0.1 C and 730 mAh g^{-1} at 5 C. After 450 cycles, the cathode can retain the specific capacity of 850 mAh g^{-1} with a high CE of 99%. Sulfur-embedded polybenzoxazine (S-BOP) with a high sulfur content of 72 wt% was prepared by Choi et al. [37]. The sulfur-embedded polymer is obtained by thermal ROP of benzoxazine in the presence of elemental sulfur (Figure 7.8b), wherein the covalent connection of sulfur and polymer is reasonably guided by the thiol groups of benzoxazine. S-BOP exhibits discharge plateaus similar to elemental sulfur at 2.3 and 2.1 V. Furthermore, S-BOP delivers a discharge capacity of 630 mAh g^{-1} at 1 C after 1000 cycles with a capacity retention rate of 92.7%.

In recent years, Fu et al. have developed a new method for preparing organosulfur polymers in the solution phase at room temperature [38–41]. They prepared polyphenylene tetrasulfide (PPTS) by stirring 1,4-benzenedithiol (1,4-BDT) and a stoichiometric proportion of the element S_8 in carbon disulfide (CS_2)/toluene mixed solution for 12 hours [38], which is composed of repeated linear unit structure with four S atoms and one benzene ring with a theoretical specific capacity of 788 mAh g^{-1} (Figure 7.9a). The PPTS-CNT electrode has an initial capacity of 633 mAh g^{-1} at 1 C and a capacity retention of 77% after 350 cycles. In addition, PPTS-CNT electrode also shows good flexibility, which has a great application prospect in flexible electronics materials. Later, a series of polyphenyl polysulfide (PPPS) were also synthesized by condensation reaction of 4,4′-thiobisbenzenethiol (TBBT) with four different molar ratios of sulfur in the same solvent at room temperature (Figure 7.9b) [40]. When the polymer is cast into Bucky paper, the formed electrode is flexible. PPPS-11 shows one reduction peak at 2.20 V from the CV curves corresponding to the formation of Li–S–Ph–S–Ph–S–Li. Then the 2.05 V voltage plateau appears, which is related to the formation of lithium polysulfide and its further transformation into Li_2S_2 and Li_2S. While PPPS-12, PPPS-13, and PPPS-14 deliver two reduction peaks at about 2.15 and 1.95 V (Figure 7.9c). The four cells exhibit the first specific capacities of 215.6, 383.9, 460.1, and 559.9 mAh g^{-1} at 0.1 C, respectively, and the retained discharge capacity of 156.3, 249.6, 280.8, and 324.1 mAh g^{-1} after 100 cycles (Figure 7.9d).

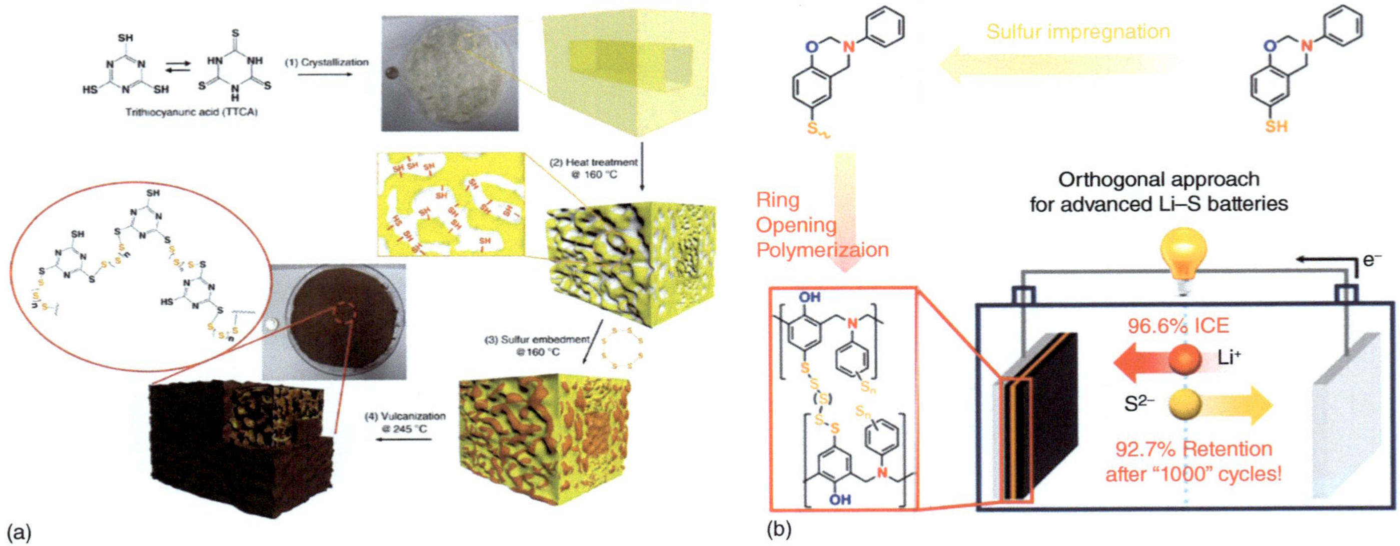

Figure 7.8 (a) Schematic drawings describing the synthetic procedures of sulfur-rich polymers with controllable morphology. Source: Adapted with permission from Kim et al. [36]/Springer Nature. (b) Preparation of S-BOPs and their application in lithium batteries. Source: Reproduced with permission from Je et al. [37]/American Chemical Society.

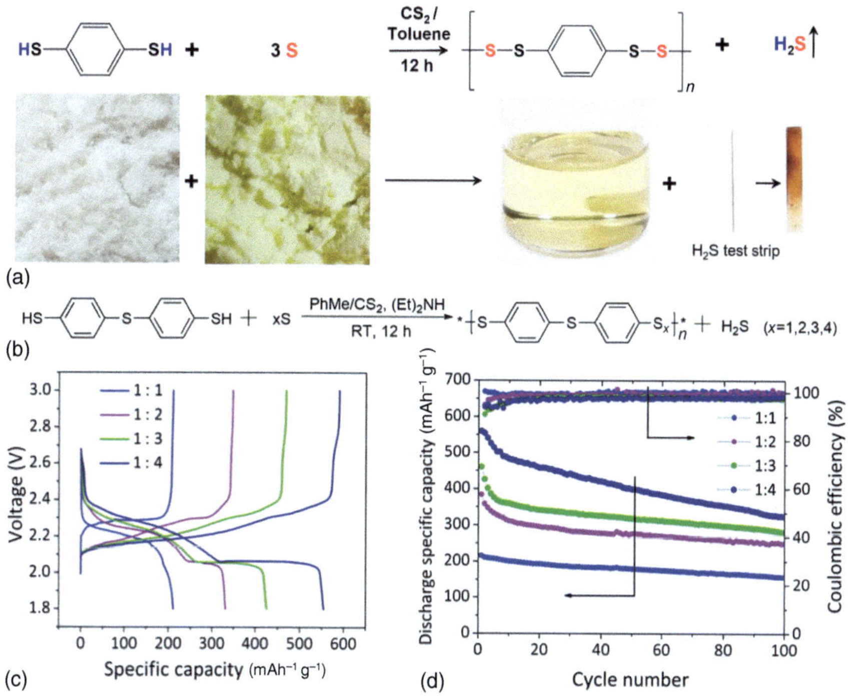

Figure 7.9 (a) The synthesis scheme and corresponding visual representation of PPTS, color change of lead acetate-based test strip verifies the production of H_2S gas. Source: Adapted with permission from Bhargav et al. [38]/American Chemical Society. (b) The synthesized polymer solutions with TBBT:S molar ratios of 1 : 1, 1 : 2, 1 : 3, and 1 : 4. (c) Discharge and recharge voltage profiles of cells with four polymer cathodes. (d) Cycling performance of the cells at 0.1 C rate. Source: Reproduced with permission from Sang et al. [40]/Royal Society of Chemistry.

7.2.3 SPAN

The SPAN was first reported by Wang et al. in 2002 [42, 43]. The polyacrylonitrile (PAN) and sulfur powder were treated at the molecular level at 300 °C (Figure 7.10a). Sulfur can be used as a dehydrogenation reagent to dehydrogenate the PAN structure, thus forming the conductive main chain configuration. The –CN functional groups of PAN are cyclized to form the stable nitrogen heterocycles in which sulfur as the stabilizer is linked. SPAN as a cathode in a lithium battery can cycle in the range of 1.0–3.0 V based on the gel composite electrolyte. The cathodic peak at 1.34 V is observed at the CV curves, which corresponds to the discharge voltage plateau at 1.47 V and stabilizes at 1.85 V in the subsequent cycle. The anodic peak appears at 2.27 V, corresponding to the charge voltage at 2.24 V. The specific capacity of SPAN can reach 850 mAh g^{-1} based on the sulfur content of 53.41 wt% at the first cycle, which indicates that most sulfur can be converted into Li_2S. Therefore, a sulfur utilization of 95.2% can be obtained. After 50 cycles, the discharge capacity can still retain above 600 mAh g^{-1}. When the sulfur content is 65 wt%, it can provide a capacity of more than 1000 mAh g^{-1}, which is necessary to achieve higher capacity. In the

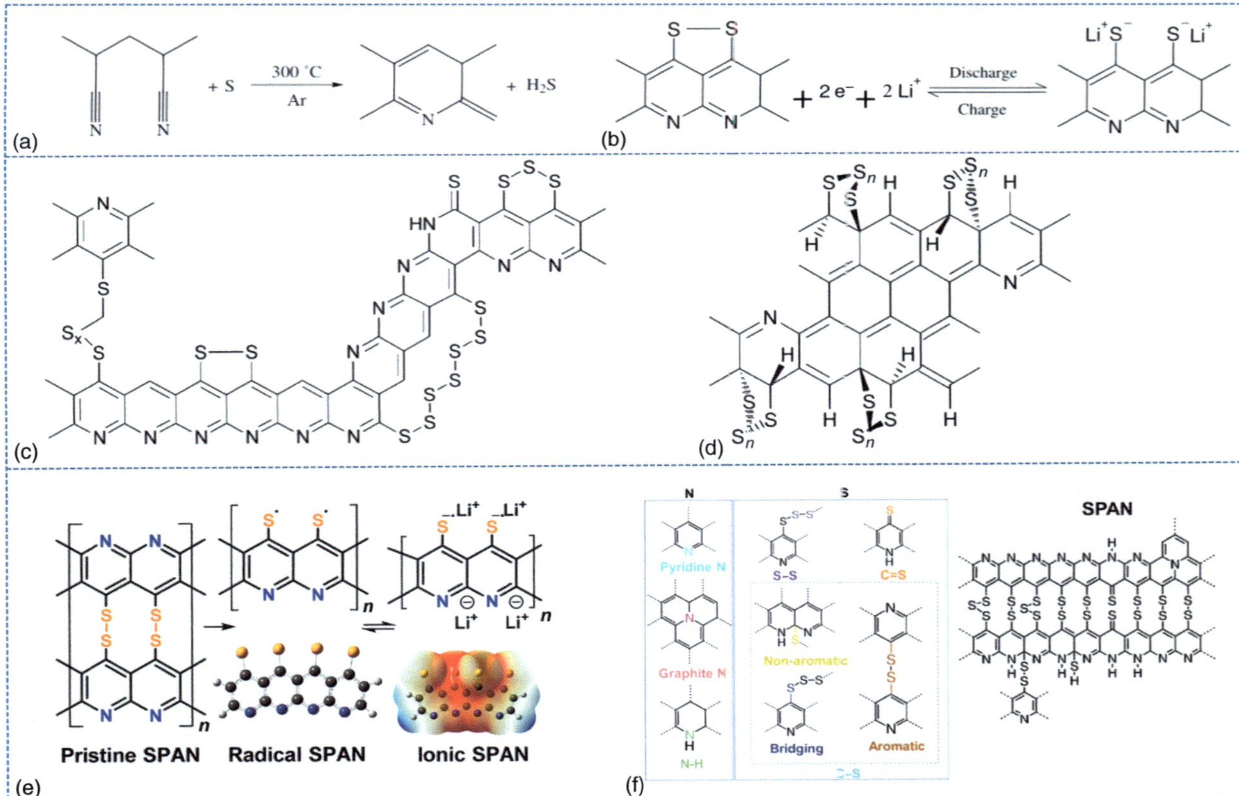

Figure 7.10 (a) Synthesis of SPAN at 300 °C by Wang et al. Source: Adapted with permission from Wang et al. [42]/Wiley-VCH. (b) The proposed lithium storage mechanism of CSM-450 by Yu et al. Source: Adapted with permission from Yu et al. [44]/Elsevier. (c) The proposed structure of SPAN by Fanous et al. Source: Adapted with permission from Fanous et al. [45]/American Chemical Society. (d) The proposed structure of SPAN by Zhang et al. Source: Adapted with permission from Zhang [46]/MDPI. (e) The lithiation mechanism of SPAN proposed by Wang et al. Source: Adapted with permission from Wang et al. [47]/American Chemical Society. (f) Summary of components and the proposed structure of SPAN by Wang et al. Source: Adapted with permission from Wang et al. [48]/American Chemical Society.

charged state, there is no obvious self-discharge phenomenon, which means that SPAN has the function of inhibiting the dissolution of active substances.

At present, there are some related research reports on the electrochemical reaction mechanism and specific structure of SPAN. In 2004, Yu et al. proposed the existence of π conjugation and C—S—S—C bonds in conductive sulfur-containing materials (CSMs) [44]. The CSM-450 prepared at a high synthesis temperature of 450 °C shows improved cycle stability, and the capacity retention rate is 92% after 240 cycles. At this synthesis temperature, the carbonization degree of elemental sulfur increases, and smaller molecules are formed, which can be better dispersed in the PAN structure. As shown in Figure 7.10b, the main chain of CSM-450 can provide the electronic pathway, and the side chain containing disulfide bonds acts as redox active sites in the process of charge and discharge. Fanous et al. used Time-of-flight secondary ion mass spectrometry (TOF-SIMS), X-ray photoelectron spectroscopy (XPS), and Fourier Transform Infrared (FTIR) spectrometer to characterize SPAN composites [45], which shows that S exists in the form ofoligo(sulfide)s, 2-pyridylthiolates, thioamides, and all the sulfur atoms are covalently attached to carbon rather than nitrogen atom (Figure 7.10c). Later, Zhang et al. put forward a more reasonable molecular structure by synthesizing various results, that is, the short $-S_x-$ chain in SPAN is covalently attached to the cyclized, partially dehydrogenated, and ribbon-like PAN backbones (Figure 7.10d) [46]. During the discharge process, both the S—S and the C—S bonds in SPAN would break resulting in a conjugated polymer backbone and Li_2S. In the subsequent charging process, Li_2S will gradually lose electrons and form a ring structure reversibly.

In recent years, the research on the molecular structure of SPAN has continued. Wang et al. continued to explore the electrochemical mechanism of SPAN [47]. They found that the sulfur radicals are generated by S—S bond cracking after the first cycle. The electron delocalization of the thiyl radical on the pyridine main chain can form the conjugated structure, which can react with lithium ion through lithium-coupled electron transfer process, thus forming an ion coordination bond reversibly (Figure 7.10e). Due to the low potential energy, the negative potential energy around the sulfur atom is easier to accommodate for Li^+ than that around the nitrogen atom. In addition, the pyridine N atom can also hold a lithium unit, thus enhancing the capacity. It shows a super-long cycle life of more than 2000 cycles with a high retention rate of 96.8%. The SPAN cathode can also react with aluminum ions (Al^{3+}) utilizing the proposed free radical electrochemical mechanism, which shows high activity, reversibility, and stability in aluminum metal batteries. The Al–SPAN cell can deliver a stable capacity of 201 mAh g^{-1}. Recently, Wang et al. synthesized three different kinds of SPAN at 300, 450, and 550 °C and analyzed the relationship between the ratio of functional groups and temperature, further clarifying the molecular structure and functional groups of SPAN (Figure 7.10f) [48]. XPS and FTIR spectroscopy show that the C–S peak in SPAN contained aromatic C=S bonds, nonaromatic C—S bonds, and C—S_x bonds combined with bridging sulfur. This study further expounds on the working mechanism of the first and subsequent cycles of SPAN and shows that aromatic functional groups contribute to maintain reversible capacity, while irreversible

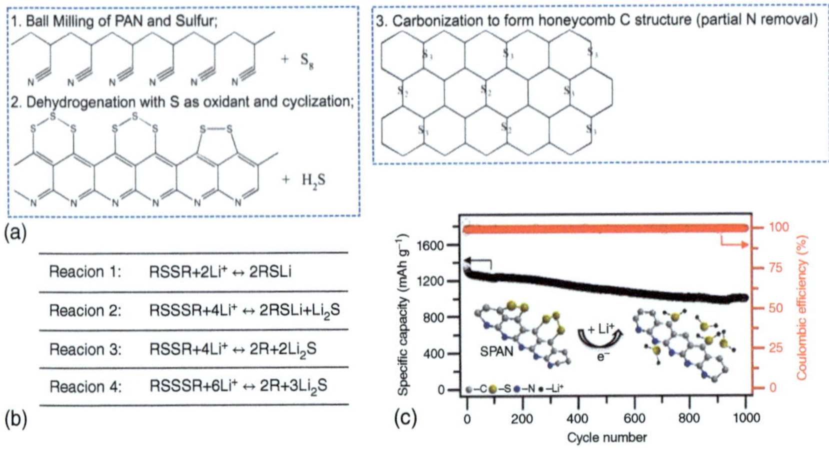

Figure 7.11 (a) Proposed synthesis route for SPAN nanocomposite cathode materials. (b) The proposed lithiation mechanisms for SPAN nanocomposites. (c) Long-cycle performance of SPAN4 electrode. Source: Reproduced with permission from Wei et al. [49]/American Chemical Society.

capacity loss is due to the transformation of nonaromatic functional groups during the first discharge.

In order to further optimize the electrochemical properties of SPAN, Wei et al. prepared a SPAN nanocomposite by a simple thermal synthesis method, which made use of the specific interaction between nitrile groups on the polymer backbone and S_8 to make PAN unstable and promote dehydrogenation and cyclization (Figure 7.11a) [49]. The sulfur in SPAN nanocomposites exists stably in the form of smaller S_2/S_3 species even after hundred cycles was proposed. The proposed lithiation mechanisms for SPAN nanocomposites are shown in Figure 7.11b, during discharge, the composite can not only break the S—S bonds but also break the C—S bonds, and finally, Li_2S is the only sulfur-containing lithiation product. SPAN can be synthesized at different temperatures of 250, 350, 450, and 600 °C. Among them, SPAN4 at 450 °C mainly exists in the form of nanospheres with an average diameter between 100 and 150 nm. SPAN4 shows excellent electrochemical performance with a discharge capacity of more than 1000 mAh g^{-1} after 1000 cycles (Figure 7.11c), which is due to the fact that the smaller molecular sulfur species are confined in the conductive host by covalent bonding and physical confinement, which completely eliminates the dissolution and shuttling of polysulfides between the lithium anode and the sulfur cathode.

In carbonate electrolyte, Li_2S_n ($n \leq 4$) shows low solubility and can react with carbonate to form a protective SEI to alleviate the dissolution problem. However, in ether electrolyte, the high solubility of Li_2S_n would cause a shuttle effect, which would show poor electrochemical performance. Chen et al. started with the unique structure of SPAN and took advantage of the fact that SPAN contains short-chain sulfur ($-S_n^{2-}-$, $2 \leq n \leq 4$). Sulfur and selenium are mixed and heat-treated according to the ratio (15 : 1, 12 : 1, 10 : 1) to form sulfur-selenium complex, and then they are dehydrogenated with PAN to form selenium-doped SPAN (Se_xSPAN, $x < 0.15$, ~50 wt% Se_xS). A small amount of selenium introduced into SPAN can accelerate the redox conversion (Figure 7.12a), making it have excellent performance in both carbonate and ether electrolytes [50]. Se can not only easily realize uniform distribution at the molecular level through Se—S bond but also contribute capacity. The lithium-ion diffusion coefficients (D_{Li+}) of the reduction peak and oxidation peak of Li–$Se_{0.06}SPAN$ battery are higher than those of Li–SPAN battery, which indicates that selenium doping contributes to rapid Li^+ transport. Based on the total mass of $Se_{0.06}SPAN$, the cathode can still provide a capacity of 546 mAh g^{-1}, which is equivalent to the excellent utilization of 75% of theoretical capacity (726 mAh g^{-1}) and retain 416 mAh g^{-1} even after 800 cycles (Figure 7.12b). Furthermore, the tellurium-doped SPAN composite was also synthesized [51, 52]. Wang et al. showed that doping tellurium with high conductivity can reduce the polarization of SPAN cathode and improve its redox kinetics [52]. The average discharge voltage of $Te_{0.045}S_{0.955}PAN$ increased slightly from 1.85–1.88 V (SPAN) to 1.88–1.91 V. The $Te_{0.045}S_{0.955}PAN$ cathode can deliver a capacity of 675 mAh g^{-1} at 0.1 A g^{-1} and retain 96.6% after 100 cycles. Li et al. introduced tellurium into SPAN as eutectic promoter to synthesize Te_xS_{1-x}@pPAN (Te-doped S@pPAN) cathode (Figure 7.12c), in which tellurium can be uniformly distributed through Te—S bond and accelerate the electrochemical reaction process (Figure 7.12d), which makes the cell show good compatibility in ether and carbonate electrolyte systems.

Furthermore, the SPAN can also be selected as cathode of sodium metal batteries [53]. Kim et al. successfully used binder-free SPAN web as cathode material for RT Na–S batteries. The charge–discharge mechanism of SPAN was characterized by XPS and Raman. During the first discharge, the electrochemical activation process was performed, and one Na per S atom (1 Na^+/S) and conjugated carbon skeleton reacted irreversibly, reducing the resistance of SPAN and not leading to the breakage of C—S and S—S bonds. After further sodiumization, a reversible charge–discharge cycle occurs. After simple activation, the reversible discharge capacity of SPAN can reach 1502 mAh $g^{-1}{}_{\text{-sulfur}}$, similar to the theoretical capacity of Na_2S. After 100 cycles, the capacity remains at 1405 mAh $g^{-1}{}_{\text{-sulfur}}$. The theoretical energy densities of the 1st and 100th cycles are 569 and 543 Wh kg^{-1}, respectively, which are higher than common LIBs (380 Wh kg^{-1}).

Magnesium–sulfur (Mg–S) battery has attracted a lot of attention due to its large theoretical volume energy density (4500 Wh l^{-1}). However, the cycle and rate stability of Mg–S battery is poor, and the self-discharge is high during the cycle, which is mainly due to the insulating properties of sulfur and its slow redox kinetics. At the same time, the volume expansion of the sulfur electrode and the shuttle effect of magnesium polysulfide (MgS_n, $4 \leq n < 8$) intermediates in the electrolyte containing ether solvent, would also cause poor battery performance. Organosulfur compounds with S—S bonds have been proposed as new cathode materials for rechargeable magnesium batteries [54–56]. The process of discharging and charging involves the fracture and remodeling of S—S bonds. NuLi et al. reported that three organosulfur cathode materials, namely 2,5-dimercapto-1,3,4-thiadiazole (DMcT), PDTDA, and CMS, are used in magnesium batteries [54]. Among them, DMcT

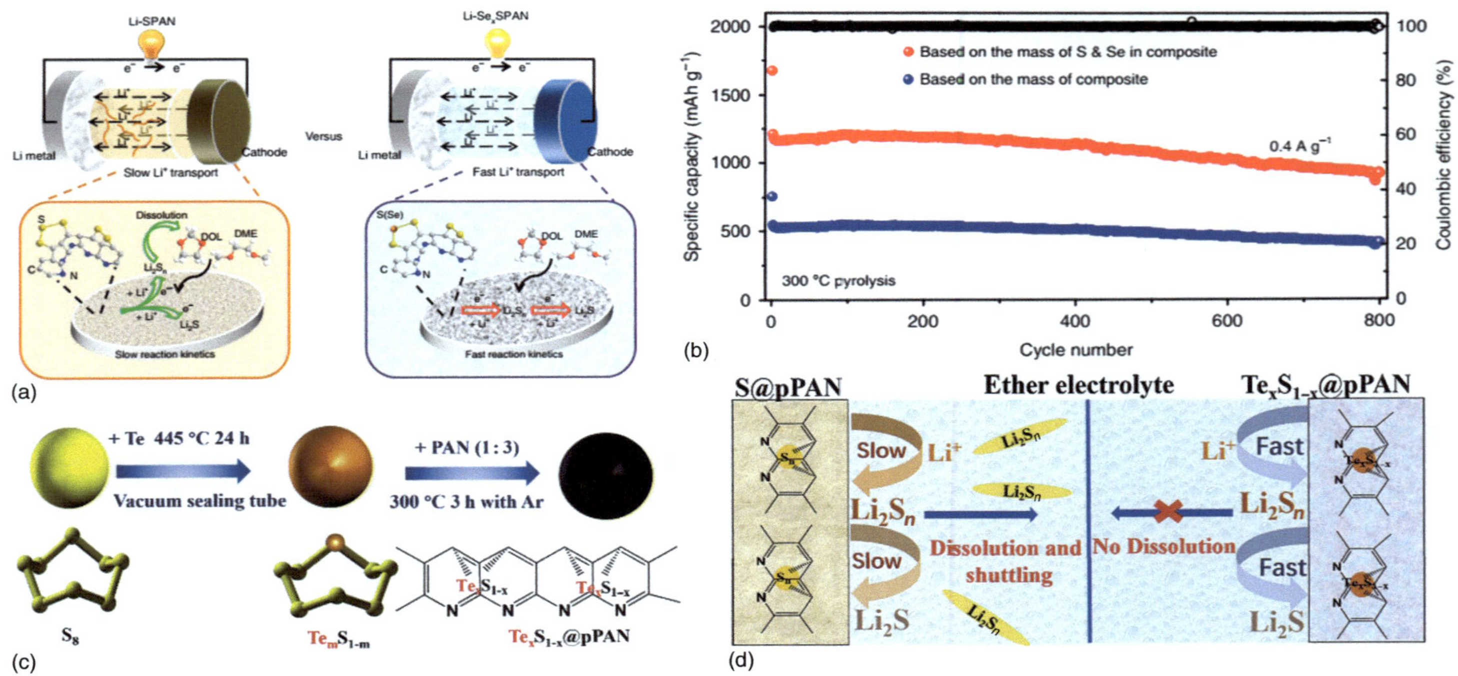

Figure 7.12 (a) The reaction process scheme of proposed Li–Se$_{0.06}$SPAN and Li–SPAN cell. (b) Cycling performance of Li–Se$_{0.06}$SPAN cell. Source: Chen et al. [50]/Springer Nature/CC BY 4.0. (c) The synthetic schematic of $Te_{0.04}S_{0.96}$@pPAN. (d) The reaction process scheme of Li–$Te_{0.04}S_{0.96}$@pPAN and Li–S@pPAN cell. Source: Reproduced with permission from Li et al. [51]/Elsevier.

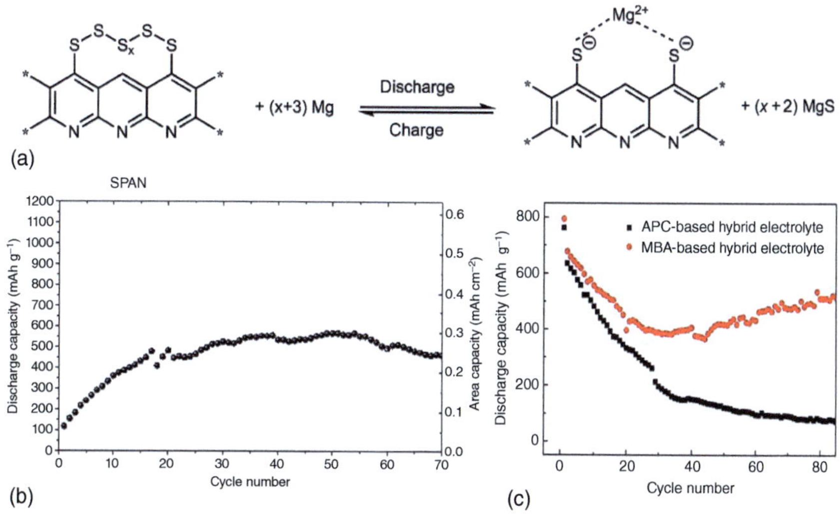

Figure 7.13 (a) Proposed overall charge and discharge reaction of SPAN used in Mg–S battery. (b) The cycle stability of Mg/SPAN-based cell cycled at C/30. Source: Reproduced with permission from Wang et al. [55]/Elsevier. (c) The cycling performance for the Mg/S@pPAN cell. Source: Reproduced with permission from Zhang et al. [56]/Elsevier.

with PAn shows a discharge plateau at 1.4 V, and PDTDA shows good dynamics and conductivity. The Mg/CMS-PAn cell exhibits an initial capacity of 120 mAh g^{-1} and good capacity retention. Wang et al. used SPAN as the cathode material of magnesium metal battery and studied its charge–discharge mechanism [55]. In the first cycle, some polysulfides (MgS_4 and MgS) may be formed during discharge (Figure 7.13a). The surface study of SPAN cathode and magnesium anode shows that MgS, MgF_2, $MgCO_3$, MgO, and $MgSO_4$ could form SEI layer on the electrode. The cell exhibits a discharge capacity of 500 mAh g^{-1} utilizing a pressed Mg anode made from Mg powder (Figure 7.13b). Furthermore, Zhang et al. used the sulfurized-pyrolyzed polyacrylonitrile (S@pPAN) composite with 47.3 wt% sulfur as the cathode electrode and matched the non-nucleophilic magnesium bis(diisopropyl)amide (MBA)–$(MgCl_2)_2$–$(AlCl_3)_2$ + LiCl/THF electrolyte for the Mg–SPAN battery [56], which not only significantly restrains the reaction between nucleophilic electrolyte and electrophilic sulfur but also effectively inhibits the dissolution and shuttle of polysulfides. In addition, a more effective cathode electrolyte interface (CEI) layer formed on the cathode with lower electrode/electrolyte internal resistance plays a key role in improving electrochemical performance. The S@pPAN cathode shows a specific capacity of 677.9 mAh g^{-1} at 0.1 C and a capacity retention of 75.9% after 85 cycles (Figure 7.13c).

7.2.4 Covalent Organosulfur Polymers

Although impregnating sulfur by forming strong covalent bonds between organic groups and sulfur can prevent the dissolution of polysulfides. However, the poor conductivity of sulfur-containing polymers still seems to be a fundamental challenge that limits their further development. At the same time, some sulfur chains connected to small organic molecules would be free in the electrolyte to form soluble polysulfides after a long-term cycle. Therefore, it is necessary to develop a special frame structure with a large void space and good electron/ion conductivity, so as to distribute as much sulfur as possible in it tightly, evenly, and regularly [57].

Covalent triazine frameworks (CTFs) are a kind of microporous polymer with a narrow pore size distribution, an extremely high specific surface area of 3000 $m^2 g^{-1}$, and medium conductivity. Coskun's group reported that CTF-1 (S-CTF-1) was synthesized from 1,4-dicyanobenzene and elemental sulfur without any catalysts or solvents at 400 °C (Figure 7.14a) [58], which realized covalent connection and uniform distribution of sulfur in CTF skeleton through in-situ vulcanization of elemental sulfur, thus effectively inhibiting the dissolution of polysulfides in electrolyte. Similar to the voltage curve of sulfur, the S-CTF-1 also shows a two-step discharge process, including the formation of lithium polysulfides at 2.35 V and the formation of Li_2S at 2.1 V. The S-CTF-1 cell with a sulfur content of 62 wt% delivers an initial capacity of 670 mAh g^{-1} with a capacity retention rate of 85.8% after 300 cycles.

Fluorinated porous triazine-based framework (FCTF-S) can play a role of physical or chemical restriction on polysulfides, thus mitigating the shuttle effect and volume expansion [59]. Adding electronegative fluorine atoms to the skeleton can provide a strong anchoring effect, inhibit dissolution, and accelerate the transformation of polysulfides. The copolymerization of perfluorinated aromatic nitrile monomers and elemental sulfur is based on the framework of triazine fluoride, the covalent attachment of sulfur, and its uniform distribution in the pores (Figure 7.14b). The Li/FCTF-S delivers a high discharge specific capacity of 1296 mAh g^{-1} at 0.1 C and a capacity of 833 mAh g^{-1} after 150 cycles at 0.5 C. Most of the methods have limited sulfur content (i.e. <80 wt%), which is a significant obstacle to the realization of high energy density in actual batteries. Coskun et al. reported the dehalogenation and nucleophilic addition reactions of perfluoroaryl units with nucleophilic sulfur radical, resulting in a high sulfur content of 86 wt% in the generated sulfur-impregnated CTF (SF-CTF) structure (Figure 7.14c) [60]. The uniform distribution of sulfur, the conjugated main chain providing electron transport, and the heteroatoms with strong binding ability to polysulfides promote the excellent electrochemical performance of SF-CTF-1 (1 : 3). The SF-CTF-1 cathode can exhibit a high discharge specific capacity of 1138.2 mAh g^{-1} at 0.05 C and retain a capacity of 520.1 mAh g^{-1} over 300 cycles at 1 C.

The mesoporous imine COF with densely arranged active vinyl groups (COF-V) can also be inversely vulcanized with sulfur to prepare the polymers with high sulfur content (S-COF-V) (Figure 7.14d) [61]. The chemical covalent bonding of sulfur dispersed in nanoporous channels with vinyl groups in COF-V and the physical limitation of mesoporous channels significantly inhibit the shuttle effect of polysulfides. The CV curve of S-COF-V shows characteristics similar to those of the sulfur cathode. Among them, the two cathodic peaks at 2.33 and 2.03 V correspond to the formation of polysulfides and Li_2S, respectively, while short-chain sulfides are oxidized to long-chain sulfides and sulfur at about 2.35 V. The S-COF-V delivers the

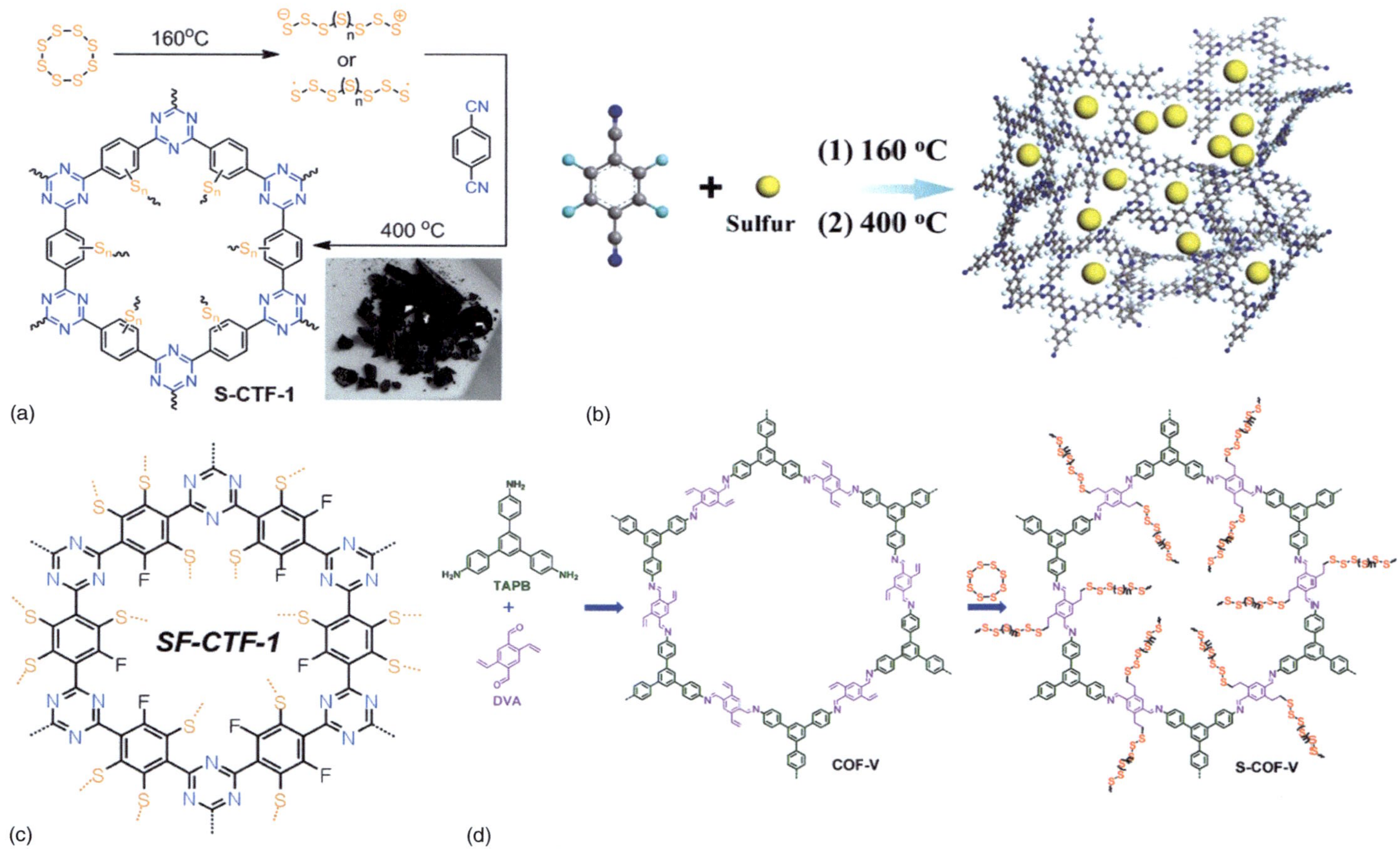

Figure 7.14 (a) Synthesis of S-CTF-1. Source: Adapted with permission from Talapaneni et al. [58]/John Wiley & Sons, Inc. (b) Schematic synthesis of the FCTF-S. Source: Adapted with permission from Xu et al. [59]/American Chemical Society. (c) The structure of SF-CTF-1. Source: Adapted with permission from Je et al. [60]/Wiley-VCH. (d) The synthetic route of S-COF-V. Source: Adapted with permission from Jiang et al. [61]/The Royal Society of Chemistry.

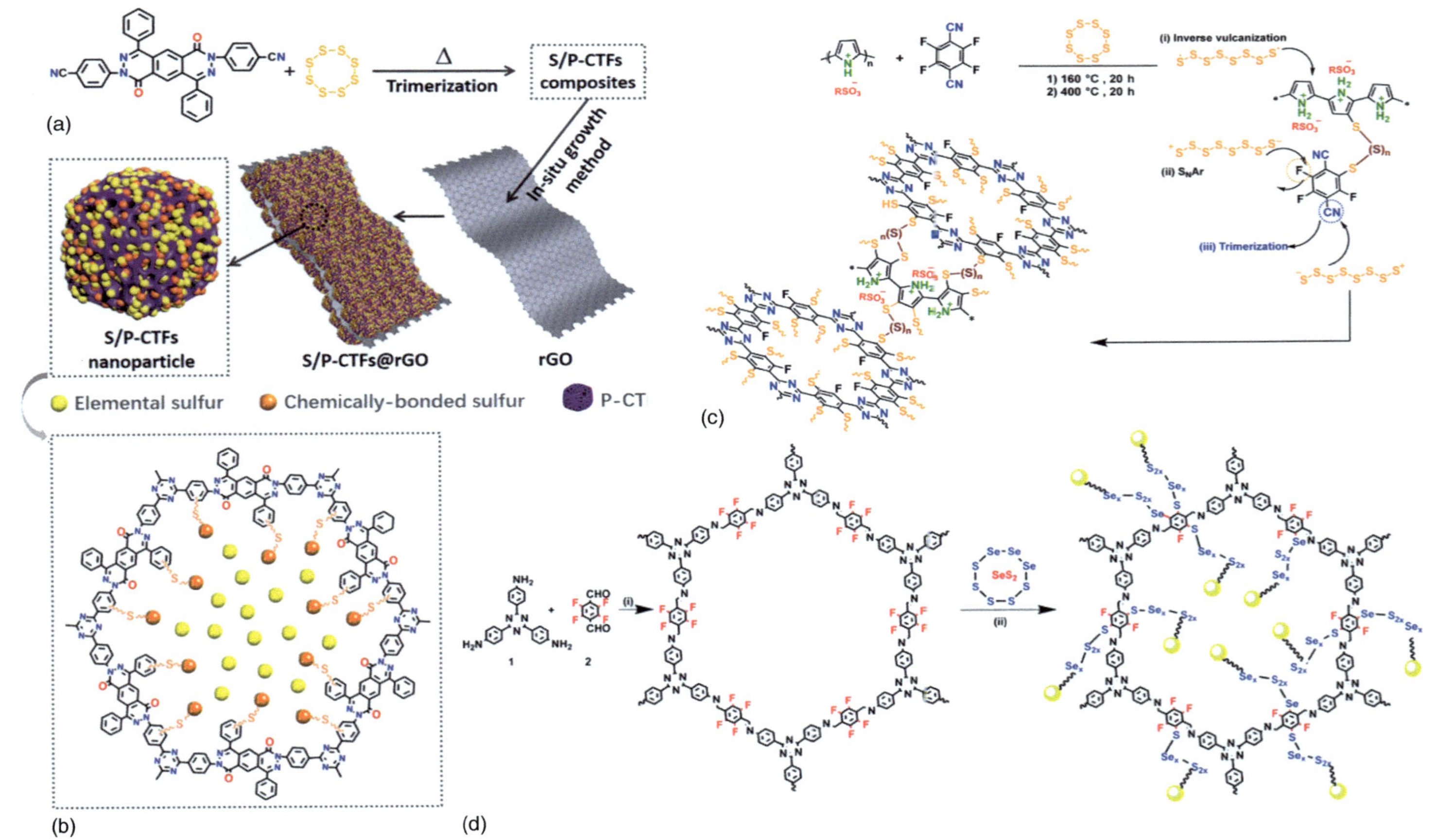

Figure 7.15 (a) Schematic synthesis of the S/P-CTFs and S/P-CTFs@rGO composites. (b) The structure of S/P-CTFs. Source: Reproduced with permission from Guan et al. [62]/American Chemical Society. (c) The synthesis of cPpy-S-CTF. Source: Adapted with permission from Kim et al. [63]/American Chemical Society. (d) Schematic illustration for the synthesis of COF-F-SeS_2. Source: Adapted with permission from Wang et al. [64]/Royal Society of Chemistry.

initial capacity of 1400 mAh g^{-1} at 0.2 C with a CE of 94.5% and presents the retained capacity of 959 mAh g^{-1} after 100 cycles.

In order to further improve the conductivity of the electrode. Guan et al. prepared S/P-CTF@rGO hybrids by in-situ growth of S/P-CTF generated by cyclization reaction of sulfur and dinitrile monomers in porous phthalazinone-based covalent triazine frameworks (P-CTFs) on conductive rGO (Figure 7.15a,b) [62]. Among them, not only the nanoporous structure can capture the polysulfide fragments in P-CTFs, but also the polar groups in P-CTFs, such as phthalazinone and triazine, can covalently bind with polysulfides, further limiting their diffusion. In addition, the introduction of conductive rGO is helpful to realize faster electron transport and electrochemical transformation. The binding energies of phthalazinone and triazine with Li_2S_4 are −2.64 and −2.51 eV, respectively, which indicates that the functional groups endow P-CTFs with strong chemical adsorption of polysulfides. The Li/S/P-CTF@rGO cell delivers the initial discharge capacity of 1130 mAh g^{-1} and retains the capacity of 920 mAh g^{-1} at 0.5 C after 500 cycles. Choi et al. incorporated one-dimensional (1D) charged polypyrrole (cPpy) into two-dimensional (2D) CTFs in the presence of elemental sulfur to synthesize 3D hybrid polymer composite cPpy-S-CTF with a high sulfur content of 83 wt%. (Figure 7.15c) [63]. The addition of cPpy in cPpy-S-CTF promotes the formation of 3D nanochannels, which not only present high affinity cationic anchoring sites for LiPSs but also provide a clear conduction path for accelerating the transmission of lithium ions and electrons. The cPpy-S-CTF delivers a capacity of 1203.4 mAh g^{-1} at 0.05 C with a capacity retention of 86.8% after 500 cycles.

Because the electronic conductivity of selenium (1×10^{-3} S cm^{-1}) is higher than that of sulfur (5×10^{-28} S cm^{-1}), it is another strategy to use sulfur-selenium solid solution material as the cathode to improve the conductivity of the battery. Wang et al. reacted the fluorine-containing covalent organic framework (COF-F) with SeS_2 at 380 °C for 8 hours to obtain the target product COF-F-SeS_2, which was used as the cathode material of lithium metal battery (Figure 7.15d) [64]. Among them, porous crystalline COF-F-SeS_2 can capture polysulfides and polyselenides through physical and chemical interactions. Moreover, the combination of high capacity of sulfur and high conductivity of selenium further makes the cathode based on porous COF show excellent electrochemical performance. The COF-F-SeS_2 cathode shows an initial discharge capacity of 1633 mAh g^{-1} at 0.1 C and retains the capacity of 970 mAh g^{-1} after 100 cycles.

7.3 Carbonyl-Derived Polymers

Carbonyl-derived polymers use carbonyl groups as active groups, which have high storage capacity and redox potential, fast reaction kinetics, and poor solubility. In order to solve the dissolution problem of small molecular compounds in organic electrolytes, one of the most effective methods is to convert them into long-chain polymer structures through copolymerization [9, 11, 65, 66]. Among them, according to the structure types of the polymers, they can be roughly divided into

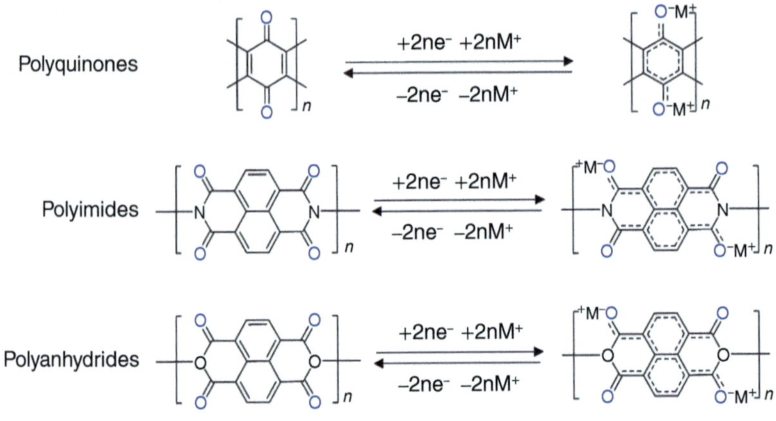

Figure 7.16 Schematic representation of the redox reactions of carbonyl-derived polymers. Source: Adapted with permission from Du et al. [6]/Wiley-VCH.

three categories, namely PQs, PIs, and PAs. They store energy through reversible tautomerism between carbonyls and enols, a process accompanied by the extraction and insertion of alkali metal ions (M^+, such as lithium, sodium, and potassium ions), as shown in Figure 7.16. Notably, although PIs and PAs contain four carbonyl groups per unit, only one carbonyl group in each imide or anhydride can undergo fully reversible redox reactions, this is due to the strong electron repulsion and steric hindrance between adjacent carbonyl groups during the reduction process, which would easily destroy the polymer structures [6, 10, 67].

7.3.1 Polyquinones

PQs are formed by the copolymerization of six-membered cyclic diketones containing two double bonds and their derivatives as monomers, that is, carbonyl groups exist in each structural unit. Among them, the two hydrogen atoms in the aromatic parent nucleus are each replaced by an oxygen atom. According to the aromatic nucleus, PQs can be simply divided into benzoquinone-based, naphthoquinone-based, anthraquinone-based, and phenanthrenequinone-based polymers [65]. We have listed some structures in Figure 7.17.

The electrochemical properties of quinone-based polymers were first reported by Manecke and Storck in 1962 [68, 69]. Later, Foos et al. synthesized poly-1,4-dimethoxybenzene (**1**) and tested its electrochemical performance when used as the cathode for lithium batteries in 1986 [70], which has relatively high theoretical specific capacity ($505\,mAh\,g^{-1}$), low solubility, and high discharge voltage (2.5–3.0 V versus Li^+/Li). However, its practically low discharge capacity ($136\,mAh\,g^{-1}$) and poor cycle performance may be due to factors such as the low purity of the synthesized products at that time. In 1999, Häringer et al. synthesized poly(5-amino-1,4-naphthoquinone) (PANQ, **10**) electrochemically, which demonstrates a discharge capacity close to the theoretical value ($290\,mAh\,g^{-1}$) [71]. In recent decades, researchers have synthesized and tested many other

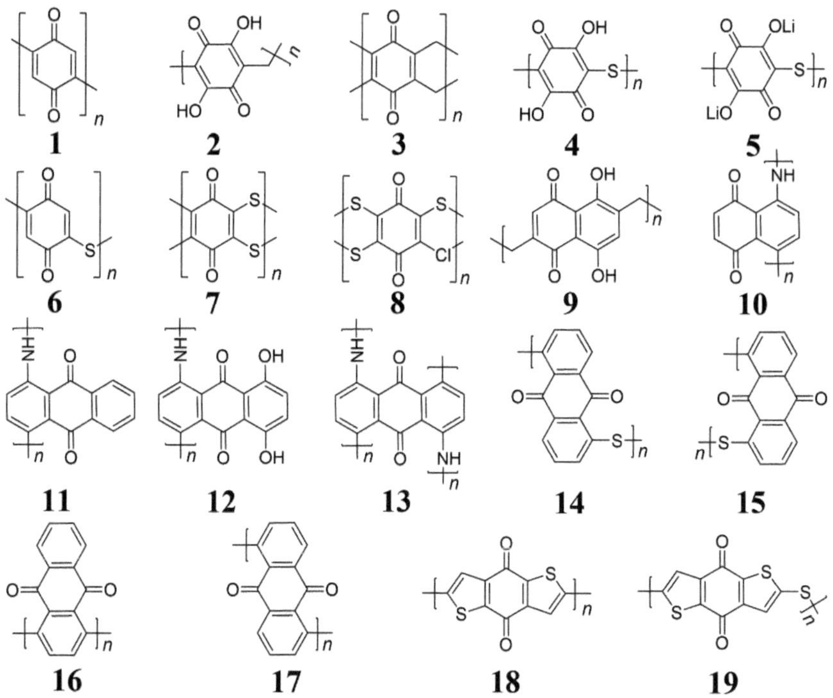

Figure 7.17 Some typical molecular structures of PQs.

quinone-based polymer cathode materials since then. Among them, the monomers are polymerized together by linkers such as methylene (–CH_2–), thioether (–S–), and amine (–NH–) [11].

The methylene (–CH_2–)-linked polymer poly(2,5-dihydroxy-1,4-benzoquinone-3,6-methylene) (PDBM, **2**) was reported by Gall et al. in early years [72], after which Oyama et al. investigated the effect of poly(3,4-ethylenedioxy-thiophene) (PEDOT) additive on the electrochemical behavior of PDBM-coated electrodes [73]. Similarly, Pirnat et al. used a simple method to synthesize a highly porous and cross-linked hydrobenzoquinone-formaldehyde polymer (HBQP) and benzoquinone-formaldehyde polymer (BQP, **3**) complex containing quinone and hydroquinone units [74]. The composite achieves good wettability (ionic wiring) and insolubility in electrolytes. However, the introduction of inactive groups increases the molecular mass and reduces the theoretical specific capacity due to the electrochemical inertness of the methylene group. In addition to this, the exact structure is still unclear as both polymers are insoluble, which makes purification and characterization in solution practically impossible.

Besides, thioether (–S–) linked PQs (**4–8, 14–15**) have also been reported. Among them, the introduction of sulfur atoms can not only further improve the redox potential and electronic conductivity but also the thioether groups formed in the polymers could provide strong intermolecular reactions, leading to excellent electrochemical performance [75, 76]. Liu et al. synthesized poly(2,5-dihydroxyl-1,4-benzoquinonyl sulfide) (PDBS, **4**) by the Phillips method with chloranilic acid (CLA) and $Na_2S{\cdot}9H_2O$ as precursors, as shown in Figure 7.18a [77]. The ^{13}C solid-state nuclear magnetic resonance (NMR) spectra (Figure 7.18b) of the PDBS electrodes at different states show that the backbone of the polymer electrodes is very stable during charging and discharging, which explains why the batteries could cycle stably. The PDBS delivers a high initial discharge capacity of 350 mAh g^{-1} and has a reversible capacity of 184 mAh g^{-1} after 100 cycles. In addition, lithiation is also an effective method to reduce the solubility of PQ. Song et al. reported the lithium salt of poly(2,5-dihydroxy-*p*-benzoquinonyl sulfide) (Li_2PDHBQS, Figure 7.18c, **5**) with the characteristics of both organic salts and polymer electrode materials by one-step polycondensation reaction [78]. The combined advantages of the O···Li···O coordination bond and increased molecular weight ($M_w = 1300\,g\,mol^{-1}$) make the Li_2PDHBQS polymer completely insoluble in the electrolyte. Li_2PDHBQS can still retain 90% of the initial capacity after 1500 cycles at a current density of 500 mA g^{-1}, and the CE can be stabilized at 99.9–100.1% after several initial cycles (Figure 7.18d).

In 2009, Song et al. synthesized poly(anthraquinonyl sulfide) (PAQS, **14**) by linking aromatic rings with sulfur atoms in a one-step reaction and used it as cathode materials for lithium batteries [79], which demonstrates excellent reversibility, cyclability, and high utilization ratio. We believe that simple PQ structures are of particular interest because they can maintain high specific capacity. Benzoquinone units with high redox potential and high theoretical capacity are the best choice. However, since the benzoquinone groups are more inclined to oxidize the sulfide anions than the anthraquinone groups, resulting in a side reaction between quinone groups and sulfide anions in addition to the polycondensation reaction between the halogenated quinone monomers and the sulfide anions. To this end, Song et al. developed a polymerization-oxidation method to synthesize poly(benzoquinonyl sulfide) (PBQS, Figure 7.19a, **6**) containing a large number of benzoquinone units by optimizing various reaction conditions [80]. PBQS delivers excellent long-term cycling stability (1000 cycles, 86%), rate capability (5000 mA g^{-1}, 72%), and high energy density (734 Wh kg^{-1}) in Li batteries. At the same time, PBQS can also be used as a cathode material for rechargeable Na batteries, which have an energy density of up to 557 Wh kg^{-1}. Similarly, Xie et al. reported a novel ladder-structured polymer poly(2,3-dithiino-1,4-benzoquinone) (PDB, Figure 7.19b, **7**), which shows an initial specific capacity of 1050 mAh g^{-1} and a capacity retention rate of 98.4% after 100 cycles [81]. After that, they successfully mixed PDB and CNTs to prepare a flexible and electrochemically stable self-supporting electrode for the all-plastic batteries (Figure 7.19c,d) [82]. The as-prepared PDB polymer cathode exhibits a high energy density of 624 Wh kg^{-1} and can be stably cycled for over 1000 cycles. Furthermore, pristine PDB and prelithiated PDB were assembled together as cathode and anode of all-plastic batteries, respectively, in order to investigate their electrochemical behavior and operational mechanisms. During discharge, the PDB accepts two Li^+ at the cathode, while the prelithiated PDB releases two Li^+ at the anode. The symmetrical full battery has a capacity of 119 mAh g^{-1} after 250 cycles at 1.0 A g^{-1}, and at the same time, the energy density can reach 276 Wh kg^{-1}. Recently, He's group

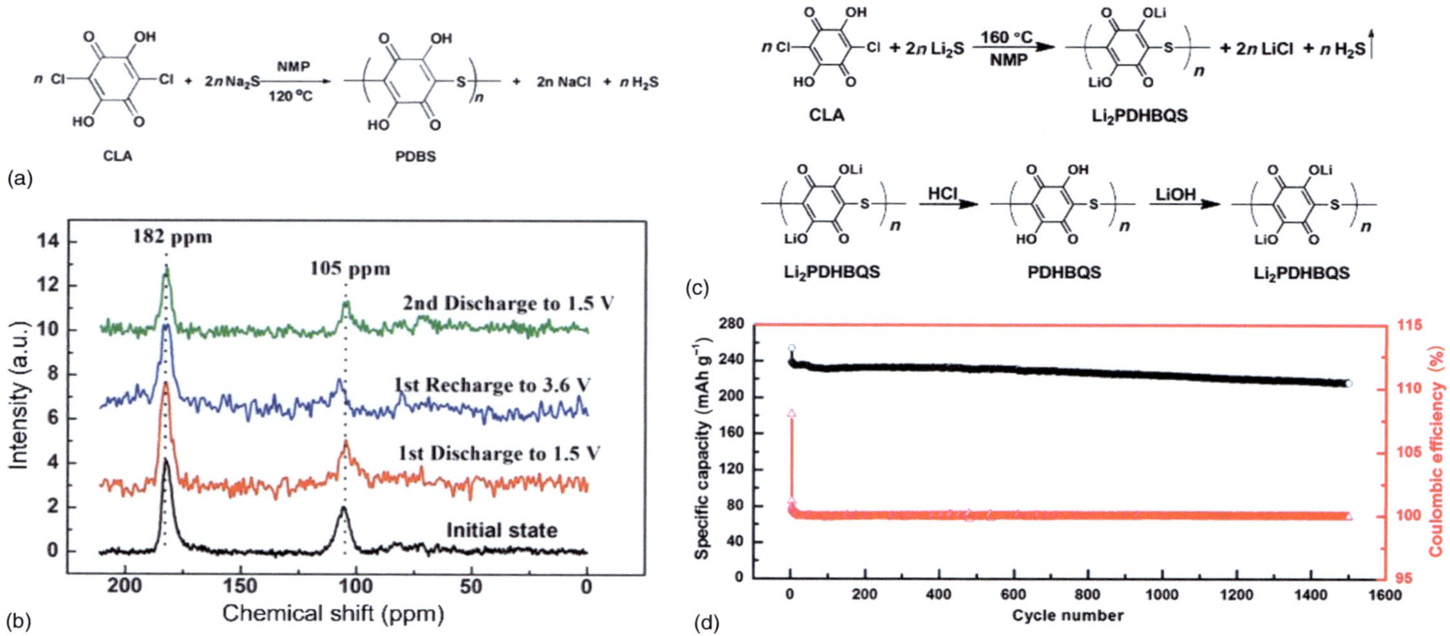

Figure 7.18 (a) The synthesis route of PDBS. (b) ^{13}C solid-state NMR spectra of the PDBS electrode at different charge or discharge states. Source: Reproduced with permission from Liu et al. [77]/Royal Society of Chemistry. (c) Synthesis of Li_2PDHBQS by one-step polycondensation of CLA and Li_2S; Purification route of Li_2PDHBQS. (d) Long-cycle performance of Li_2PDHBQS electrode with absorbed water. Source: Reproduced with permission from Song et al. [78]/Royal Society of Chemistry.

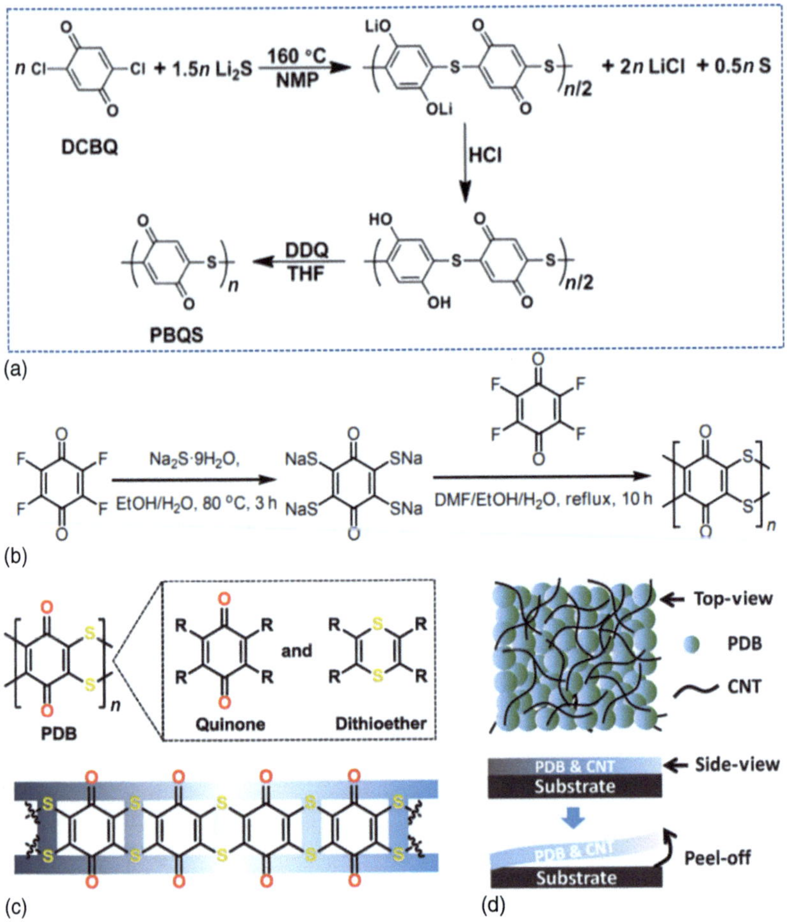

Figure 7.19 (a) The synthesis route of PBQS. Source: Adapted with permission from Song et al. [80]/Wiley-VCH. (b) The synthesis route of PDB. Source: Adapted with permission from Xie et al. [81]/Science China Press. (c) Chemical structure of the ladder-structured PDB. (d) Schematic procedures for preparing a free-standing PDB electrode film. Source: Reproduced with permission from Xie et al. [82]/John Wiley & Sons.

investigated the electrochemical performance of the chlorine-substituted polymer poly(2-chloro-3,5,6-trisulfide-1,4-benzoquinone) (PCTB, **8**) as electrode material for Li-organic batteries [83, 84]. Among them, the existence of electron-withdrawing groups increases the ability of organic cathode to accept electrons, making the electrodes exhibit a high potential for lithium storage.

In addition to benzoquinone-derived polymers, other quinone-derived polymers are also widely used for organic electrodes. Kassam et al. used lithiated 1,4,5,8-naphthalenetetraol formaldehyde polymer (**9**) with naphthoquinone-based structure as cathode materials for all organic Li-ion batteries in 2011 [85]. Häringer et al. synthesized a conducting copolymer PANQ (**10**) bridged by an amine (–NH–) moiety of PAn backbone, in which the –NH– units only exhibit a redox activity at high potential without affecting the redox process of the quinone centers [71]. The battery with the porous $LiMn_2O_4$/carbon counter electrode exhibits a redox potential of about 2.6 V and a discharge-specific capacity close to the theoretical value (313 mAh g^{-1}) of about 290 mAh g^{-1} based on 2e$^-$ transfer. Similarly, some anthraquinone-based organic polymers containing the –NH– units have attracted more attention in recent years due to their special structures. Compared with the above naphthoquinone-based polymers, they exhibit extended conjugated structures and enhanced conductivity to some extent. Tang et al. used the synthesized poly-1-amino-anthraquinone (PAAQ, **11**) as cathode material for lithium batteries and analyzed the physicochemical and electrochemical properties of the materials [86]. The research shows that not only the quinone groups can undergo redox reaction, but also the conductive PAn skeleton π-conjugated system can undergo doping/de-doping during the charge–discharge process of PAAQ. It can achieve a discharge capacity of 218.3 mAh g^{-1} in the first cycle and still maintain relatively high CE after 25 cycles. After that, Zhao et al. investigated the electrochemical behavior of poly(5-amino-1,4-dyhydroxy anthraquinone) (PADAQ, **12**) synthesized by a simple oxidation process for rechargeable lithium batteries [87]. This polymer exhibits low solubility and non-crystallinity in the electrolyte, which is different from its monomer, thus maintaining the stability of the electrode structure well. The electrochemical mechanism of PADAQ is shown in Figure 7.20a. The analogous PAn chain undergoes doping/dedoping reactions between 3.3 and 3.7 V, and redox reactions of carbonyl and hydroxyl groups occur at 3.3–1.5 V and 1.5–1.0 V, respectively. Compared with the monomer, the cycling performance of PADAQ is improved. At a cut-off voltage of 1.5–3.7 V and a current density of 400 mA g^{-1}, the initial discharge capacity of PADAQ is 101 mAh g^{-1}, which can be maintained at 126 mAh g^{-1} after 50 cycles. It can reach a discharge capacity of 95 mAh g^{-1} even at a high current density of 1400 mA g^{-1}. Furthermore, Liu et al. used 1,5-diaminoanthraquinone (DAAQ, **13**) electrochemically polymerized on conductive CNTs/NTs as the organic cathode materials for LIBs [88]. The relationship between the redox properties and the structures of the polymerized DAAQ and 1-aminoanthraquinone (AAQ) was investigated by experiments and theoretical calculations. Potential-dependent CV measurements reveal a bipolar charge storage mechanism of polymerized DAAQ. Namely, an n-type doping/dedoping mechanism for the binding of quinone groups and Li^+ in the low potential window of 1.5–3 V and a p-type doping/dedoping mechanism for the interaction between secondary amines and the PF_6^- anion in the high potential window of 3.0–4.5 V, as shown in Figure 7.20b. The poly-DAAQ-NT cathode with a mass fraction of 60 wt% polymer provides a capacity of up to 311 mAh g^{-1}, much higher than AQ-NT (Figure 7.20c).

It has been found that the different substitution positions of the linker on the carbonyl-based polymers or the different kinds of linker can also greatly affect the electrochemical properties of the materials. Xu et al. investigated the effect of two polymers poly(1,5-anthraquinonyl sulfide) (P15AQS, **14**) and poly(1,8-anthraquinonyl sulfide) (P18AQS, **15**) containing linkers at different positions as organic electrode materials on electrochemical behavior [89]. The P15AQS

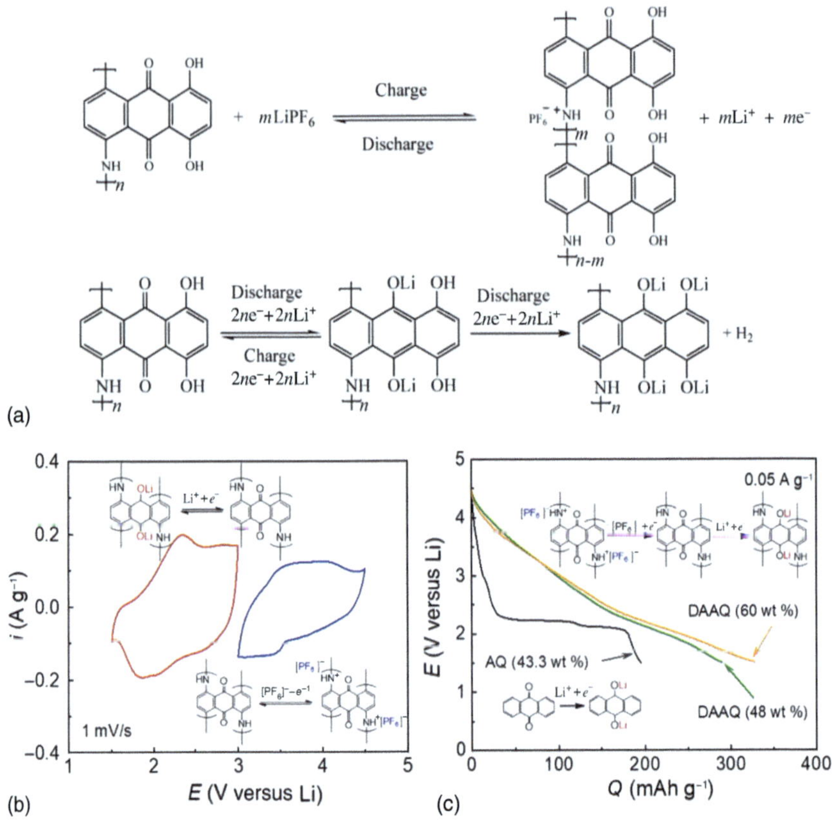

Figure 7.20 (a) The electrochemical mechanism scheme of PADAQ. Source: Adapted with permission from Zhao et al. [87]/Elsevier. (b) Potential-dependent CV scans of the poly-DAAQ-NT. (c) The discharge profiles of the AQ-NT and poly-DAAQ-NT. Source: Reproduced with permission from Liu et al. [88]/American Chemical Society.

with less steric hindrance exhibits higher discharge capacity, rate performance, and longer cycle stability. It can achieve a discharge specific capacity of 250 mAh g^{-1} in the first cycle, which is slightly higher than the theoretical specific capacity of 225 mAh g^{-1}, while P18AQS with the same theoretical specific capacity can only release a discharge capacity of 90 mAh g^{-1} under the same conditions. After that, Song et al. synthesized two polymers poly(1,4-anthraquinone) (P14AQ, **16**) and poly(1,5-anthraquinone) (P15AQ, **17**) through a simple one-step polycondensation and explored the relationship between the structures and properties of these materials in 2015 [90]. These two polymers directly couple the electroactive rings together without any additional inactive connecting units, thus increasing theoretical capacity, structural conjugation effect, and electron transport capability. Surprisingly, P14AQ has a theoretical energy density (2.14 V * 263 mAh g^{-1} = 563 Wh kg^{-1}) slightly higher than the inorganic material $LiCoO_2$ (3.8 V * 140 mAh g^{-1} = 530 Wh kg^{-1}), ultrahigh rate performance, and cycle stability. It has a discharge capacity of 263 mAh g^{-1} at 0.2 C, and the capacity retention rates are 98%, 96%, 91%, 84%, 78%,

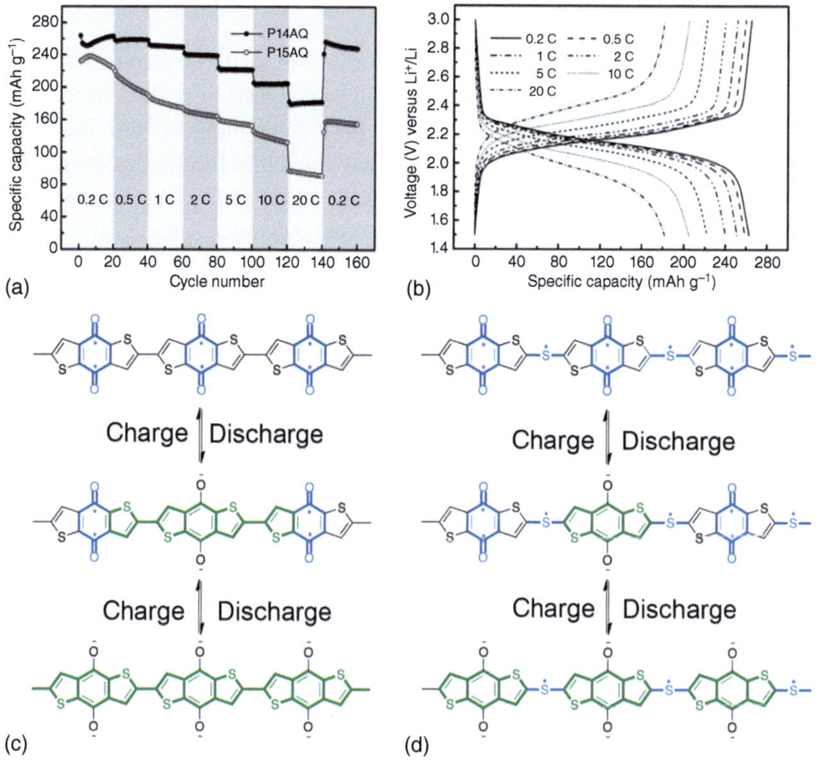

Figure 7.21 (a) Rate performance and (b) the corresponding voltage profiles of P14AQ and P15AQ at different current rates. Source: Reproduced with permission from Song et al. [90]/John Wiley & Sons. (c–d) Conjugation evolution of PBDTD and PBDTDS during the discharge/charge process. Source: Adapted with permission from Jing et al. [91]/Elsevier.

and 69% at current rates of 0.5, 1, 2, 5, 10, and 20 C, respectively (Figure 7.21a,b). Meanwhile, the reversible capacity can reach 248 and 235 mAh g^{-1} at 1 C and 2 C, and the capacity retention rate is 98.1% and 99.4% after 1000 cycles, respectively. Jing et al. designed and synthesized two low-priced cross-conjugated polymers poly(benzo[1,2-*b*:4,5-*b*′]dithiophene-4,8-dione-2,6-diyl) (PBDTD, **18**) and poly(benzo[1,2-*b*:4,5-*b*′]dithiophene-4,8-dione-2,6-diyl sulfide) (PBDTDS, **19**) as cathode materials for LIBs (Figure 7.21c,d) [91]. The cross-conjugated PBDTD evolves into a through-conjugation during lithiation, whereas PBDTDS maintains the cross-conjugated state and changes the molecule from a planar conformation to a helical conformation due to the presence of ether bonds that disrupt the conjugation of the polymer. PBDTD exhibits better rate performance than PBDTDS due to the fact that the through-conjugated structure formed after reduction is more conducive to intra-chain electron conduction. Furthermore, the planar conformation is more favorable for extended electron delocalization than the helical one. These works have shed new light on the relationship between the structure and electrochemical properties of polymers.

7.3.2 Polyimides

PIs are a kind of carbonyl polymer with high thermal stability, electrochemical activity, and good mechanical strength, which can be prepared by polycondensation reaction between diamines and dianhydrides [92]. Depending on the aromatic nucleus attached to the imine groups, PIs can be divided into three types: pyromellitic diimide-based (PMDI, I), naphthalene diimide-based (NDI, II), and perylene diimide-based (PDI, III) (R: different organic groups) polymers. Among them, with the increase in the volume of molecular aromatic nuclei, its theoretical specific capacity will decrease, while the redox potential will increase. Therefore, how to optimize the capacity and voltage reasonably is an important issue. In recent years, researchers have developed many PI-derived cathode materials [6]. Figure 7.22 shows the structures of some representative PIs.

Song et al. designed and synthesized five typical PIs (PI-**1**, **2**, **5**, **6**, **7**) by selecting different dianhydrides and diamines as the starting materials and compared the differences between their electrochemical properties [93]. These imide-based polymers share a similar redox reaction process with the main active sites at the oxygen atoms. As shown in Figure 7.23a,b, it is found that the average discharge voltage of the cells increased from 2.05 to 2.50 V from PI-**1**, **2**, and **6**, **7** to **5**, while the voltage gap became smaller. Of these, PI-**5**, which has the best conjugate structure, demonstrates the best electrochemical performance. This research opens the way for further research on the application of PIs as cathodes in energy storage. Nevertheless, the search for PI cathodes with better electrochemical properties is still on. PMDI-based polymers have attracted much attention due to their high theoretical specific capacity. In 2014, Wei's group prepared binder-free flexible electrode material PI-**2**/SWNTs (single-wall carbon nanotube) by in situ polymerization of PI-**2** on SWNTs, in which the porous structures of SWNTs film achieve good contact between the electrode material and the collector, thus improving the electronic conductivity of PI-**2**/SWNTs and achieving ultrahigh rate performance [96]. Similarly, the group further polymerized pyromellitic dianhydride (PMDA) and 2,6-diaminoanthraquinone (AQ) to produce a new PI (PMAQ, **3**) with high capacity and excellent thermal stability [94], in which the PMAQ particles and the SWNTs network are interwoven to form a self-supporting electrode without any binder. Due to the high conductivity and flexibility of SWNTs, the flexible electrode of PMAQ-30% SWNTs achieves extremely high rate performance and long-term stable cycling, maintaining the discharge capacity of 120 mAh g^{-1} at 20 C (Figure 7.23c) and the capacity retention of 91.5% after 300 cycles (Figure 7.23d). In addition to some main-chain types of PMDI-based polymers reported above, Zindy et al. copolymerized PMDI and 1,4-dibromobenzene by a nontraditional direct (hetero)arylation polymerization (DHAP) method to synthesize a series of PMDI-based polymers with different side-chain substituents (Figure 7.23e, **4**), which were used as active materials for Li^+ and Na^+ batteries [95]. The effect of side-chain engineering by varying the length of the alkyl side-chains and using cross-linked and thermocleavable chains was investigated experimentally and by density functional theory (DFT) calculations. This study provides ideas for the rational design of new organic electrode materials.

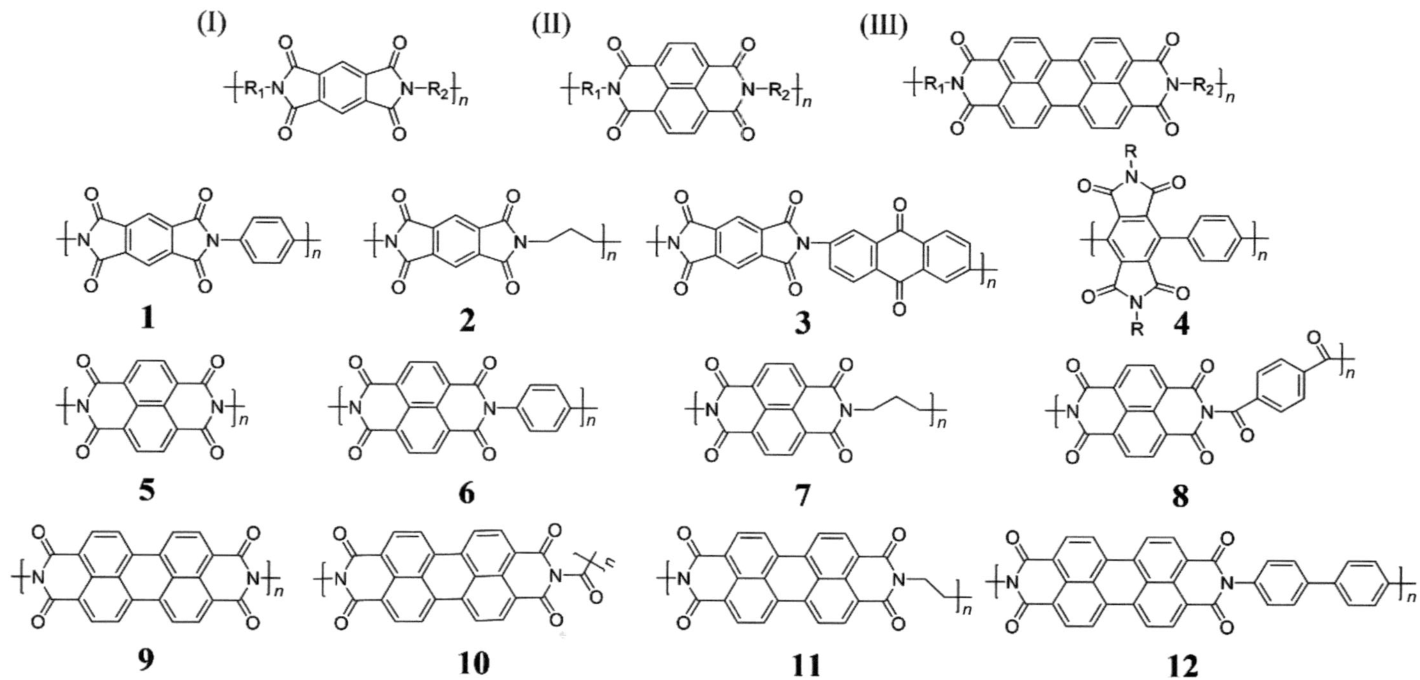

Figure 7.22 Some typical molecular structures of PIs.

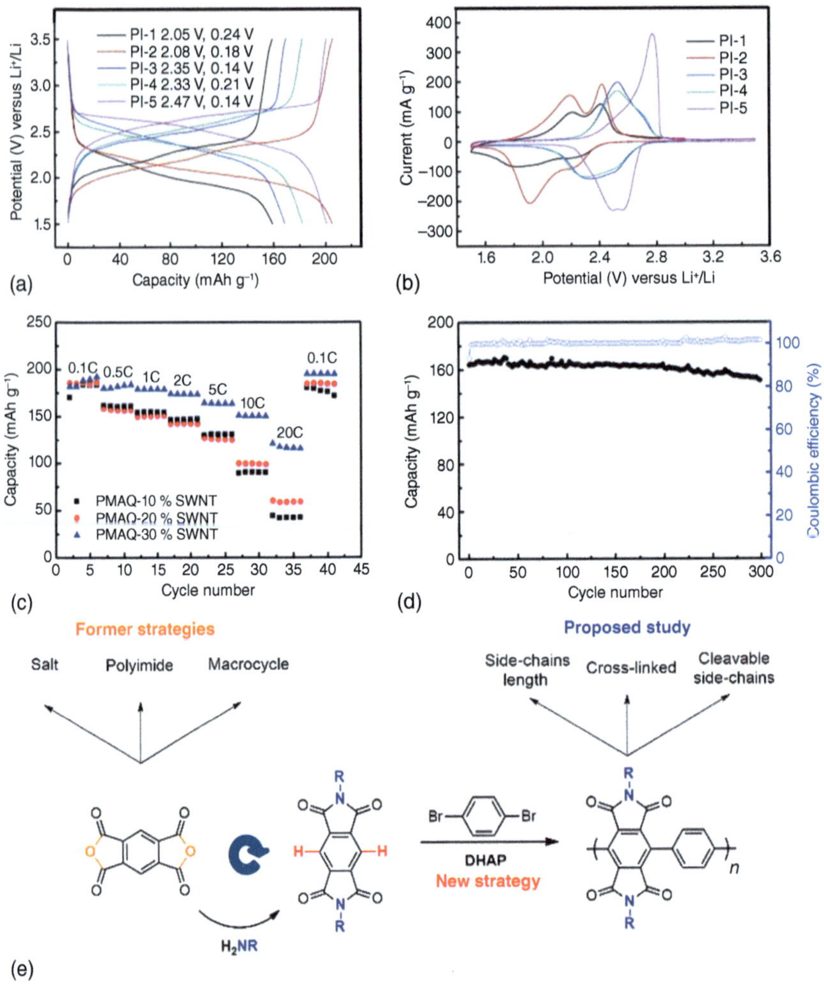

Figure 7.23 (a) Voltage profiles and (b) CV curves of the five polyimide samples. Source: Reproduced with permission from Song et al. [93]/John Wiley & Sons. (c) Rate performance and (d) cycling performance of the PMAQ-30% SWNT composite. Source: Adapted with permission from Wu et al. [94]/The Royal Society of Chemistry. (e) Diagram of the present DHAP method compared with the earlier traditional method. Source: Reproduced with permission from Zindy et al. [95]/Royal Society of Chemistry.

In addition to PMDI-based polymers, NDI and PDI-based polymers have received considerable attention for their improved electrical conductivity due to their expanded conjugate structure. However, the expansion of molecular structure will inevitably lead to a decrease in theoretical specific capacity. Therefore, some researchers have investigated the electrochemical properties of PIs with different aromatic core structures. Wang et al. systematically synthesized a series of dianhydride-based PIs containing a variety of commercially available dianhydrides such as PMDA, 1,4,5,8-naphthalenetetracarboxylic dianhydride (NTCDA), and perylene 3,4,9,10-tetracarboxylic dianhydride (PTCDA) by a simple one-pot reaction

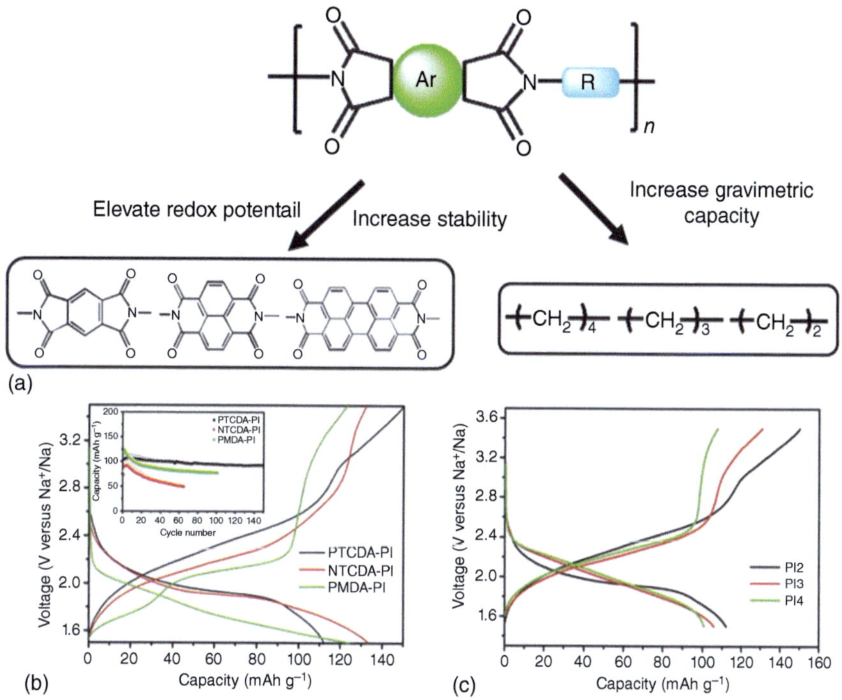

Figure 7.24 (a) Structure of different polyimides. Charge–discharge curves of different (b) benzene-ring-based and (c) alkyl-chain-based PIs electrodes. Source: Reproduced with permission from Wang et al. [97]/John Wiley & Sons.

and used in sodium-organic batteries (Figure 7.24a) [97]. The average discharge voltage plateau gradually increases from 1.73 to 1.89 and 1.94 V with the changes of the connecting skeleton from PMDA to NTCDA and PTCDA (Figure 7.24b), which is due to the introduction of additional electron-withdrawing groups resulting in a difference of the lowest unoccupied molecular orbital (LUMO) energy. At the same time, the PTCDA-based polymers demonstrate better thermal stability and more stable cycling performance, with a discharge capacity of 93.4 mAh g^{-1} after 150 cycles. The PTCDA-based PIs with different alkyl chain lengths are named PI2, PI3, and PI4. Among them, PI2 exhibits a relatively high gravimetric capacity, which is due to the lower molecular weight of molecules with shorter alkyl chain lengths (Figure 7.24c).

The different substituents attached to the aromatic nucleus of PIs also give rise to different electrochemical behaviors. Wang et al. reported two aromatic dianhydride-derived PIs (PI1-**7**, PI2-**6**) combined with CNTs as organic cathode materials and matched with ionic an liquid-modified non-nucleophilic organic electrolyte for rechargeable Mg^{2+} batteries and investigated their electrochemical properties [98]. The π-conjugated PIs can provide an abundance of redox-active sites and a highly reversible multi-electron redox process, the 3D cross-linked conductive network of CNTs can improve electronic conductivity, and the non-nucleophile

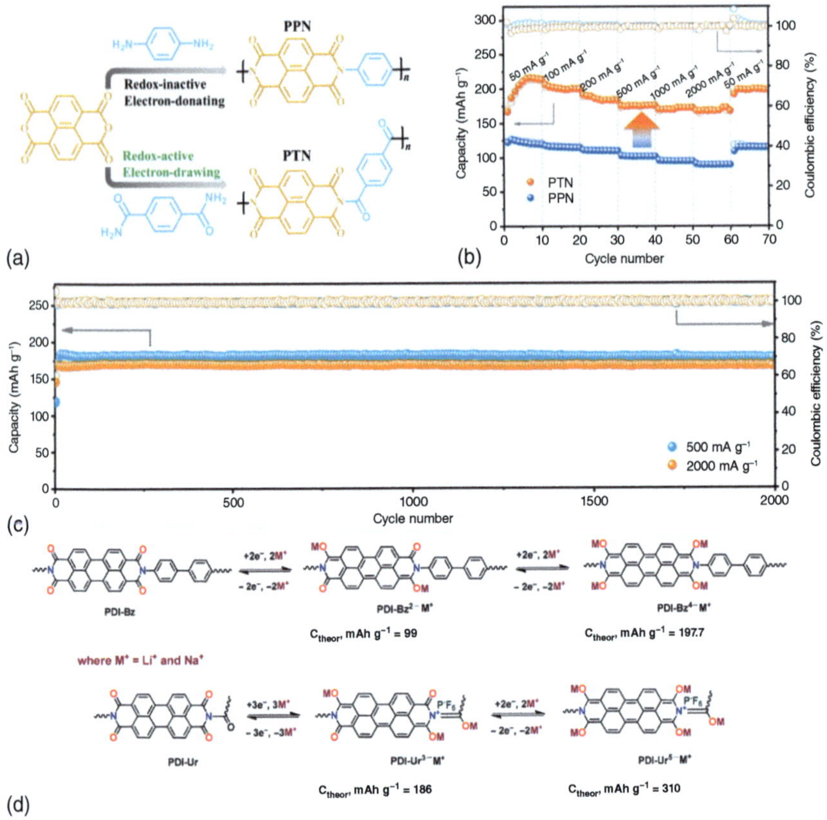

Figure 7.25 (a) Schematic diagram of the synthesis of PPN and PTN. (b) Rate the performance of the PPN and PTN. (c) Long-term cycling performance of PTN for 2000 cycles. Source: Reproduced with permission from Wang et al. [99]/Elsevier. (d) The redox reaction mechanism of PDI-Bz and PDI-Ur-based cathode for rechargeable metal batteries. Source: Adapted with permission from Ruby et al. [100]/American Chemical Society.

electrolyte also contributed to the large voltage window and good interfacial compatibility. PI2 can be stably cycled for 8000 cycles at a high rate of 20 C due to its lower HOMO-LUMO gap than PI1, resulting in better electronic conductivity and improved utilization ratio for Mg^{2+} storage. After that, Wang et al. synthesized PPN (**6**) and PTN (**8**) by copolymerization of NTCDA with p-phenylenediamine (PA) and terephthalamide (TA), respectively (Figure 7.25a), and compared their electrochemical behaviors [99]. The introduction of electron-withdrawing groups and special π-conjugated structure in PTN could not only improve the specific capacity and working potential but also achieve high energy density. In addition, it could also effectively facilitate charge transport and achieve high-rate performance. The PTN-based cathode shows high rate performance and long-cycle stability, with a capacity retention of 78.2% at 2000 mA g^{-1} compared to 50 mA g^{-1} (Figure 7.25b) and a capacity retention of 97.3% after 2000 cycles at 2000 mA g^{-1} (Figure 7.25c).

Sharma et al. designed and synthesized the PI derivatives of PTCDA by the reaction of PTCDA and hydrazine, ethylene diamine, and carbonyl diamine, namely poly(3,4,9,10-perylenetetracarboxylic dianhydride hydrazine) (HP, **9**), poly (3,4,9,10-perylenetetracarboxylic dianhydride urea) (UP, **10**), and poly(3,4,9,10-perylenetetracarboxylic dianhydride ethylene diamine) (EDP, **11**) in 2013, respectively [101]. The structures are verified by various characterization techniques such as FTIR, X-ray diffraction (XRD), and NMR spectroscopy. Although the structures of these molecules are slightly different, the discharge voltage plateaus are all around 2.5 V versus Li^+/Li, while the coulomb efficiency is close to 100%. Among them, the UP electrode demonstrates excellent electrochemical performance in terms of capacity, cycling stability, and rate capability, which can be attributed to the presence of additional carbonyl groups, enhanced conductivity, and favorable band gap. Similarly, Ruby Raj et al. reported two PTCDA-based PIs (perylenediimide-urea (PDI-Ur, **10**) and perylenediimide-benzidine (PDI-Bz, **12**)) synthesized by the reactions between non-bay-substituted PTCDA units and bifunctional amine compounds and used as organic electrodes for rechargeable Li^+ and Na^+ batteries [100]. In the voltage window range of 1.5–3.5 V versus Li^+/Li and 1.2–3.5 V versus Na^+/Na, PDI-Bz undergoes a four-electron reaction during the reduction process. First, a carbonyl group gains one electron to form a radical anion (PDI-$Bz^{\cdot-}$) as an intermediate, and then another electron to produce a dianion species with metal-ions (PDI-$Bz^{2-}M^+$), and finally a tetra-anion species with metal ions (PDI-$Bz^{4-}M^+$) is formed. In the case of PDI-Ur, a trianion species with metal ions (PDI-$Ur^{3-}M^+$) will be first produced due to the presence of an additional carbonyl group (the C=O group in carbonyl diamine moiety), and finally a penta-anion species (PDI-$Ur^{5-}M^+$). At the same time, the N^+ positively charged at imide position could interact with a PF_6^- to form a stable compound. A total of five electrons can be reversibly exchanged in this process, that is, there are four reaction sites on the conjugated carbonyl groups presented in the perylene structure and one on the urea group presented in the imide region of PDI-Ur (Figure 7.25d).

7.3.3 Polyanhydrides

On the basis of stable enolization, one carbonyl group in each anhydride group of PAs can also undergo reversible redox processes, which is similar to the redox mechanism of PIs. Therefore, PAs are also potential cathode materials for rechargeable batteries [6]. However, up to now, there are few reports on such.

Han et al. prepared 3,4,9,10-perylene-tetracarboxylicacid-dianhydride sulfide (PTCDA/S, P500) polymers by mixing PTCDA with excess sulfur and calcining them at 400–500 °C in a flowing argon atmosphere, and characterized the structures by FTIR, XRD, NMR, etc. [102] Not only can electron transfer in PTCDA/S be achieved by π–π stacking between layers, but the thioether bonds therein provide an alternative pathway for electron transfer that is caused by π-electron delocalization of the perylene rings with a much lower energy barrier than π–π stacking, which can lead to a great increase in the conductivity and total electron transfer efficiency. The P500 has a discharge capacity of 131 mAh g^{-1} at the high current density of 100 mA g^{-1} and a charge/discharge voltage platform close to 2.6/2.4 V, which gradually increases to a larger value of 148 mAh g^{-1} after 250 cycles, and a discharge capacity of 180 mAh g^{-1} at the low current density of 25 mA g^{-1}.

7.4 Covalent Organic Frameworks-Derived Polymers

COFs are crystalline organic porous polymers composed of light elements such as hydrogen, boron, carbon, nitrogen, and oxygen, which have the advantages of low density, large specific surface area, high porosity, good physical and chemical stability, adjustable structure, and easy functionalization, and have received a lot of attention in recent years in the field of energy conversion and storage [92].

Yang et al. synthesized PI-based COF materials Tp-DANT-COF and Tb-DANT-COF by the reaction of 2,7-diaminobenzo[*lmn*][3,8]phenanthroline-1,3,6,8(*2H,7H*)-tetraone (DANT) with 1,3,5-triformylphloroglucinol (Tp) and 1,3,5-triformylbenzene (Tb) via a simple solvothermal route (Figure 7.26a) [103]. Both COF materials have high thermal and structural stability, and their ordered porous networks and strong covalent interaction could facilitate the transport of ions and charges and avoid the dissolution of organic active molecules. In addition, the slight difference in

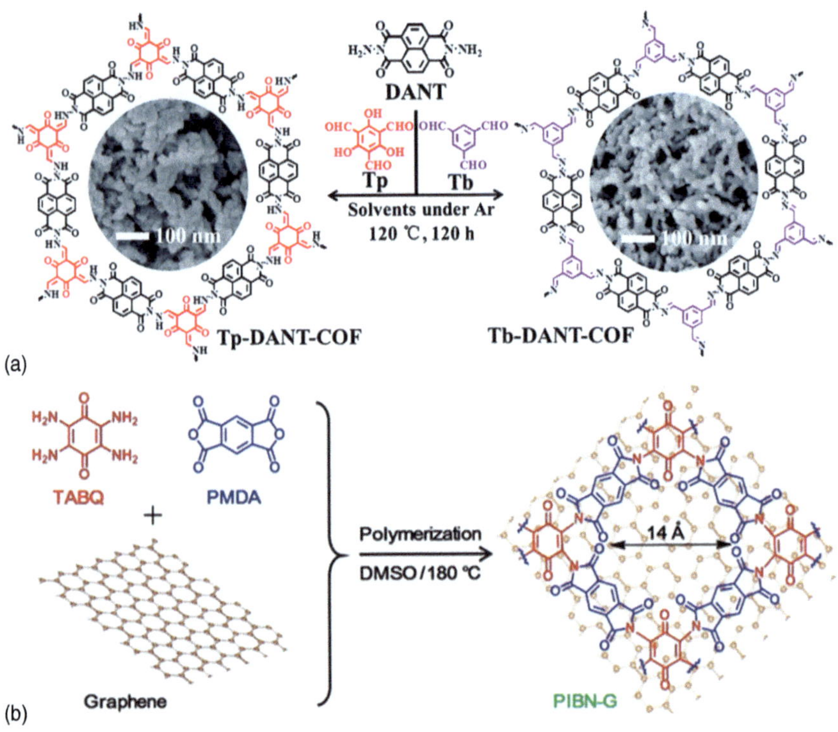

Figure 7.26 (a) Synthesis of Tp-DANT-COF and Tb-DANT-COF. Source: Adapted with permission from Yang et al. [103]/Royal Society of Chemistry. (b) Synthesis route of PIBN and PIBN-G. Source: Reproduced with permission from Luo et al. [104]/John Wiley & Sons.

molecular structure also affects its electrochemical behavior, and Tp-DANT-COF and Tb-DANT-COF with different linkers show different electrochemical properties. In order to increase the redox reaction sites on the molecule to achieve a higher theoretical specific capacity, Luo et al. synthesized the COF, poly(imide-benzoquinone) (PIBN), with a large number of carbonyl reaction sites by in-situ polymerization of tetramino-benzoquinone and PMDA on graphene, and used it as cathode material for Li^+ batteries (Figure 7.26b) [104]. Thanks to its special microporous structure and the strong interaction between graphene and PIBN, PIBN-G exhibits high electron conduction and Li^+ diffusion, which is beneficial to the redox reaction of carbonyl groups and therefore has a large reversible capacity. This unique design makes the carbonyl groups in each repeating unit of PIBN involve two consecutive $8Li^+$ and $2Li^+$ reactions during lithiation and delithiation. PIBN-G has a discharge capacity of 271 and 193.1 mAh g^{-1} at 0.1 and 10 C, respectively, and the capacity retention rate of 86% can still be maintained after 300 cycles.

In recent years, some novel polymer COF materials with ingenious designs have been reported continuously, which not only have multiple active sites and therefore high theoretical specific capacities but also demonstrate unique redox mechanisms. It is believed that the simplest way to increase the theoretical specific capacity is to increase the redox component of the molecular unit and reduce the inactive component as much as possible. Wu et al. prepared BQ1-COF by simple polycondensation of hexaketocyclohexane (HKH) and tetramino-benzoquinone in *N*-methylpyrrolidone (NMP) (Figure 7.27a) [105]. The abundant carbon–oxygen (C=O), carbon–nitrogen (C=N) groups, and a small amount of inactive groups in this molecule make it have a theoretical specific capacity of up to 773 mAh g^{-1}, which is almost the highest among polymer-based cathode materials. Figure 7.27b illustrates the single-layer conjugated skeleton of the BQ1-COF. The optimal lithiation mechanism of BQ1-COF was analyzed by theoretical and DFT experiments. The N and O atoms in this system act as collaborative centers that can combine with lithium to form stable five-membered chelate rings (C–N–Li–O–C). In addition, electroactive pyridine N and pyrazine N atoms in π-conjugated C=N systems can strongly coordinate with lithium ions to form N—Li—N bond. Although each repeating unit can accept 18 e^- in theory, the binding energy increases significantly after the insertion of 12 Li^+ due to steric hindrance, making subsequent lithiation more difficult. Therefore, BQ1-COF can store up to 12 lithium ions in the operating voltage window of 3.5–1.2 V during the actual lithiation process. The stable 2D-conjugated network structure and delicate channels in the BQ1-COF can promote rapid Li^+ transport, resulting in ultrahigh rate performance and long-term stable cycling, with a discharge capacity of 81% after 1000 cycles and a discharge capacity of 170.7 mAh g^{-1} even at a current density of 7.73 A g^{-1}. Similarly, Yang et al. reported the use of a novel[3+3] truxenone (TRO)-based COFs synthesized by truxenone-based triamine (TRO) and 1,3,5-triformylphloroglucinol (TFP) as cathode materials for all-solid-state Li^+ batteries [106]. According to electrochemical and theoretical calculations based on the model compound, it is clear that COF-TRO would involve $6Li^+$ in the lithiation/delithiation process (Figure 7.27c), corresponding to a theoretical capacity of 274 mAh g^{-1}, which is the almost highest capacity reported

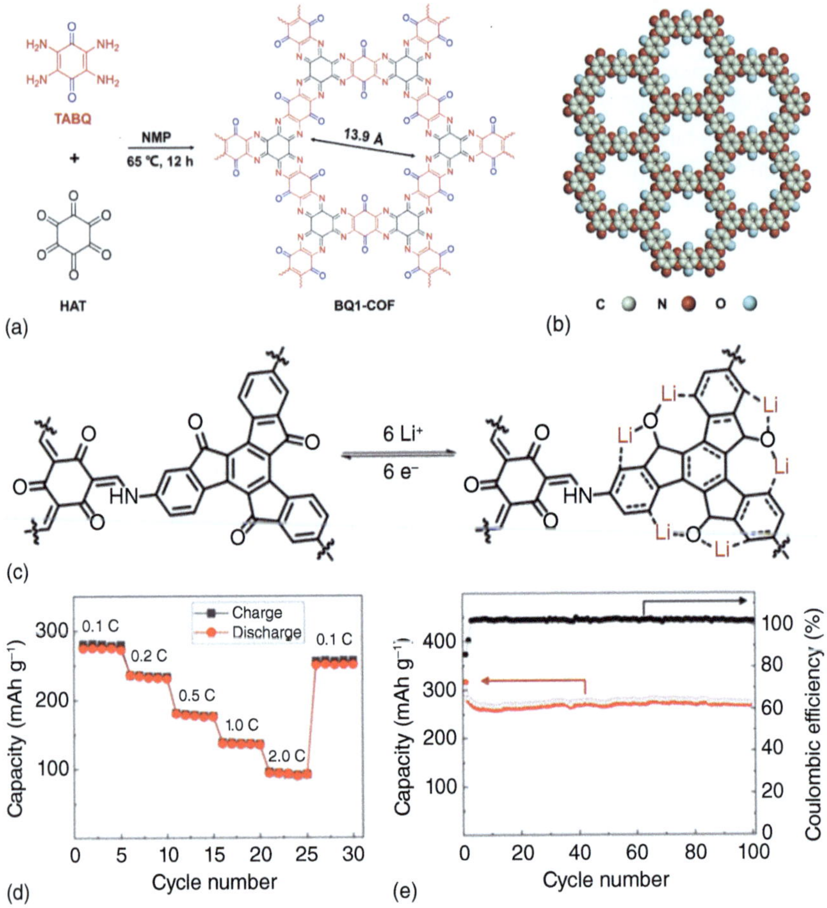

Figure 7.27 (a) Synthesis of BQ1-COF. (b) Diagram of the single-layer conjugated skeleton of BQ1-COF. Source: Adapted with permission from Wu et al. [105]/Elsevier. (c) Lithiation/delithiation mechanism of COF-TRO. (d) Rate performance of COF-TRO. (e) Long-term cycling performance of COF-TRO for 100 cycles. Source: Adapted with permission from Yang et al. [106]/Wiley-VCH.

for COF materials in the cathode of all-solid-state LIBs. The combination of ordered crystalline COF structure and high-density carbonyl groups greatly facilitates Li^+ storage for redox reactions. As a result, a discharge capacity of up to 268 mAh g^{-1} is delivered, which is almost 97.5% of the theoretical capacity (Figure 7.27d). In addition, the capacity retention rate is still 99.9% after 100 cycles at 0.1 C (Figure 7.27e).

In response to the problem that extended polygons in early typical 2D COF materials often tend to pack tightly in an eclipsed manner due to strong π–π interactions, internal active sites deeply buried within the 1D channels are difficult to reach due to the long ion transport pathways, which would inevitably lead to under-utilization of redox active sites, thus reducing their capacity and rate performance. To this end, Wang et al. proposed to speed up the diffusion-controlled

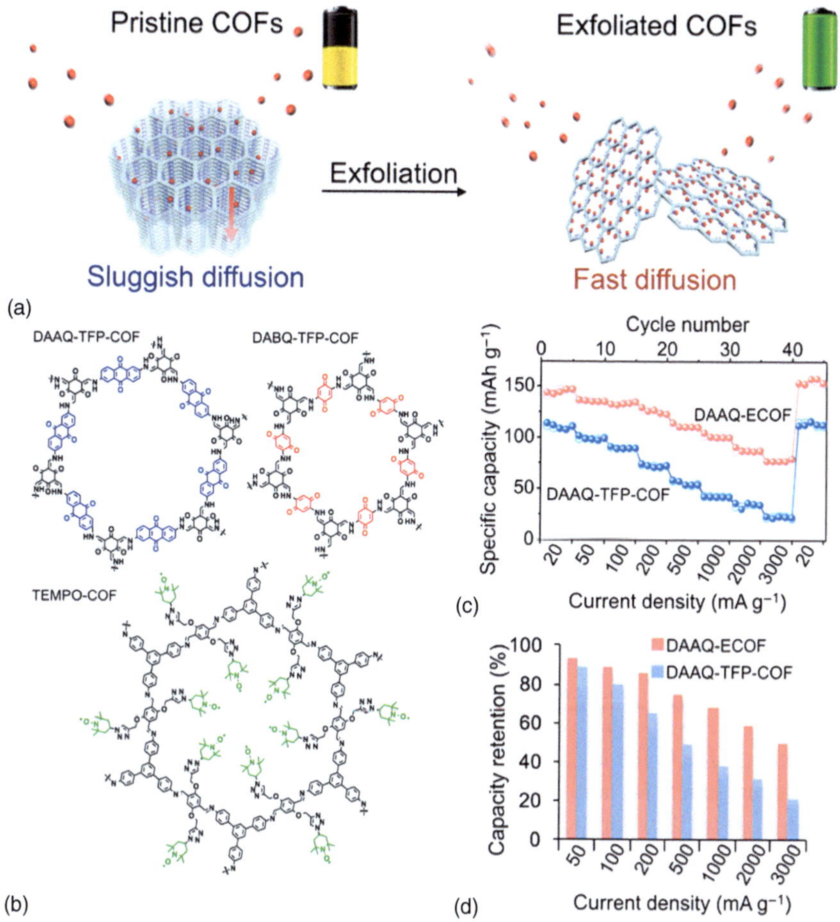

Figure 7.28 (a) Schematic diagram for the exfoliation of 2D redox-active COFs into exfoliated COFs as cathodes for lithium-ion batteries. (b) Structures of DAAQ-TFP-COF, DABQ-TFP-COF, and TEMPO-COF. (c) Rate performance and (d) capacity retention of DAAQ-ECOF and DAAQ-TFP-COF. Source: Reproduced with permission from Wang et al. [107]/American Chemical Society.

redox reactions in COFs by stripping the layered redox-active COFs into 2D few-layer nanosheets and prepared three exfoliated COFs (ECOFs) materials based on benzoquinone (DABQ-TFP-COF), anthraquinone (DAAQ-TFP-COF), and nitroxide (TEMPO-COF) (Figure 7.28a,b) [107]. Compared with pristine COFs, ECOFs have a shorter Li^+ diffusion pathway, which significantly improves the utilization efficiency of redox sites and realizes fast Li^+ storage capacity. As a result, the electrochemical process controlled by ion diffusion in the pristine COF-based cathode becomes a process dominated by charge transfer in the ECOFs. The DAAQ-ECOF demonstrates excellent rate performance with a discharge capacity of 145 mAh g^{-1} at a current density of 20 mA g^{-1} and a capacity retention rate of 93%, 89%, 86%, 74%, 68%, 60%, and 50% with increasing current rates from 50, 100, 200, 500, 1000, 2000, and 3000 mA g^{-1}, respectively (Figure 7.28c,d).

7.5 Organic Radical-Derived Polymers

Organic radical polymers (ORPs) are a kind of polymer material with stable radical groups on the side chains, which are composed of the radical active centers and the polymer skeleton bonded with radicals. Radical polymers have attracted much attention due to their advantages, such as stable and diverse structures, fast electrode reaction rate, high reversibility, and good flexibility. Among them, the most widely studied are the nitroxide radical polymers, which can combine with anions (ClO_4^-, PF_6^-) by p-doping/p-dedoping at high potentials and with metal ions by n-doping/n-dedoping at low potentials, while 2,2,6,6-tetramethyl-1-piperidinyloxy (TEMPO)-based organic compounds are the most widely studied nitroxide radical polymers [6, 7, 108–110]. Therefore, our discussion here mainly focuses on some TEMPO-based ORPs. Here are some representative structures listed. (Figure 7.29).

Nakahara et al. synthesized a stable nitroxide radical polymer, poly(2,2,6,6-tetramethylpiperidinyloxy methacrylate) (PTMA), and applied it to the cathode materials of rechargeable batteries for the first time in 2002 [111]. PTMA shows an average discharge voltage of 3.5 V and a discharge capacity of 77 mAh kg^{-1}, which corresponds to 70% of the theoretical capacity. In particular, the capacity retention rate is almost constant after 500 cycles. This has led to the development of TEMPO-based electrode active materials. In the following years, they further optimized the PTMA electrode materials [112, 113], which is the most common structure in ORPs containing TEMPO units. However, the PTMA exhibits high self-discharge due to the high solubility of the polymers in the organic electrolyte. At the same time, the addition of a large amount of carbon materials in the electrode reduces the energy density of the batteries. Therefore, Kim et al. prepared a PTMA-impregnated CNT composite electrode containing 68 wt% PTMA and 32 wt% CNT by dissolution–diffusion–washing process, in which the active polymer is encapsulated in CNT (Figure 7.30a) [114]. The CNT with a large specific surface area can not only hold a large amount of active material, but its strong absorption effect can also bind and prevent the polymer from dissolving into the electrolyte. At the same time, the CNT also provides an effective electron conduction path, and the electrolyte can easily penetrate into the electrode, thus realizing rapid ion diffusion. PTMA-impregnated CNT delivers excellent electrochemical performance, with an initial discharge capacity of 222 mAh g^{-1} at 0.1 C, and a retention rate of about 93% after 100 cycles at 0.5 C. When the current density is increased from 0.1, 0.5,

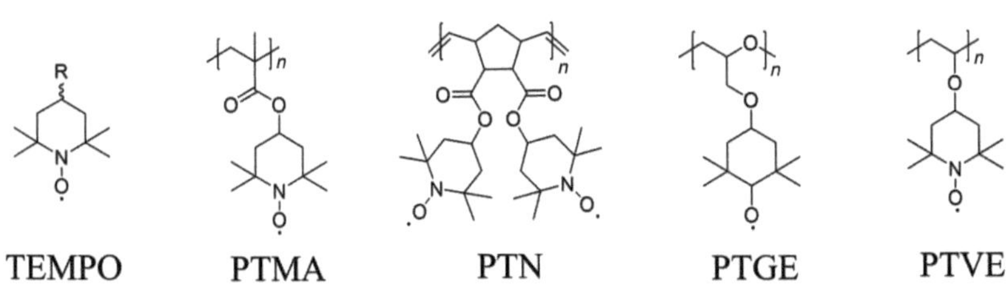

Figure 7.29 Some typical molecular structures of TEMPO-based radical polymers.

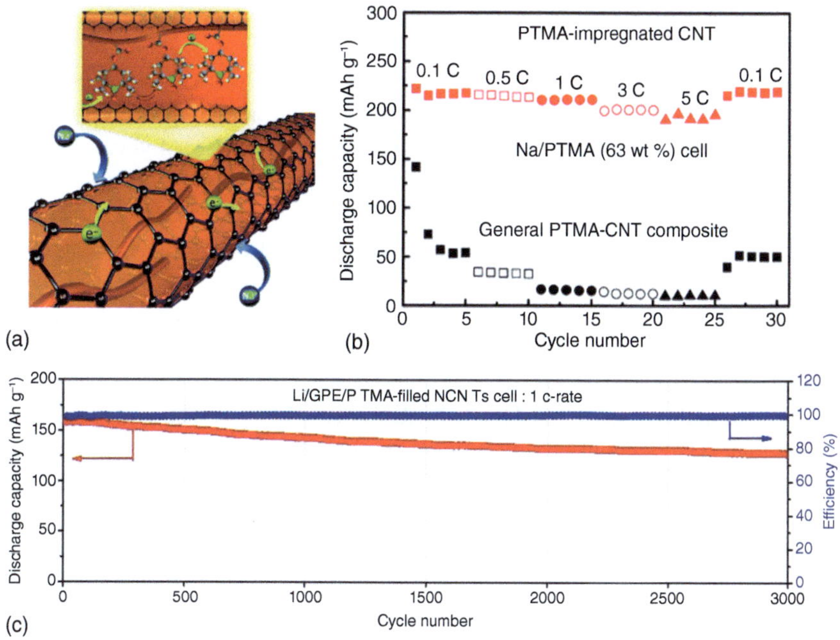

Figure 7.30 (a) Schematic illustration of the PTMA-impregnated CNT structure. (b) Rate performance of PTMA-CNT composite and PTMA-impregnated CNT electrodes. Source: Reproduced with permission from Kim et al. [114]/Royal Society of Chemistry. (c) Long-term cycle performance of the Li/GPE/PTMA-filled NCNTs cell. Source: Reproduced with permission from Byeon et al. [115]/Elsevier.

1, 3 to 5 C, the discharge capacity can be maintained at about 222, 216, 210, 200, and 190 mAh g^{-1}, respectively (Figure 7.30b). Then, in order to further improve the conductivity and stability of the electrodes, Byeon et al. further prepared nitrogen-doped carbon nanotubes (NCNTs) filled with PTMA polymers in the central pore, in which the active materials accounted for more than 70 wt% [115]. PTMA can undergo a unique redox reaction in the charge–discharge process. In the initial state, PTMA exists as the nitroxide radical in the side chain of TEMPO, which can be converted into an oxoammonium cation at 3–4 V by p-type reaction or an aminoxy anion at 1.5–3 V by n-type reaction. This study determines the different properties of each reaction for the first time. The PTMA-filled NCNTs cells with a PI-based GPE can achieve high cycle stability, with a capacity retention rate of more than 80% after 3000 cycles (Figure 7.30c), and can achieve reversible capacities of 199.8 and 89.6 mAh g^{-1} at current densities of 0.2 C and 20 C, respectively. Similarly, Zhu et al. prepared PTMA-GO composite by grafting PTMA onto graphene oxide (GO) by click chemistry [116]. The preparation method is completed at room temperature, which can not only preserve the oxygen-containing functional groups on the surface of GO but also be simple and practical for large-scale production. The PTMA-GO electrode has very good charge and discharge reversibility and is almost twice the conductivity of the PTMA electrode. Moreover, the discharge-specific capacity of PTMA-GO is maintained at around 141 mAh g^{-1} after 300 cycles, which is almost 2.3 times higher than that of the PTMA electrode.

In addition to PTMA, there are other types of TEMPO-based polymers, such as poly(TEMPO-norbornene) (PTN), poly(2,2,6,6-tetramethylpiperidinyl-*N*-oxyl glycidyl ether) (PTGE), and poly(2,2,6,6-tetramethylpiperidinyloxy-4-yl vinyl ether) (PTVE). In the early stage, Katsumata et al. synthesized a polymer containing two TEMPO moieties in each unit by direct ring-opening metathesis polymerization (ROMP) of TEMPO-containing acetylene monomers and norbornene monomers in the presence of Grubbs' catalysts and used it in rechargeable batteries [117]. In order to better inhibit the dissolution of polymers, cross-linking is a good method. Subsequently, Suga et al. prepared patterned cross-linked films **1b** containing TEMPO-substituted polynorbornenes by spin-coating a toluene-ethyl lactate solution **1a** and a small amount of the bis(azide) derivatives **3** onto an ITO glass substrate by ultraviolet (UV) irradiation and a photo-mask for the first time (Figure 7.31a) [118], which inhibited the dissolution of the active material into the electrolyte due to the insolubility of the films in common organic solvents and provided high mechanical stability. It can be seen from SEM that the radical polymers and carbon nanofibers (RP-CNFs) obtained by photocrosslinking are well dispersed, and the electrode has almost no significant capacity attenuation after 1000 cycles at the cut-off voltages of 3.0–4.2 V (Figure 7.31b).

Sukegawa et al. synthesized a cross-linked poly(4-glycidyloxy TEMPO) (CL-PTGE) with high radical concentrations and molecular weights via anionic ring-opening copolymerization of 4-glycidyloxy TEMPO and a small amount of the diglycidyl ether using pentaerythritol/*t*-BuP_4 initiator (Figure 7.32a) [119]. This poly(ethylene oxide) (PEO)-like structure not only allows ions to be transported throughout the polymer layers but also has good swelling properties in conventional electrolyte solutions. At the same time, the SWCNT conductive network enhances the electronic conductivity (Figure 7.32b). The CL-PTGE/SWCNT composite electrode containing only 10% SWCNT exhibits fast and reversible charge storage capacity

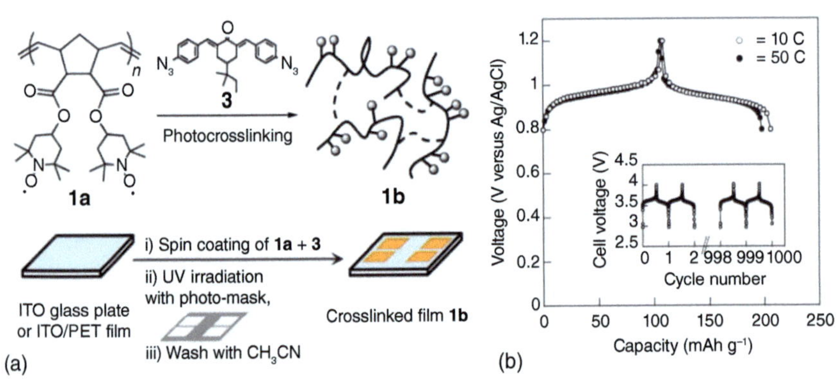

Figure 7.31 (a) The synthesis route of the TEMPO-substituted polynorbornenes. (b) Electrochemical performance of the RP-CNFs composite electrode. Source: Suga et al. [118]/Royal Society of Chemistry.

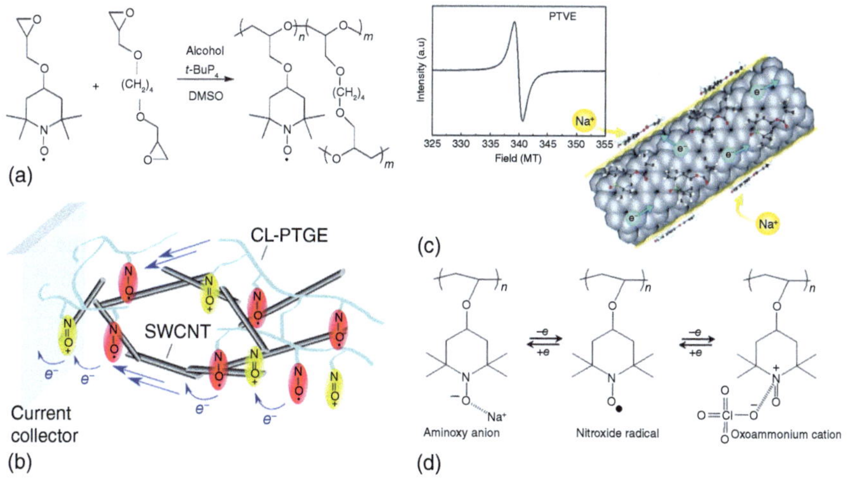

Figure 7.32 (a) The synthesis of CL-PTGE. (b) Schematic diagram of the charge transport of CL-PTGE/SWCNT composite layer swelling in electrolyte solution. Source: Sukegawa et al. [119]/Royal Society of Chemistry. (c) Schematic and (d) reaction formula of PTVE-functionalized CNTs composite cathode during redox reaction. Source: Kim et al. [120]/ Royal Society of Chemistry.

and cycling stability. The charging capacity is 80% of the theoretical capacity, and the capacity retention rate is over 70% after 200 cycles. In order to further improve the conductivity of ORPs, Joo et al. made a breakthrough in 2018 [121]. The research team demonstrated a nonconjugated conductive polymer with a subambient glass transition temperature that can undergo rapid solid-state charge transfer reactions as well as having a conductivity of up to 28 siemens per meter and a channel length of 0.6 μm. The polyether backbone provides the polymer with excellent flexibility and crystallinity, which allows the thermal annealing at 80 °C to form open core-shell sites in electronic communications without worrying about the degradation of nitroxide radicals. This percolation network of nitroxide radicals has greatly facilitated the development of electronic communications with an unprecedented electrical conductivity of about 10 S m^{-1}. Subsequently, Lutkenhaus et al. have commented on this work [122].

Koshika et al. designed a full cell containing a PTVE polymer cathode, an aqueous electrolyte, and a zinc anode [123]. The hydrophilic polymer has high affinity with the aqueous electrolyte, high electrical conductivity, and a theoretical specific capacity of 131 mAh g^{-1}, which is higher than other TEMPO-based polymers PTMA (112 mAh g^{-1}) and PTN (109 mAh g^{-1}) reported before [124]. The full battery exhibits an output voltage of 1.7 V, which can be reversibly cycled for more than 500 cycles. At almost the same time, PTVE polymer film is also used as the electrode material for half-cells [125], demonstrating high rate performance and excellent cycling stability. It can realize 97% of its theoretical capacity (127 mAh g^{-1}) even when charged at 1200 C (equivalent to full charging for 3 seconds). In addition, Kim et al. designed advanced Na-ion organic full cells using PTVE-functionalized CNTs composite cathode (Figure 7.32c), microporous GPE based on an electrospun PI membrane, and a hard carbon anode [120]. As we can see from Figure 7.32d, PTVE-CNT composite will undergo a two-step electron transfer process during charging and discharging, which is similar to the redox mechanism of most ORPs. The PTVE in the composite acts as the active material and binder, occupying the interior and exterior of the CNTs, which improves the actual energy density of the battery. At the same time, the π–π* interaction (The peak of 289.6–290.0 eV can be seen in the XPS spectrum of PTVE-CNT belonging to π–π* shake-up.) between the aromatic rings of the PTVE and the CNTs reduces the dissolution of the electrodes. In addition, the use of PI GPE can avoid the dissolution of the active material in the electrolyte and improve thermal stability. The Na/PTVE-CNTs deliver a discharge capacity of 128.6 and 68.2 mAh g^{-1} at 0.5 C and 10 C, respectively, with a CE of nearly 100%.

7.6 Other Polymers

In addition, some compounds with active sites on the N atoms, *e.g.* triphenylamine, hexaazatrinaphthalene-based polymers, have also received wide attention in recent years [7]. They can lose/gain electrons and combine/remove anions from the electrolyte during charge and discharge, which is different from the redox process of the polymers mentioned above. These electrode materials typically exhibit fast reaction kinetics and excellent cycling stability in conventional Li^+ batteries and even in other metal-ion batteries such as Na^+ and potassium ions (K^+).

7.6.1 Triphenylamine-Based Polymers

Fan et al. designed a potassium-based dual-ion full battery (PDIB) with poly triphenylamine (PTPAn) as cathode, graphite as anode, and KPF_6 in EC/DEC as the supporting electrolyte for the first time and analyzed the electrochemical mechanism of PTPAn by XPS [126]. In the pristine state, the full survey XPS spectrum of the PTPAn electrode shows three peaks at around 285, 400, and 534 eV, corresponding to peaks of C 1s, N 1s, and O 1s, respectively. When charged to 4.0 V, three new peaks appeared near 135, 192, and 687 eV, corresponding to P 2p, P 2s, and F 1s, respectively, indicating that PF_6^- ions have been successfully incorporated into the PTPAn cathode. In contrast, when the PDIBs discharged to 1.0 V, the XPS peaks of P 2p and P 2s disappeared and the F 1s weakened significantly, indicating that the PF_6^- ions were released into the electrolyte during the discharge (Figure 7.33a). As shown in Figure 7.33b, during the charging process, the nitrogen atoms on the cathode electrode of the PTPAn would lose electrons to form positively charged N^+, and combine with PF_6^- ions in the electrolyte to form a metastable compound $N^+PF_6^-$. During the discharge process, N^+ ions in $N^+PF_6^-$ compounds would accept electrons from the cathode and become natural N atoms again, and the PF_6^- ions would also leave the cathode and dissolve back into the electrolyte. At the same time, K^+ in the electrolyte could be intercalated back and forth in the graphite anode. In this process, the cathode and anode gain and lose the same number of electrons. PDIBs deliver a high median discharge voltage of 3.23 V, superior rate performance, and

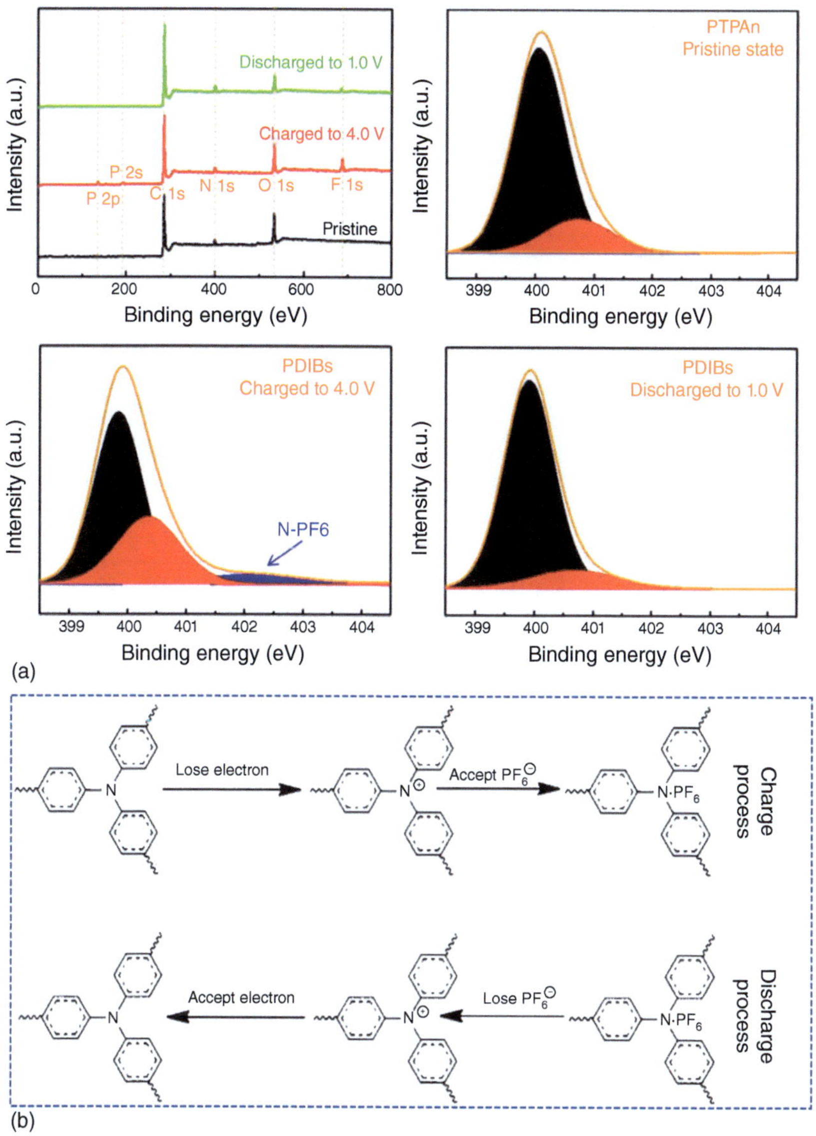

Figure 7.33 (a) XPS of the PTPAn electrode at different states in the PDIB half cells. (b) The reaction mechanism of the PDIBs during charging and discharging. Source: Reproduced with permission from Fan et al. [126]/American Chemical Society.

excellent cycling stability, with a capacity retention rate of 75.5% after 500 cycles, corresponding to a capacity decay of 0.049% per cycle.

7.6.2 Hexaazatrinaphthalene-Based Polymers

Mao et al. applied an environmentally friendly, abundant, and sustainable polymer poly(hexaazatrinaphthalene) (PHATN) as a universal cathode for rechargeable

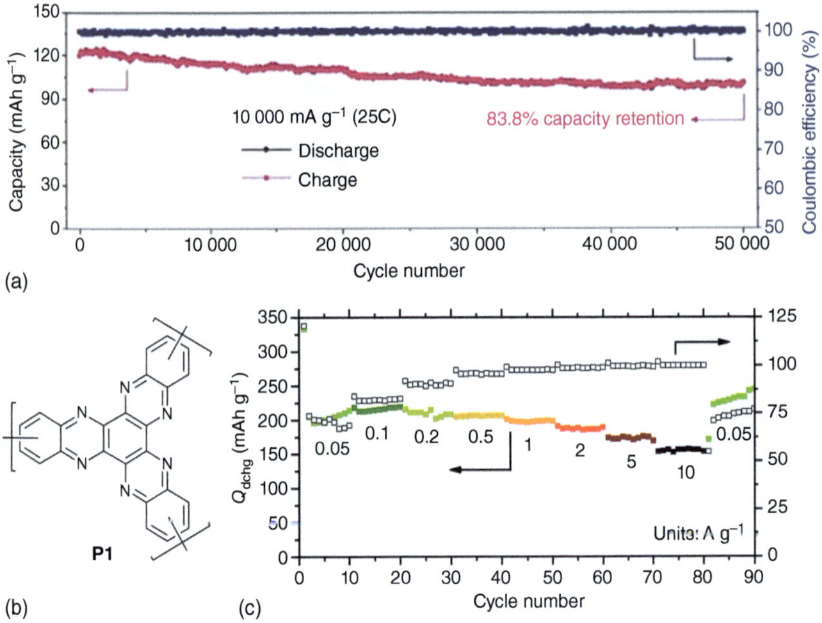

Figure 7.34 (a) Long-term cycling performance of the Na/PHATN cell at 25 C for 50 000 cycles. Source: Reproduced with permission from Mao et al. [127]/John Wiley & Sons. (b) Structures of the P1. (c) The rate performance of P1 in potassium-ion batteries. Source: Reproduced with permission from Kapaev et al. [128]/Royal Society of Chemistry.

Na-ion (NIBs), Mg-ion (RMBs), and Al-ion (RABs) batteries, where the redox active centers are based on multi-electron-deficient pyrazine sites [127]. The redox mechanism of PHATN is analyzed by ex situ XPS, FTIR, and Raman. The N 1s peak of pristine PHATN has two characteristic peaks at 398.8 and 400.4 eV, which represent −C=N− and −C−N−, respectively. When discharged to 1.0 V, a new peak representing Na−N appears at 398.6 eV, while the peak of −C=N− decreases. After charging to 3.5 V, the peak of −C=N− recovered and the peak of Na−N disappeared, which indicated that −C=N− could be reversibly sodium and de-sodium in PHATN electrode. Similar phenomena are observed in FTIR and Raman spectra, which prove that the reversible electrochemical reaction between imine groups and Na ions occurs in the process of charge and discharge. The PHATN cathode demonstrates almost the best performance in NIBs with fast reaction kinetics, ultra-stable cycling, and high energy density. At the ultrahigh current density of 10 A g^{-1} (25 C), PHATN still has a discharge capacity of about 100 mAh g^{-1} with a capacity retention rate of 83.8% after 50 000 cycles (Figure 7.34a). In addition, Kapaev et al. used the hexaazatriphenylene (HAT)-based polymers (P1, Figure 7.34b) synthesized from 3,3′-diaminobenzidine and triquinoyl as cathode materials and matched with the optimized DME-LiTFSI electrolyte for fast and stable Li^+, Na^+, and K^+ batteries [128]. In particular, the K^+ batteries with P1 electrode exhibit excellent electrochemical performance with high specific capacity, energy density, and excellent rate performance, which have a discharge specific capacity of 169 mAh g^{-1}

at the current density of $10\ A\ g^{-1}$ (Figure 7.34c) and almost no capacity attenuation after 4600 cycles, indicating that polymer P1 is the most durable cathode material for nonaqueous K^+ batteries reported to date.

7.7 Summary and Outlook

Polymer-based materials have shown fascinating advantages as organic electrodes, they have the characteristics of environmental friendliness, adjustable structure, and abundant resources. In recent years, a wide variety of polymer materials have been used in the field of energy storage, showing great development prospects. In this chapter, we describe the applications of different kinds of polymer materials as cathode in rechargeable batteries, including organosulfur-derived, carbonyl-derived, COFs-derived, ORPs-derived, and other polymers (Figure 7.35), which exhibit different reaction mechanisms and interesting electrochemical phenomena.

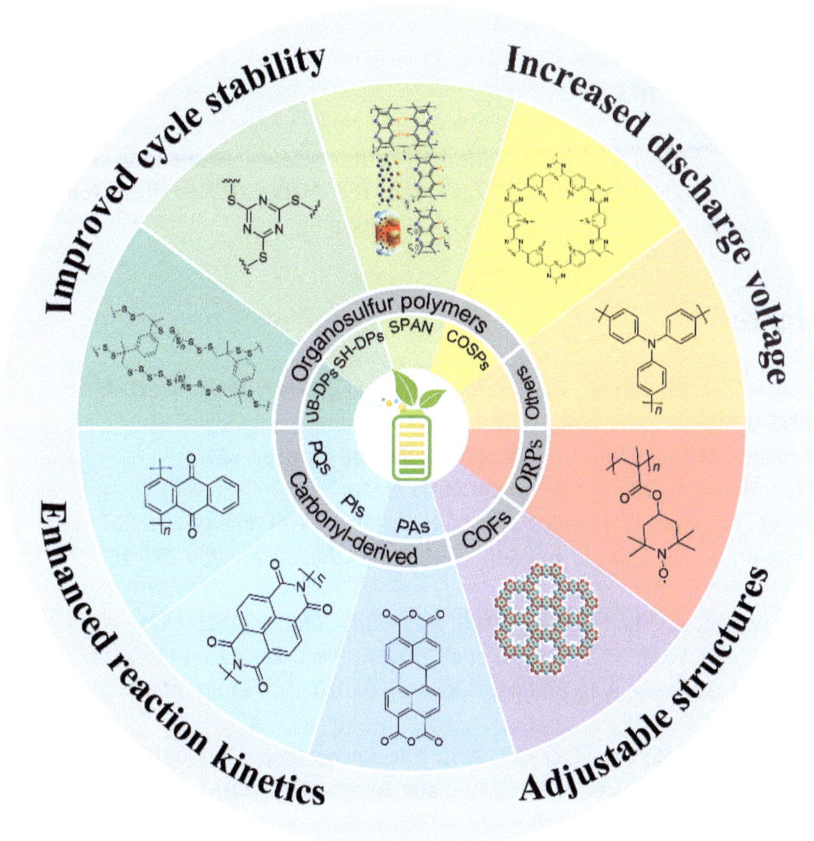

Figure 7.35 Summary illustration of polymer-based cathode materials in rechargeable batteries.

Nevertheless, there is still a long way to go before polymer-based organic electrode materials are ready for practical application. Therefore, the future development needs to focus on solving the following problems: (i) Novel polymer molecules with relatively high theoretical specific capacity and discharge voltage should be designed to achieve higher energy density; (ii) Most of the currently reported polymer electrode materials are mainly used in Li^+ batteries, more new polymer materials should be used in other energy storage devices such as Mg^{2+}, K^+, and redox flow batteries; (iii) Development of more advanced in-situ characterization tools and new techniques to better investigate the redox mechanisms; (iv) The introduction of self-supporting conductive materials such as graphene or carbon nanotubes to increase the conductivity of polymer-based materials in electrodes; (v) Deeply understand the matching relationship between electrode materials with different functional groups and electrolyte, and optimize electrolyte to make the compatibility between polymer electrode materials and electrolyte better; (vi) Exploring simpler, more efficient, environmentally friendly and mass-produced routes for the synthesis of organic polymer materials; (vii) Solid electrolytes or flame-retardant materials should be used in polymer batteries to improve their safety performance.

Acknowledgments

The work was supported by the National Natural Science Foundation of China (Grant Nos. U2004214 and 21975225) and the National Key Research and Development Program of China (2022YFB2402200).

References

1 Etacheri, V., Marom, R., Elazari, R. et al. (2011). *Energy Environ. Sci.* 4: 3243–3262.

2 Armand, M. and Tarascon, J.-M. (2008). *Nature* 451: 652–657.

3 Manthiram, A. (2020). *Nat. Commun.* 11: 1550.

4 Li, W., Erickson, E.M., and Manthiram, A. (2020). *Nat. Energy* 5: 26–34.

5 Scrosati, B., Hassoun, J., and Sun, Y.-K. (2011). *Energy Environ. Sci.* 4: 3287–3295.

6 Du, W., Du, X., Ma, M. et al. (2022). *Adv. Funct. Mater.* 32: 2110871.

7 Muench, S., Wild, A., Friebe, C. et al. (2016). *Chem. Rev.* 116: 9438–9484.

8 Schon, T.B., Tilley, A.J., Bridges, C.R. et al. (2016). *Adv. Funct. Mater.* 26: 6896–6903.

9 Wu, Y., Zeng, R., Nan, J. et al. (2017). *Adv. Energy Mater.* 7: 1700278.

10 Schon, T.B., McAllister, B.T., Li, P.F., and Seferos, D.S. (2016). *Chem. Soc. Rev.* 45: 6345–6404.

11 Zhang, Q., Cui, X., Hao, S. et al. (2021). *Mater. Today* 50: 170–198.

12 Ren, S., Sang, P., Guo, W., and Fu, Y. (2022). *Polym. Chem.* 13: 5676–5690.

13 Liu, J., Wang, M., Xu, N. et al. (2018). *Energy Storage Mater.* 15: 53–64.

14 Liu, K., Zhao, H., Ye, D., and Zhang, J. (2021). *Chem. Eng. J.* 417: 129309.
15 Chung, W.J., Griebel, J.J., Kim, E.T. et al. (2013). *Nat. Chem.* 5: 518–524.
16 Simmonds, A.G., Griebel, J.J., Park, J. et al. (2014). *ACS Macro Lett.* 3: 229–232.
17 Wei, Y., Li, X., Xu, Z. et al. (2015). *Polym. Chem.* 6: 973–982.
18 Chang, C.-H. and Manthiram, A. (2018). *ACS Energy Lett.* 3: 72–77.
19 Hu, G., Sun, Z., Shi, C. et al. (2017). *Adv. Mater.* 29: 1603835.
20 Wu, F., Chen, S., Srot, V. et al. (2018). *Adv. Mater.* 30: 1706643.
21 Li, X., Yuan, L., Liu, D. et al. (2020). *Energy Storage Mater.* 26: 570–576.
22 Sun, Z., Xiao, M., Wang, S. et al. (2014). *J. Mater. Chem. A* 2: 9280–9286.
23 Fu, C., Li, G., Zhang, J. et al. (2016). *ACS Energy Lett.* 1: 115–120.
24 Wang, Z., Shen, X., Li, S. et al. (2022). *J. Energy Chem.* 64: 496–502.
25 Kang, H., Kim, H., and Park, M.J. (2018). *Adv. Energy Mater.* 8: 1802423.
26 Guo, W. and Fu, Y. (2020). *Chem. Eur. J.* 26: 13322–13331.
27 Dong, Y., Lu, P., Ding, Y. et al. (2021). *SusMat* 1: 393–412.
28 Du, H., Feng, S., Luo, W. et al. (2020). *J. Mater. Sci. Technol.* 55: 1–15.
29 Zhou, J., Qian, T., Xu, N. et al. (2017). *Adv. Mater.* 29: 1701294.
30 Dong, P., Han, K.S., Lcc, J.I. et al. (2018). *ACS Appl. Mater. Interfaces* 10: 29565–29573.
31 Gomez, I., Mantione, D., Leonet, O. et al. (2018). *ChemElectroChem* 5: 260–265.
32 Liu, X., Li, Y., Xu, X. et al. (2021). *J. Energy Chem.* 61: 104–134.
33 Ghosh, A., Shukla, S., Monisha, M. et al. (2017). *ACS Energy Lett.* 2: 2478–2485.
34 Zhou, D., Chen, Y., Li, B. et al. (2018). *Angew. Chem. Int. Ed.* 57: 10168–10172.
35 Katsuhiko Naoi, K.-I.K., Mori, M., and Komiyama, M. (1997). *J. Electrochem. Soc.* 144: L173–L175.
36 Kim, H., Lee, J., Ahn, H. et al. (2015). *Nat. Commun.* 6: 7278.
37 Je, S.H., Hwang, T.H., Talapaneni, S.N. et al. (2016). *ACS Energy Lett.* 1: 566–572.
38 Bhargav, A., Bell, M.E., Cui, Y., and Fu, Y. (2018). *ACS Appl. Energy Mater.* 1: 5859–5864.
39 Bhargav, A., Chang, C.H., Fu, Y., and Manthiram, A. (2019). *ACS Appl. Mater. Interfaces* 11: 6136–6142.
40 Sang, P., Si, Y., and Fu, Y. (2019). *Chem. Commun.* 55: 4857–4860.
41 Sang, P., Song, J., Guo, W., and Fu, Y. (2021). *Chem. Eng. J.* 415: 129043.
42 Wang, J., Yang, J., Xie, J., and Xu, N. (2002). *Adv. Mater.* 14: 963–965.
43 Wang, J., Yang, J., Wan, C. et al. (2003). *Adv. Funct. Mater.* 13: 487–492.
44 Yu, X.-G., Xie, J.-Y., Yang, J. et al. (2004). *J. Electroanal. Chem.* 573: 121–128.
45 Fanous, J., Wegner, M., Grimminger, J. et al. (2011). *Chem. Mater.* 23: 5024–5028.
46 Zhang, S. (2014). *Energies* 7: 4588–4600.
47 Wang, W., Cao, Z., Elia, G.A. et al. (2018). *ACS Energy Lett.* 3: 2899–2907.
48 Wang, S., Lu, B., Cheng, D. et al. (2023). *J. Am. Chem. Soc.* 145: 9624–9633.
49 Wei, S., Ma, L., Hendrickson, K.E. et al. (2015). *J. Am. Chem. Soc.* 137: 12143–12152.
50 Chen, X., Peng, L., Wang, L. et al. (2019). *Nat. Commun.* 10: 1021.
51 Li, S., Han, Z., Hu, W. et al. (2019). *Nano Energy* 60: 153–161.

52 Wang, K., Guan, Y., Jin, Z. et al. (2019). *J. Energy Chem.* 39: 249–255.
53 Kim, H., Sadan, M.K., Kim, C. et al. (2021). *Chem. Eng. J.* 426: 130787.
54 NuLi, Y., Guo, Z., Liu, H., and Yang, J. (2007). *Electrochem. Commun.* 9: 1913–1917.
55 Wang, P., Kappler, J., Sievert, B. et al. (2020). *Electrochim. Acta* 361: 137024.
56 Zhang, S., Ren, W., NuLi, Y. et al. (2022). *Chem. Eng. J.* 427: 130902.
57 Zhu, D., Xu, G., Barnes, M. et al. (2021). *Adv. Funct. Mater.* 31: 2100505.
58 Talapaneni, S.N., Hwang, T.H., Je, S.H. et al. (2016). *Angew. Chem. Int. Ed.* 55: 3106–3111.
59 Xu, F., Yang, S., Jiang, G. et al. (2017). *ACS Appl. Mater. Interfaces* 9: 37731–37738.
60 Je, S.H., Kim, H.J., Kim, J. et al. (2017). *Adv. Funct. Mater.* 27: 1703947.
61 Jiang, Q., Li, Y., Zhao, X. et al. (2018). *J. Mater. Chem. A* 6: 17977–17981.
62 Guan, R., Zhong, L., Wang, S. et al. (2020). *Appl. Mater. Interfaces* 12: 8296–8305.
63 Kim, J., Elabd, A., Chung, S.-Y. et al. (2020). *Chem. Mater.* 32: 4185–4193.
64 Wang, D.-G., Wang, Y., Song, M. et al. (2019). *Chem. Commun.* 55: 13247–13250.
65 Han, C., Li, H., Shi, R. et al. (2019). *J. Mater. Chem. A* 7: 23378–23415.
66 Xie, J. and Zhang, Q. (2016). *J. Mater. Chem. A* 4: 7091–7106.
67 Häupler, B., Wild, A., and Schubert, U.S. (2015). *Adv. Energy Mater.* 5: 1402034.
68 Manecke, G. and Storck, W. (1963). *J. Polym. Sci. Part C* 4: 1457–1466.
69 Manecke, G. and Storck, W. (1962). *Angew. Chem. Int. Ed.* 1: 659–660.
70 Foos, J.S., Erker, S.M., and Rembetsy, L.M. (1986). *J. Electrochem. Soc.* 133: 836–841.
71 Häringer, D., Novák, P., Haas, O. et al. (1999). *J. Electrochem. Soc.* 146: 2393–2396.
72 Le Gall, T., Reiman, K.H., Grossel, M.C., and Owen, J.R. (2003). *J. Power Sources* 119–121: 316–320.
73 Oyama, N., Sarukawa, T., Mochizuki, Y. et al. (2009). *J. Power Sources* 189: 230–239.
74 Pirnat, K., Mali, G., Gaberscek, M., and Dominko, R. (2016). *J. Power Sources* 315: 169–178.
75 Ma, T., Zhao, Q., Wang, J. et al. (2016). *Angew. Chem. Int. Ed.* 55: 6428–6432.
76 Kolek, M., Otteny, F., Schmidt, P. et al. (2017). *Energy Environ. Sci.* 10: 2334–2341.
77 Liu, K., Zheng, J., Zhong, G., and Yang, Y. (2011). *J. Mater. Chem.* 21: 4125–4131.
78 Song, Z., Qian, Y., Liu, X. et al. (2014). *Energy Environ. Sci.* 7: 4077–4086.
79 Song, Z., Zhan, H., and Zhou, Y. (2009). *Chem. Commun.* 4: 448–450.
80 Song, Z., Qian, Y., Zhang, T. et al. (2015). *Adv. Sci. (Weinh)* 2: 1500124.
81 Xie, J., Wang, Z., Gu, P. et al. (2016). *Sci. China Mater.* 59: 6–11.
82 Xie, J., Wang, Z., Xu, Z.J., and Zhang, Q. (2018). *Adv. Energy Mater.* 8: 1703509.
83 Wei, W., Li, L., Zhang, L. et al. (2018). *Electrochem. Commun.* 90: 21–25.
84 Wei, W., Li, L., Zhang, L. et al. (2018). *Mater. Lett.* 213: 126–130.

85 Kassam, A., Burnell, D.J., and Dahn, J.R. (2011). *Electrochem. Solid-State Lett.* 14: A22–A23.

86 Tang, Z. and Xu, G. (2003). *Acta Phys. Chim. Sin.* 19: 307–310.

87 Zhao, L., Wang, W., Wang, A. et al. (2013). *J. Power Sources* 233: 23–27.

88 Liu, T., Kim, K.C., Lee, B. et al. (2020). *ACS Appl. Energy Mater.* 3: 3728–3735.

89 Xu, W., Read, A., Koech, P.K. et al. (2012). *J. Mater. Chem.* 22: 4032–4039.

90 Song, Z., Qian, Y., Gordin, M.L. et al. (2015). *Angew. Chem. Int. Ed.* 54: 13947–13951.

91 Jing, Y., Liang, Y., Gheytani, S., and Yao, Y. (2017). *Nano Energy* 37: 46–52.

92 Wang, H., Yao, C.-J., Nie, H.-J. et al. (2020). *J. Mater. Chem. A* 8: 11906–11922.

93 Song, Z., Zhan, H., and Zhou, Y. (2010). *Angew. Chem. Int. Ed.* 49: 8444–8448.

94 Wu, H.P., Yang, Q., Meng, Q.H. et al. (2016). *J. Mater. Chem. A* 4: 2115–2121.

95 Zindy, N., Blaskovits, J.T., Beaumont, C. et al. (2018). *Chem. Mater.* 30: 6821–6830.

96 Wu, H., Shevlin, S.A., Meng, Q. et al. (2014). *Adv. Mater.* 26: 3338–3343.

97 Wang, H.-G., Yuan, S., Ma, D.-L. et al. (2014). *Adv. Energy Mater.* 4: 1301651.

98 Wang, Y., Liu, Z., Wang, C. et al. (2020). *Energy Storage Mater.* 26: 494–502.

99 Wang, J., Liu, H., Du, C. et al. (2022). *Chem. Eng. J.* 444: 136598.

100 Ruby, R.M., Mangalaraja, R.V., Contreras, D. et al. (2019). *ACS Appl. Energy Mater.* 3: 240–252.

101 Sharma, P., Damien, D., Nagarajan, K. et al. (2013). *J. Phys. Chem. Lett.* 4: 3192–3197.

102 Han, X., Chang, C., Yuan, L. et al. (2007). *Adv. Mater.* 19: 1616–1621.

103 Yang, D.-H., Yao, Z.-Q., Wu, D. et al. (2016). *J. Mater. Chem. A* 4: 18621–18627.

104 Luo, Z., Liu, L., Ning, J. et al. (2018). *Angew. Chem. Int. Ed.* 57: 9443–9446.

105 Wu, M., Zhao, Y., Sun, B. et al. (2020). *Nano Energy* 70: 104498.

106 Yang, X., Hu, Y., Dunlap, N. et al. (2020). *Angew. Chem. Int. Ed.* 59: 20385–20389.

107 Wang, S., Wang, Q., Shao, P. et al. (2017). *J. Am. Chem. Soc.* 139: 4258–4261.

108 Xie, Y., Zhang, K., Yamauchi, Y. et al. (2021). *Mater. Horiz.* 8: 803–829.

109 Wilcox, D.A., Agarkar, V., Mukherjee, S., and Boudouris, B.W. (2018). *Annu. Rev. Chem. Biomol. Eng.* 9: 83–103.

110 Zhang, H., Gao, Y., Liu, X.H. et al. (2021). *Adv. Funct. Mater.* 32: 2107718.

111 Nakahara, K., Iwasa, S., Satoh, M. et al. (2002). *Chem. Phys. Lett.* 359: 351–354.

112 Nishide, H., Iwasa, S., Pu, Y.-J. et al. (2004). *Electrochim. Acta* 50: 827–831.

113 Nakahara, K., Iriyama, J., Iwasa, S. et al. (2007). *J. Power Sources* 165: 870–873.

114 Kim, J.-K., Kim, Y., Park, S. et al. (2016). *Energy Environ. Sci.* 9: 1264–1269.

115 Byeon, H., Gu, B., Kim, H.-J. et al. (2021). *Chem. Eng. J.* 413: 127402.

116 Zhu, J., Zhu, T., Tuo, H. et al. (2020). *Macromol. Chem. Phys.* 221: 2000160.

117 Katsumata, T., Satoh, M., Wada, J. et al. (2006). *Macromol. Rapid Commun.* 27: 1206–1211.

118 Suga, T., Konishi, H., and Nishide, H. (2007). *Chem. Commun.* 17: 1730–1732.

119 Sukegawa, T., Sato, K., Oyaizu, K., and Nishide, H. (2015). *RSC Adv.* 5: 15448–15452.

120 Kim, H.W., Kim, H.-J., Byeon, H. et al. (2020). *J. Mater. Chem. A* 8: 17980–17986.
121 Joo, Y., Agarkar, V., Sung, S.H. et al. (2018). *Science* 359: 1391–1395.
122 Lutkenhaus, J. (2018). *Science* 359: 1334–1335.
123 Koshika, K., Sano, N., Oyaizu, K., and Nishide, H. (2009). *Macromol. Chem. Phys.* 210: 1989–1995.
124 Song, Z. and Zhou, H. (2013). *Energy Environ. Sci.* 6: 2280–2301.
125 Koshika, K., Sano, N., Oyaizu, K., and Nishide, H. (2009). *Chem. Commun.* 836–838.
126 Fan, L., Liu, Q., Xu, Z., and Lu, B. (2017). *ACS Energy Lett.* 2: 1614–1620.
127 Mao, M., Luo, C., Pollard, T.P. et al. (2019). *Angew. Chem. Int. Ed.* 58: 17820–17826.
128 Kapaev, R.R., Zhidkov, I.S., Kurmaev, E.Z. et al. (2019). *J. Mater. Chem. A* 7: 22596–22603.

8

Organic Anode

8.1 Introduction

The anode is one of the indispensable components of the battery. Graphite, due to its high electronic conductivity, low cost, and abundant raw materials, has served as anode material in commercial lithium-ion batteries (LIBs) since they were introduced by Sony in 1991 [1]. To date, graphite still occupies the vast majority of the anode market share and will remain so in the coming decades. However, as the demand for battery energy density increases, the limited specific capacity of graphite inhibits its further development. Lithium metal anode has a high theoretical specific capacity of 3860 mAh g^{-1} and a low negative electrochemical potential of −3.04 V versus the standard hydrogen electrode, which endows lithium metal batteries (LMBs) with high discharging voltage and energy density [2]. But the high electrochemical and chemical reactivity of lithium metal leads to the continuous side reaction and low utilization of lithium metal, as well as the safety hazards caused by the formation of lithium dendrites, hindering the commercial application of lithium metal anodes. Although great progress has been made in the protection of lithium metal anode in recent years, it is still not possible to use them on a large scale [3]. Another representative anode is lithium titanate ($Li_4Ti_5O_{12}$, LTO), which was regarded as "zero strain" electrode material due to the negligible volume changes during the redox process [4]. In addition, the charge–discharge voltage of LTO is about 1.55 V versus Li/Li^+, such voltage can avoid the decomposition of electrolytes and the formation of lithium dendrites, enabling stable electrochemical performance and safety. While its poor electrical conductivity (10^{-8}–10^{-13} S cm^{-1}), high redox voltage, and low specific capacity (175 mAh g^{-1}) limit its rate capability and energy density.

Due to their structural diversity, environmental friendliness, tunable molecular structure, and low cost, organic anodes are gaining more and more attention [5]. In this chapter, they were divided into conjugated carboxylates, Schiff bases, azo compounds, covalent organic frameworks (COFs), and thiophene compounds according to the redox process. We systematically review the development of organic anodes in rechargeable batteries from the perspective of molecular engineering, introduce their redox mechanism and molecular structure-electrochemical behavior relationship in detail, and provide guidance for future developments.

Rechargeable Organic Batteries: Materials, Mechanisms, and Prospects, First Edition.
Yongzhu Fu, Xiang Li, Shuai Tang, and Wei Guo.

8.2 Conjugated Carboxylates

Conjugated carboxylates have been regarded as one of the promising anode materials in rechargeable metal-ion batteries due to their low redox potential (0.7 ~ 1.4 versus Li/Li^+, 0.3 ~ 0.8 versus Na/Na^+, and 0.3 ~ 0.9 versus K/K^+). In 2009, Armand et al. first reported two conjugated dicarboxylates, lithium terephthalate ($Li_2C_8H_4O_4$, Li_2-BDC) and lithium *trans–trans*-muconate ($Li_2C_6H_4O_4$) as anode materials in lithium batteries [6]. Later, several carboxylate-based anode materials are reported, and they can be divided into aromatic dicarboxylates and aliphatic carboxylates.

8.2.1 Aromatic Dicarboxylates

For aromatic dicarboxylates, their redox mechanism is shown in Figure 8.1a. During the discharge process, the carbonyl groups of the terephthalates are reduced and

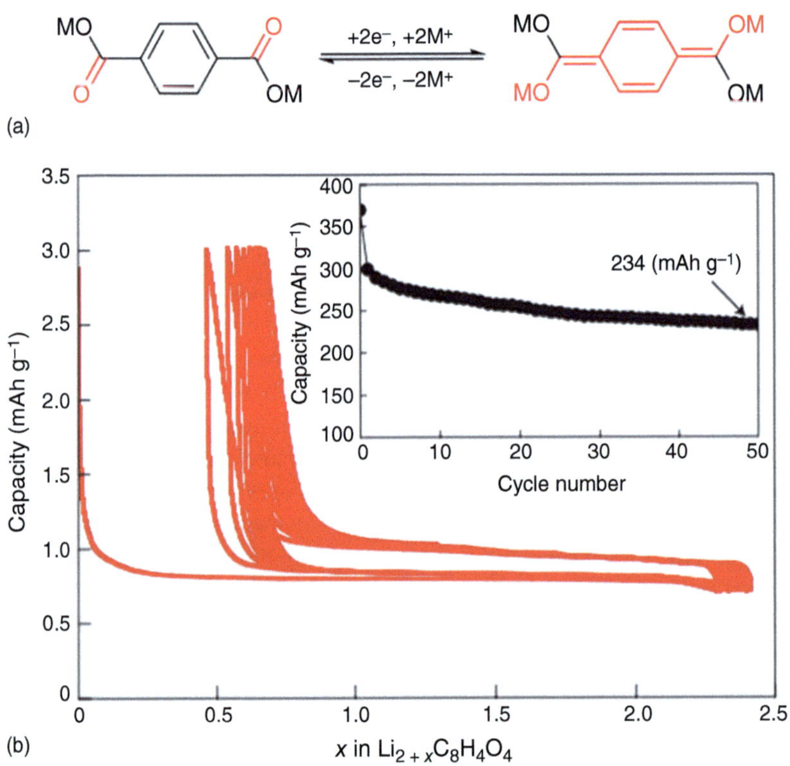

Figure 8.1 (a) Redox mechanism of aromatic dicarboxylates. (b) Charge–discharge profiles of $Li/Li_2C_8H_4O_4$ cell at 0.1 C rate; inset is the corresponding cycling performance. Source: Reproduced with permission from Armand et al. [6]/Springer Nature.

combine with the metal ions (Li^+, Na^+, or K^+) in the electrolyte to form M—O bonds, and the conjugated structure of aromatic dicarboxylate changes.

Li_2-BDC, as the simplest aromatic dicarboxylate, is prepared via the acid–base-type reaction of terephthatic and $LiOH{\cdot}H_2O$ in ethanol solution and possesses a theoretical specific capacity of 301 mAh g^{-1} based on the transfer of 2 e^-[6]. The $Li/Li_2C_8H_4O_4$ cell delivers stable electrochemical performance of 50 cycles at 0.1 C rate in carbonate electrolyte with a low discharge voltage of 0.8 V versus Li/Li^+ and small polarization of 0.1–0.2 V (Figure 8.1b). Meanwhile, the good reversibility of $Li_2C_8H_4O_4$ is also proved by infrared (IR) spectroscopy, X-ray diffraction (XRD), and electron paramagnetic resonance (EPR) spectra.

In addition, metal terephthalates such as sodium terephthalate ($Na_2C_8H_4O_4$, Na_2-BDC) [7, 8] and potassium terephthalate ($K_2C_8H_4O_4$, K_2-BDC) [9, 10] have also been explored in Na-ion and K-ion batteries. Zhao et al. reported Na_2-BDC, which is ball-milled with Ketjin black (KB) for reduced particle size and increased electronic conductivity and covered with an Al_2O_3-protected layer for suppressed decomposition of electrolytes during initial discharge [7]. The composite electrode shows low discharge voltage at 0.29 V versus Na/Na^+ and a stable reversible capacity of 250 mAh g^{-1} (theoretical specific capacity: 255 mAh g^{-1}) at 0.1 C rate. Lei et al. reported the application of K_2-BDC as anode materials in 1,2-dimethoxyethane (DME)-based electrolytes for K-ion batteries [9]. The theoretical specific capacity of K_2-BDC is 221 mAh g^{-1}, and the redox voltage is around 0.6 V versus K/K^+. Benefiting from the stable solid electrolyte interphase (SEI) formed in the ether electrolyte, the K_2-BDC electrode achieves capacity retention of 94.6% after 500 cycles at a current of 200 mA g^{-1}. Meanwhile, they also assembled an all-organic cell with $K_2(CO)_6$ as cathode and $K_2C_8H_4O_4$ as anode, which shows an output voltage of 1.8 V and a discharge capacity of 145 mAh g^{-1}, corresponding to the energy density of 52 Wh kg^{-1}. Wang et al. prepared graphene oxide-wrapped K_2-BDC hollow microrods, which benefited from improved K^+ ion transfer kinetics, enhanced electronic conductivity, and reduced diffusion distance. The composite electrode shows better electrochemical performance than bulk K_2-BDC [10].

Based on these pioneering researches, several studies have been proposed to improve the electrochemical performance of dicarboxylates via molecular structure modification. These strategies are shown in Figure 8.2 and discussed in the following.

8.2.1.1 Effect of Metal Cation

Basically, the stable layered structure of conjugated dicarboxylates is formed via the electrostatic interaction between metal cations and carboxylate groups. The type of metal cation affects the physical (e.g. solubility and thermal stability) and electrochemical properties of dicarboxylates; several metal cations such as K^+ [11], Ag^+ [12], Ca^{2+} [13–15], Al^{3+} [13], Ba^{2+}, Sr^{2+} [14], Zn^{2+} [16], Ni^{2+} [17, 18], Fe^{2+} [17], and Co^{2+} [19, 20] have been used in carboxylate electrode materials for rechargeable batteries.

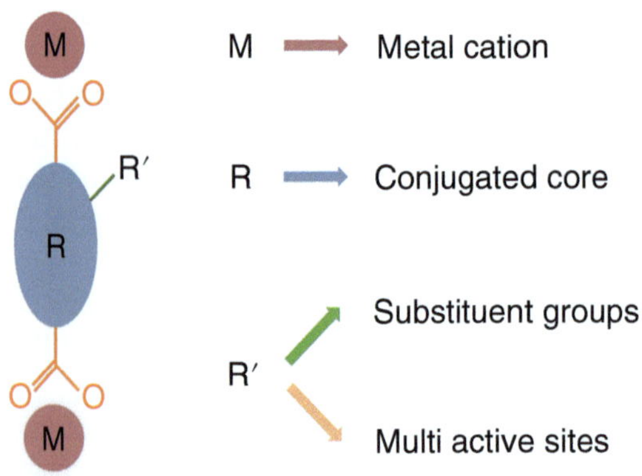

Figure 8.2 The modified strategies of dicarboxylates.

Most of the metal cations do not contribute capacity during the redox process, they only improve the structure and thermal stability. Wang et al. found that Ca-BDC shows lower solubility (1.8 ppm) compared with that of Li_2-BDC (19 ppm) in carbonate electrolyte, which is beneficial to suppress the loss of active materials [13]. In addition, they also prove that the Ca-BDC electrode exhibits excellent structure stability even under a low discharge voltage, leading to a high initial discharge capacity of 399 mAh g^{-1} and stable electrochemical performance of 40 cycles at a voltage range of 0.005–3.0 V. It is obvious that the Ca-BDC electrode delivers a higher capacity than the theoretical value of 263 mAh g^{-1} (based on the conversion reaction of $2e^-$); however, the authors attributed it to the capacity contribution of carbon black, which locks further exploration. Later, they reported three alkaline earth metal terephthalates (Ca-BDC, Sr-BDC, and Ba-BDC) and found that the metal cation with a smaller radius tends to exhibit the ionic bond behavior, contributing to lower discharge voltage, higher discharge capacity, and more stable cycling performance [14]. Deng et al. reported that in the application of K_2-BDC as anode materials in lithium batteries, the replacement of Li ions in the structure of Li_2-BDC with K ions enables a more stable lattice architecture due to the better size matching between anions and cations [11]. Meanwhile, they also introduce graphene matrix into the electrode to improve the electronic conductivity and particle dispersity, such composite electrode achieves a stable discharge capacity of 122 mAh g^{-1} at 8 C rate after 500 cycles.

Different from the above studies, Lee et al. reported that the transition metal ions Ni^{2+} and Fe^{2+} in Ni-BDC and Fe-BDC are also involved in the redox process [17]. The lithiation process consists of two parts: first, the divalent transition metal ions are reduced to transition metal nanoparticles at high potential, followed by further lithiation of unsaturated dicarboxylates into saturated organic scaffolds at low potential (Figure 8.3). And the delithiation process is reversible. The Li/Ni-BDC and Li/Fe-BDC cells deliver discharge capacities of 1402 and 1474 mAh g^{-1}, respectively, corresponding to 10 e^- transfer after subtracting the capacity contribution from conductive Super P. A similar redox mechanism was also reported by Kim et al., who prepared self-standing heteromat nickel 2,6-naphthalenedicarboxylate (Ni-NDC) electrodes via spatially embedding NiNDC into heteromat frameworks of single-walled carbon nanotubes (SWCNTs) and polyacrylonitrile (PAN) nanofibers

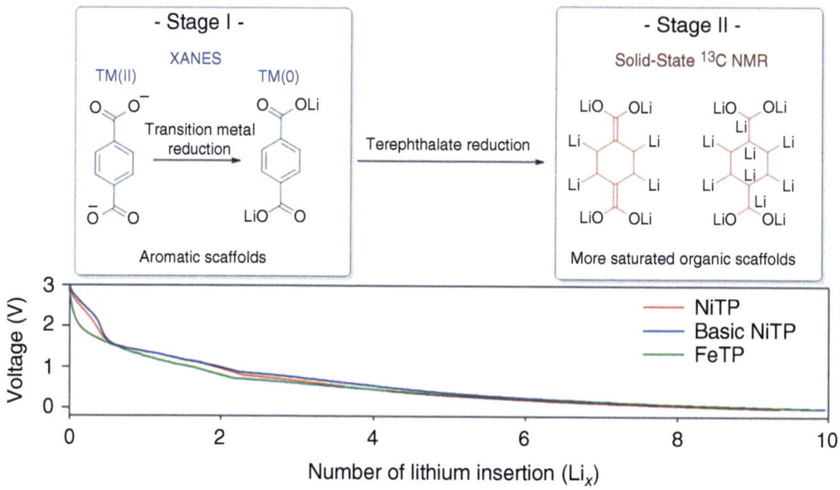

Figure 8.3 Proposed lithiation mechanism of transition metal-tetrephthalate. Source: Reproduced with permission from Lee et al. [17]/John Wiley & Sons.

(Figure 8.4a) [18]. Such structure provides sufficient electron/ion transport pathways for improved redox kinetics, enabling a reversible capacity of 360 mAh g^{-1} at 5 C rate. The differential capacity and the redox process of the NiNDC anode are shown in Figure 8.4b,c, the three lithiation stages correspond to the reduction of Ni^{2+} (stage I) and organic ligand transformations (stage II and stage III). In addition, Xue et al. reported another kind of redox process [12]. They prepared Ag_2-BDC and found that the Ag^+ ion would be reduced to conductive Ag particles in the first discharge process, which would not be oxidized in the following cycles. Such a mechanism is proved by XRD, and the formed Ag particles are detected. The Ag_2-BDC electrode both shows stable electrochemical performance in Li-ion and Na-ion batteries.

8.2.1.2 Effect of Conjugated Core

The conjugated core is an important component in the structure of dicarboxylates, which directly affects the electrochemical performance of dicarboxylates. So far, the modification of conjugated core mainly includes three aspects: (i) aromatic extension, (ii) linking with double/triple bond, (iii) utilization of heterocyclic.

For aromatic extension, Fédèle et al. first reported dilithium 2,6-naphthalene di-carboxylate (Li_2-NDC) synthesized via a freeze-drying process using the acid and lithium hydroxide [21]. Although the extension of the aromatic structure reduces the theoretical specific capacity to 235 mAh g^{-1}, the increased π-conjugation enhances the electronic delocalization and redox kinetics. The redox mechanism of Li_2-NDC is similar to that of Li_2-BDC, corresponding to the transfer of 2 e^- and 2 Li^+ (Figure 8.5a). And the Li/Li_2-NDC cell delivers a discharge capacity of 200 mAh g^{-1} with an output voltage of 0.88 V versus Li/Li^+. In addition, Yasuda et al. prepared Li_2-NDC with narrowed π-stacked naphthalene interlayer distance via molecular self-assembly [25]. They found that such a structure is beneficial to

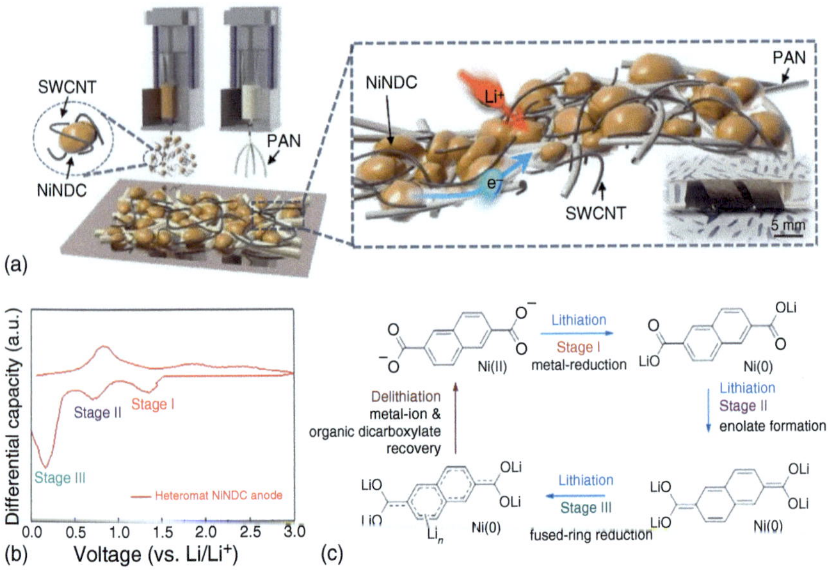

Figure 8.4 (a) Schematic illustration of the preparation of heteromat Ni-NDC. (b) Differential capacity (dQ/dV) and (c) the redox process of NiNDC anode. Source: Adapted with permission from Kim et al. [18]/Elsevier.

the formation of more efficient ion and electron pathways in the framework, which enables low internal resistance and a high discharge capacity of 220 mAh g^{-1}.

The Na_2-NDC [26–28] and K_2-NDC [29] are explored in Na-ion and K-ion batteries, respectively. Madabalmi et al. reported the Na/Na_2-NDC cell with a discharge voltage of 0.33 V versus Na/Na^+, giving a reversible discharge capacity of 145 mAh g^{-1} after 70 cycles at a current density of 40 mA g^{-1} [26]. Meanwhile, they also assembled a full cell coupling Na_2-NDC anode with $Na_3V_2O_2(PO_4)_2F$/reduced graphene oxide (rGO) cathode, which delivers discharge capacity of 150 mAh g^{-1} (weight based on anode material) with an average output voltage of 3.2 V. Na_2-NDC is also reported by Desai et al. to be synthesized via an energy-efficient and rapid microwave-assisted method, showing improved electrochemical performance with high initial Coulombic efficiency (ICE) and being more suitable for large-scale preparation [28]. K_2-NDC reported by Li et al. exhibits a redox potential of 0.55 V versus K/K^+ and a discharge capacity of 139 mAh g^{-1} at 100 mA g^{-1} with a capacity retention of 93% after 300 cycles [29].

4,4′-Biphenyldicarboxylate (BPDC) salt exhibits a longer separation distance between the two end carboxylate units than that of BDC and NDC, e.g. Li_2-BPDC (10.30 Å) > Li_2-NDC (8.314 Å) > Li_2-BDC (5.69 Å) (Figure 8.5b) [22], which enables improved rate capability. As shown in Figure 8.5c, the Li/Li_2-BPDC cell delivers a discharge capacity of 183 mAh g^{-1} after 25 cycles at 2 e^-/h rate, higher than that of Li_2-NDC (134 mAh g^{-1}) and Li_2-BDC (54 mAh g^{-1}) even though the Li_2-BPDC has the lowest theoretical specific capacity of 211 mAh g^{-1} among them. The same good rate capability also exists in Na_2-BPDC [30] and K_2-BPDC [24]. Na/Na_2-BPDC cell

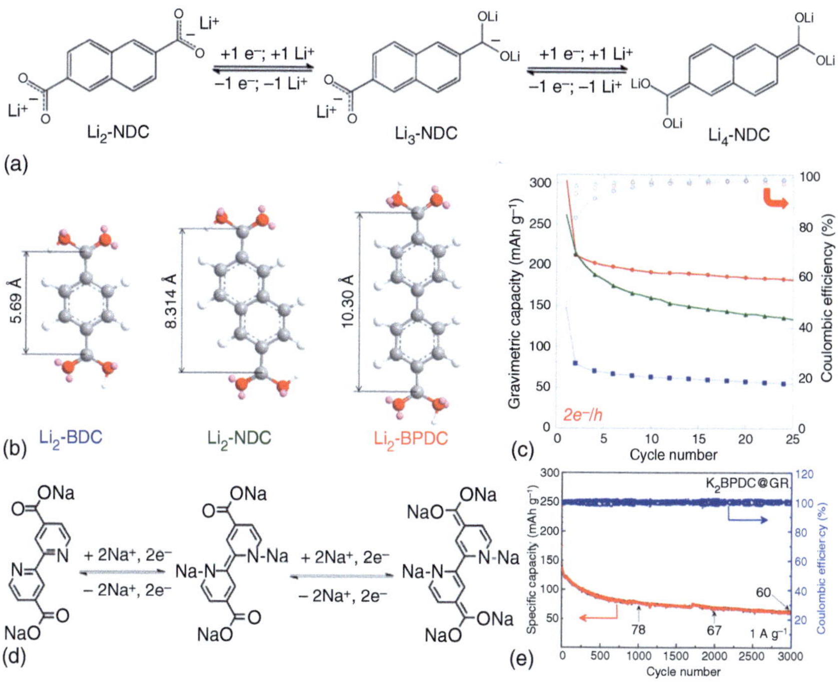

Figure 8.5 (a) The redox mechanism of Li_2-NDC. Source: Adapted with permission from Fedele et al. [21], The Electrochemical Society. (b) Comparison of the distance between the two end carboxylates in Li_2-BDC, Li_2-NDC, and Li_2-BPDC. (c) Cycling performance of Li_2-BDC (blue curve), Li_2-NDC (green curve), and Li_2-BPDC (red curve) at 2 e^-/h rate. Source: Reproduced with permission from Fédèle et al. [22]/American Chemical Society. (d) Redox mechanism of Na-DCA. Source: Adapted with permission from Holguin et al. [23], Elsevier. (e) Cycling performance of K_2-BPDC/graphene composite electrode at 1 A g^{-1}. Source: Reproduced with permission from Li et al. [24]/American Chemical Society.

reported by Choi et al. shows a reversible capacity of 200 mAh g^{-1} with a discharge voltage of 0.5 V versus Na/Na^+ and stable performance over 150 cycles, and a capacity of 100 mAh g^{-1} even at a high rate of 20 C [30]. Later, Jia et al. replaced the biphenyl group with the phenylpyridine group to design two isomers, namely sodium 5-(4-carboxyphenyl) picolinate (Na-CPP) and sodium 6-(4-carboxyphenyl) nicotinate (Na-CPN), and compared the different electrochemical properties brought by the different substitution positions [31]. Interestingly, they found that Na-CPP exhibits better capacity retention and rate performance than Na-CPN due to its smaller particle size, polarization, and charge-transporting resistance. The Na-CPP electrode delivers a discharge capacity of 197 mAh g^{-1} with an output voltage of 0.47 V versus Na/Na^+ and a high capacity retention of 99.1% after 350 cycles at 100 mA g^{-1}. Furthermore, it achieves a capacity retention of 89.9% after 1200 cycles at a 5 C rate. They also applied Na-CPP anode in an all-organic Na-ion full cell coupling with commercial disodium rhodizonate (DSR) cathode, which shows a discharge capacity of about 100 mAh g^{-1}. Recently, Holguin et al. reported that 2,2′-bipyridine-4,4′-dicarboxylic acid disodium salt (Na-DCA) with a bipyridine group,

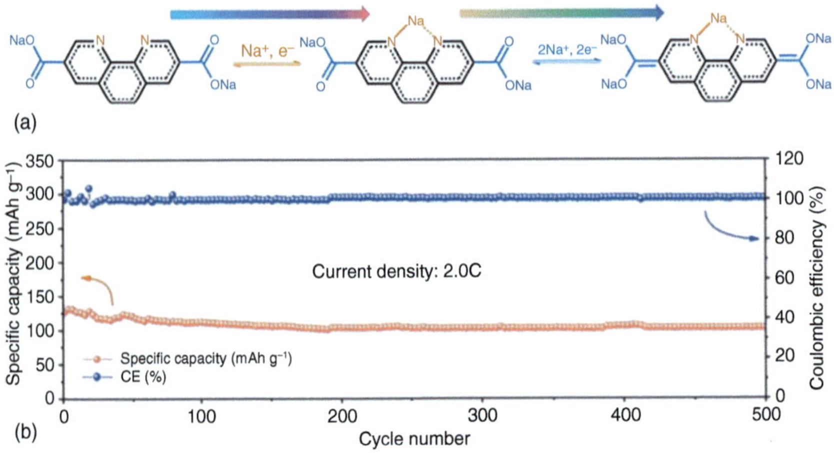

Figure 8.6 (a) Redox mechanism and (b) long cycling performance of Na-PD electrode. Source: Reproduced with permission from Jia et al. [33]/Royal Society of Chemistry.

the bipyridine group also can serve as available active sites to provide additional capacity contribution (Figure 8.5d) [23]. When combined with N-doped rGO, the composite electrode shows a discharge capacity of 300 mAh g^{-1} at 50 mA g^{-1} with a pair of redox plateaus at about 0.48 V. For K_2-BPDC, the K_2-BPDC/graphene composite electrode reported by Li et al. shows a specific capacity of 75 mAh g^{-1} after 3000 cycles at a high current density of 1 A g^{-1} (Figure 8.5e) [24].

Lithium 2,6-anthracene dicarboxylate (Li_2-ADC), featuring a long conjugation length and strong electronic delocalization, is reported by Fédèle et al., delivering a discharge capacity of 133 mAh g^{-1} with a capacity retention of 77% after 50 cycles at an ultrahigh current density of 10 C rate (1 C = 193 mA g^{-1}) [32]. Recently, Jia et al. prepared sodium 1,10-phenanthroline-3,8-dicarboxylate (Na-PD) with a large conjugated structure and improved electronic conductivity [33]. Interestingly, the N atoms on phenanthroline could give additional redox active sites for Na^+ storage (Figure 8.6a), leading to a three-electron transfer redox mechanism. The Na-PD electrode delivers a reversible capacity of 126 mAh g^{-1} after 500 cycles at 2 C rate (Figure 8.6b).

A π-conjugated N-containing heteroaromatic hexacarboxylate (Li_6-HAT) with four aromatic rings is reported by Xia et al.; the π-conjugated N-containing heteroaromatic center enables excellent reversibility of (de)lithiation process, and the π-conjugated nitrogen also serves as electrochemically active sites during the redox process [34]. The proposed redox mechanism is shown in Figure 8.7a. And the Li_6-HAT electrode delivers a capacity of 294.5 mAh g^{-1} after 3000 cycles at 800 mA g^{-1}, corresponding to a capacity retention of 84.1% (Figure 8.7b).

Perylen-3,4,9,10-tetracarboxylate (PTC) salts with hyper-conjugated core unit of perylene have been prepared and used as anode materials in rechargeable batteries [35–39]. Although the significantly increased relative molecular mass reduces their theoretical specific capacity, their excellent rate capability still enables them to have good application prospects.

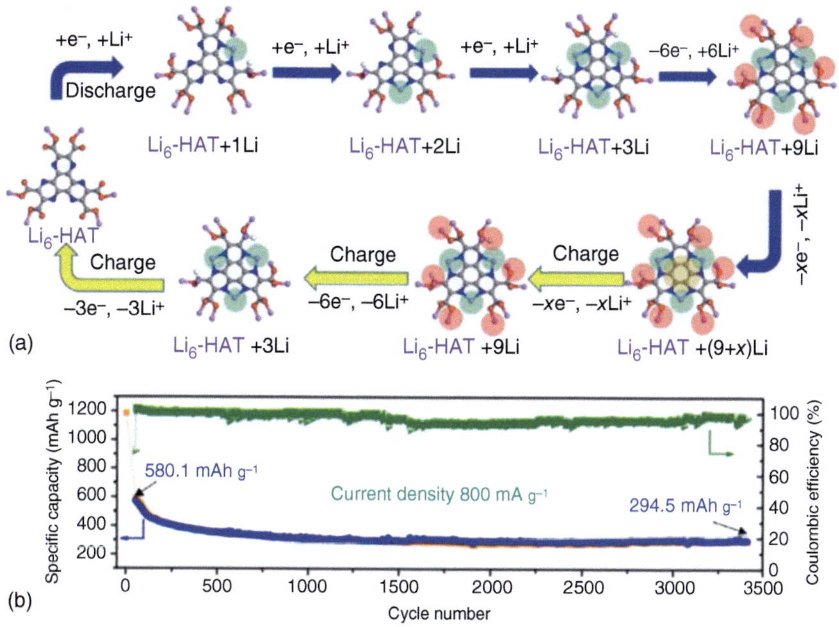

Figure 8.7 (a) The structural evolution of Li_6-HAT during the charge–discharge process. (b) Cycling performance of Li_6-HAT electrode at 800 mA g^{-1}. Source: Reproduced with permission from Xia et al. [34]/Elsevier.

The second type of modification of conjugated core is linking with a double/triple bond. Renault et al. explored the conjugated dicarboxylates linking with C=C and C≡C outside the aromatic unit, which are dilithium benzenediacrylate (Li_2-BDA) [40] and dilithium benzenedipropiolate (Li_2-BDP) [41], respectively. The carbon-coating Li_2-BDA electrode shows a reversible capacity of 170 mAh g^{-1} at 0.05 C rate with a discharge voltage of 1.2 V versus Li/Li^+. And the Li_2-BDP electrode delivers a stable performance of 100 cycles with a capacity of about 130 mAh g^{-1} at a voltage range of 0.5–3.5 V. When Li_2-BDP is cycled at 0–3.5 V, it delivers an ultrahigh specific capacity of 1363 mAh g^{-1} due to the further reduced unsaturated carbon–carbon bonds. However, rapid capacity losses are observed owing to intrinsic instability, side reactions between carbonate electrolyte and reactive C—Li nucleophilic species, and severe volume expansion.

Walker et al. prepared lithium 4,4′-tolane-dicarboxylate (Li_2-TDC) linked with C≡C inside the aromatic unit using different solvents (e.g. methanol, ethanol) [42]. They found that the crystallization of the materials depends on the type of solvent used in the synthesis, thus affecting their electrochemical properties. Li_2-TDC synthesized in methanol solvent shows a discharge capacity of 200 mAh g^{-1} with a stable electrochemical performance of 50 cycles, and the discharge voltage is 0.65 V versus Li/Li^+. Later, Wang et al. synthesized sodium 4,4′-stilbene-dicarboxylate (Na_2-SDC) and applied it as anode material in sodium batteries [43]. The Na_2-SDC electrode delivers a discharge capacity of 204 mAh g^{-1} after 50 cycles at 50 mA g^{-1} with a discharge voltage of about 0.45 V versus Na/Na^+ (Figure 8.8a), corresponding

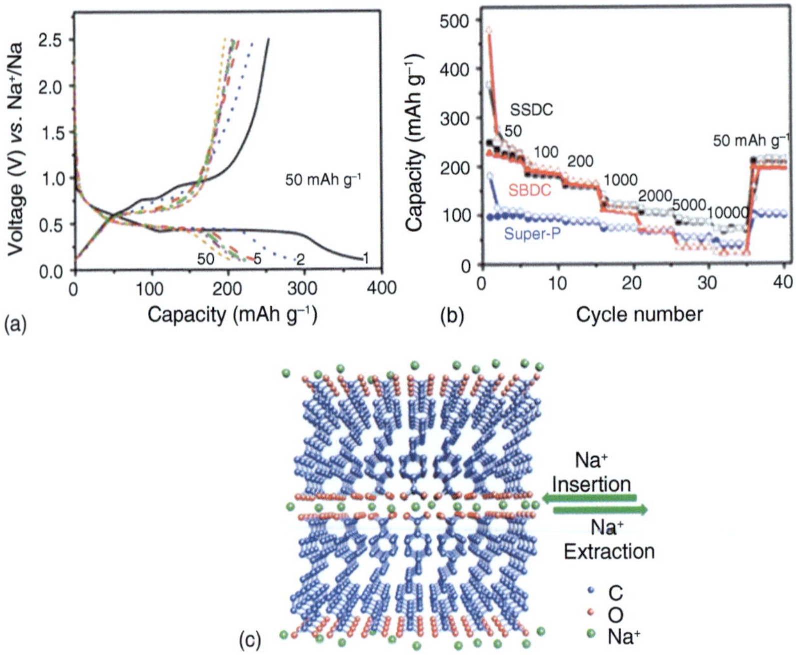

Figure 8.8 (a) Voltage profiles and (b) rate cyclability of Na_2-SDC electrode. (c) Schematic molecular packing of Na_2-SDC and the formed channel at the active center for insertion/extraction of Na^+ ions. Source: Reproduced with permission from Wang et al. [43]/American Chemical Society.

to a capacity retention of more than 90%. In addition, the Na_2-SDC electrode also shows excellent rate capability; it delivers a discharge capacity of $100\,mAh\,g^{-1}$ at $2\,A\,g^{-1}$ and $72\,mAh\,g^{-1}$ at an ultrahigh rate of $10\,A\,g^{-1}$, and the capacity can be fully recovered after being cycled at $10\,A\,g^{-1}$ as shown in Figure 8.8b. The authors believe that such high-rate performance can be attributed to the large *d*-spacing distance and the lamellar packing of the molecules, which constructs a fast diffusion channel at the active center for Na^+ insertion/extraction between the layers (Figure 8.8c).

The third type uses heterocyclic chemistry to regulate the electronic density of the conjugated core, which leads to a significant difference in redox potential. For example, thiophene [44], pyridine [45–47], and pyrazine [45] based dicarboxylate derivatives have been studied. The discharge voltage of dilithium thiophene-2,5-dicarboxylate (Li_2-ThDC) is 1.0 V versus Li/Li^+, as shown in Figure 8.9a, a little higher than that of Li_2-BDC (0.8 V) [44]. And the theoretical specific capacity of Li_2-ThDC should be $291\,mAh\,g^{-1}$ based on the transfer of 2 e^- per molecule. However, it delivers an actual reversible specific capacity of $850\,mAh\,g^{-1}$ at a current density of $30\,mA\,g^{-1}$ under the voltage range of 0–3.0 V (corresponding to 5.8 e^- transfer), which is due to the excessive lithiation process at low reduction potential.

Disodium 2,5-pyridinedicarboxylate ($Na_2C_7H_3NO_4$, Na_2-PDC) and disodium 2,5-pyrazinedicarboxylate ($Na_2C_6H_2N_2O_4$, Na_2-PzDC) are reported as anode materials in lithium batteries by Wu et al. [45]. They both show high discharge

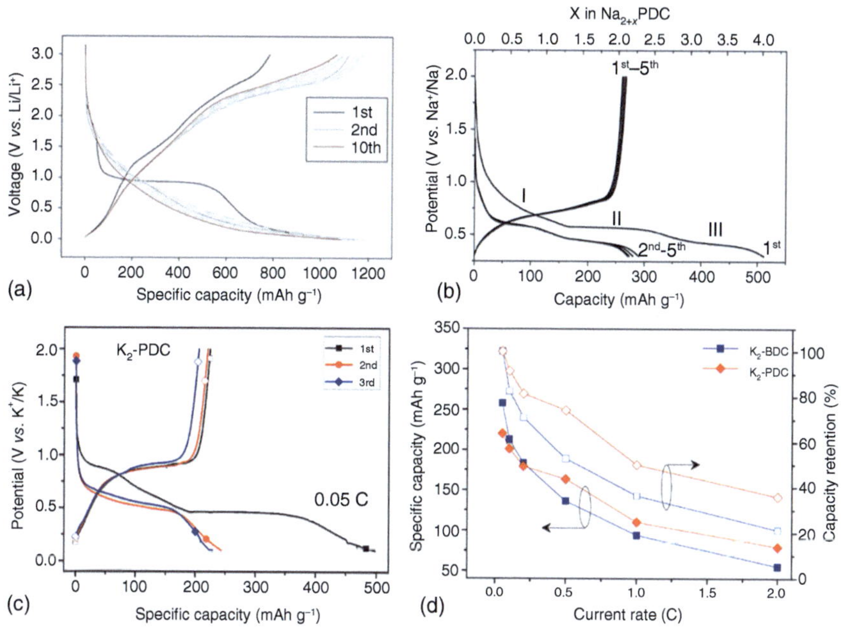

Figure 8.9 Charge–discharge profiles of (a) Li_2-ThDC. Source: Reproduced with permission from Lee et al. [44]/American Chemical Society. (b) Na_2-PDC. Source: Reproduced with permission from Padhy et al. [46]/John Wiley & Sons. (c) K_2-BDC. (d) The rate performance of K_2-BDC and K_2-PDC. Source: Reproduced with permission from Deng et al. [47]/Elsevier.

voltage at ~1.2 V versus Li/Li^+ (Na_2-PDC) and ~1.5 V versus Li/Li^+ (Na_2-PzDC), respectively, which is detrimental for anode materials. And Na_2-PDC electrode delivers a discharge specific capacity of 168 mAh g^{-1} after 20 cycles at 2 C rate, while Na_2-PzDC electrode delivers a discharge specific capacity of 180 mAh g^{-1} after 200 cycles at 0.1 C rate. Later, Na_2-PDC was also reported as anode material in sodium batteries [46], which shows a reversible capacity of 236 mAh g^{-1} at 0.1 C rate. It is worth noting that two distinct plateaus at 0.57 V (marked as region II) and 0.41 V (marked as region III) versus Na/Na^+ are observed during the discharge process, as shown in Figure 8.9b. The authors proposed that the first sodiation happens through bond formation between the inserted Na^+ and the N atom on the pyridine, and the second sodiation happens via conventional carboxylates. Similarly, K_2-PDC is reported as anode material in potassium batteries by Deng et al. and compared with K_2-BDC [47]. The authors found that K_2-PDC shows higher discharge voltage (K_2-PDC: ~0.6 V versus K/K^+, K_2-BDC: ~0.5 V versus K/K^+) but better rate capability than K_2-BDC, as shown in Figure 8.9c and d. K_2-PDC electrode delivers a discharge capacity of 176 mAh g^{-1} after 100 cycles at 0.2 C rate, while K_2-BDC electrode delivers 158 mAh g^{-1}.

8.2.1.3 Effect of Substituent Groups

The output voltage of the cells can directly affect the energy density (E_{cell} = V_{cell} × Capacity), which depends on the voltage of the cathode and anode

($V_{cell} = V_{cathode} - V_{anode}$). By reducing the output voltage of anode materials, the output voltage of the full cell can increase, achieving a high energy density under a given specific capacity. The introduction of substituent groups into the aromatic core of dicarboxylates can effectively regulate the electronic density and the potential; generally, electron-donating groups are proposed to be able to decrease the discharge voltage, and electron-withdrawing groups are able to increase the discharge voltage. Several substituent groups have been reported, e.g. methyl groups [48], amino groups [49], methoxy groups [50], and halogenated groups [51]. Lakraychi et al. reported dilithium 2,5-dimethylterephthalate (Li_2-DMT) [48]. Due to the electro-donating effect of methyl groups, Li_2-DMT shows a discharge voltage of 0.65 V versus Li/Li^+ and delivers a discharge capacity of 160 mAh g^{-1} (corresponding to 61% of the theoretical capacity). Renault et al. reported dilithium 2-aminoterephthalate (Li_2-ATP) with a primary amine (-NH_2) group [49]. The compound presents an average reduction potential of 0.9 V versus Li/Li^+ and a discharge capacity of 180 mAh g^{-1} (corresponding to 65% of the theoretical capacity), which also shows a stable performance during the 50 cycles, benefiting from the stabilized local environment around lithium during reduction.

In sodium batteries, Park et al. prepared three monosubstituted dicarboxylates using $-NO_2$ (NO_2–Na_2–BDC), –Br (Br–Na_2–BDC), and $-NH_2$ (NH_2–Na_2–BDC) as substituents and reported their electrochemical performances in Na metal cells [52]. These compounds display different redox voltages depending on the substituent, as shown in Figure 8.10. Recently, Huang et al. synthesized three disodium halogenated dicarboxylates to explore the impact of halogen atoms on electrochemical performance [51]. Among them, the authors found that disodium 2,5-difluoroterephthalate (Na_2-DFTP) shows the best performance compared with 2,5-dichloroterephthalate (Na_2-DCTP), disodium 2,5-dibromoterephthalate (Na_2-DBTP), and Na_2-BDC, such as a high discharge capacity of 212 mAh g^{-1}, a long cycle life of 300 cycles, and a high rate performance of 5 A g^{-1}, which is due to the F ions in Na_2-DFTP interacting with Na^+ during the redox process, reducing the solubility in the electrolyte, and enhancing the cyclic stability.

8.2.1.4 Multi Active Sites

Increasing the redox-active sites is one of the effective ways to improve the theoretical specific capacity. Maiti et al. prepared trilithium 1,2,4-benzenetricarboxylates (Li_3-BTC) with a theoretical specific capacity of 352 mAh g^{-1} based on the transfer of three e^-/Li^+ [53]. However, the Li_3-BTC electrode only delivers a discharge capacity of 162 mAh g^{-1}, which is less than half of the expected theoretical capacity, expressing low utilization of the reactive carboxylates. Similarly, Gu et al. reported trisodium 1,2,4-benzenetricarboxylates (Na_3-BTC) [54]. They think Na_3-BTC still undergoes a two-electron transfer reaction with a theoretical specific capacity of 194 mAh g^{-1}. The Na_3-BTC delivers a discharge capacity of 157.3 mAh g^{-1} after 50 cycles at 25 mA g^{-1}. In addition, the electron-withdrawing –COONa group is beneficial to accelerated charge transfer and improved structural stability, achieving an outstanding rate capability of 2 A g^{-1} and a long cycle life of 500 cycles.

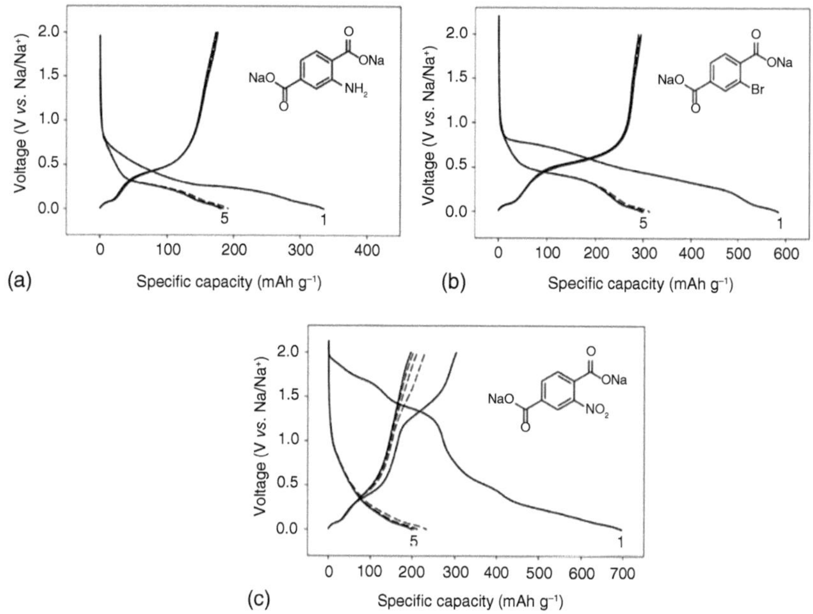

Figure 8.10 Charge–discharge profiles of (a) NH_2-Na_2-BDC, (b) Br-Na_2-BDC, and (c) NO_2-Na_2-BDC. Source: Reproduced with permission from Park et al. [52]/John Wiley & Sons.

Tetralithium 1,2,4,5-benzenetetracaboxylate (Li_4-BTC) with a further increased theoretical specific capacity of 385 mAh sg^{-1} based on the transfer of four e^-/Li^+ was reported [55, 56]. Experimentally, Li_4-BTC electrode delivers an increased specific capacity of 1016 mAh g^{-1} and is then stabilized at 600 mAh g^{-1} after 100 cycles. The much higher capacity is due to the excessive lithiation mentioned above at low voltage. Qin et al. synthesized tetrasodium 1,2,4,5-benzenetetracaboxylate (Na_4-BTC) and applied it as anode material in sodium batteries [57]. Similar to Na_3-BTC, the authors suppose that only the two carboxylate groups at the para positions are electrochemically active to enable two electrons to transfer and reversibly react with two e^-/Na^+ (Figure 8.11a). The Na_4-BTC electrode exhibits a redox potential of ~0.65 V versus Na/Na^+ and delivers a reversible capacity of 192.7 mAh g^{-1} at a current density of 20 mA g^{-1} (Figure 8.11b). Furthermore, Na_4-BTC also achieves a long cycle life of 1000 cycles at 2 A g^{-1}. Pan et al. reported tetrapotassium 1,2,4,5-benzenetetracaboxylate (K_4-BTC) in potassium batteries [58]. Interestingly, K_4-BTC shows four active sites for K^+ ion storage, and the corresponding mechanism is shown in Figure 8.11c. K_4-BTC delivers a high capacity of 290 mAh g^{-1} with a capacity retention of 83% after 1000 cycles (Figure 8.11d).

8.2.2 Aliphatic Dicarboxylates

For aliphatic dicarboxylates, which mainly refer to vinylic compounds. In 2009, Armand et al. first reported this kind of compound (namely dilithium

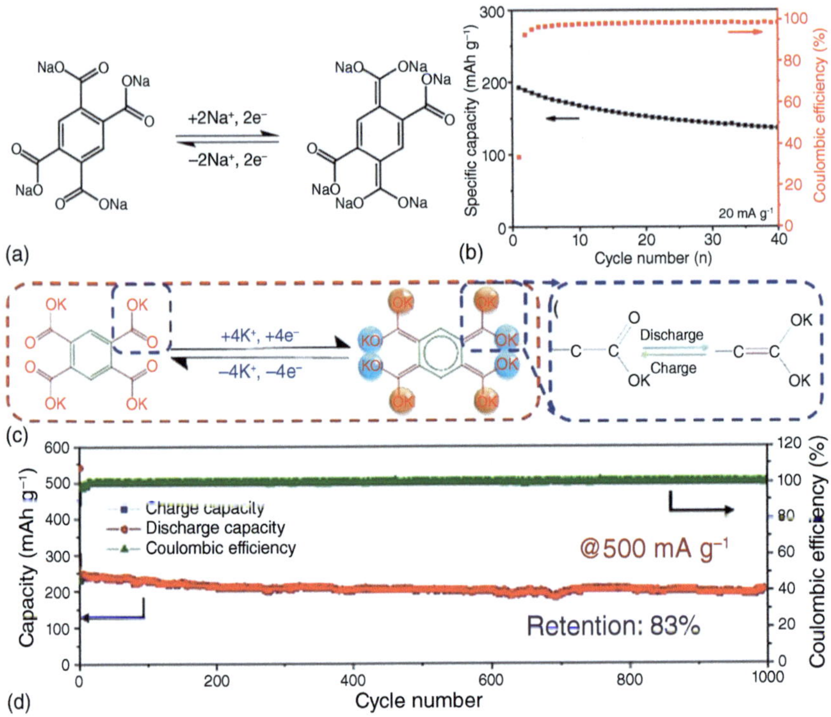

Figure 8.11 The redox mechanism (a), (c), and cycling performance (b), (d) of Na_4-BTC and K_4-BTC electrodes, respectively. Source: (a and c) Reproduced with permission from Qin et al. [57]/Royal Society of Chemistry. (b and d) Reproduced with permission from Pan et al. [58]/John Wiley & Sons.

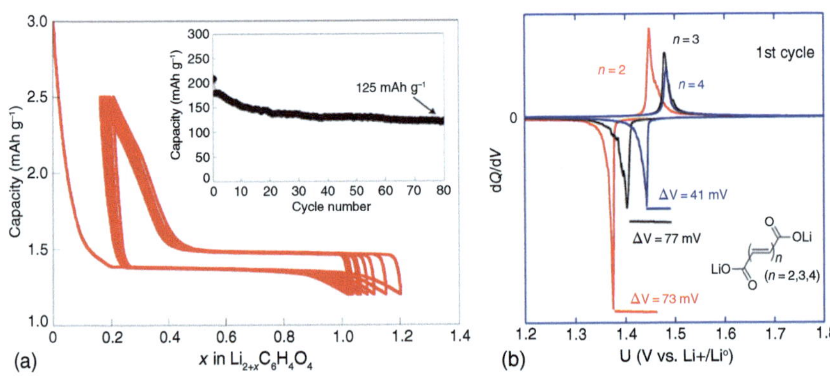

Figure 8.12 (a) Charge–discharge profiles of $Li/Li_2C_6H_4O_4$ cell at 0.1 C rate; inset is the corresponding cycling performance. Source: Reproduced with permission from Armand et al. [6]/Springer Nature. (b) Comparison of the derivative curves (d*Q*/d*V*) of aliphatic dicarboxylates with different numbers of vinylic groups. Source: Reproduced with permission from Walker et al. [59]/Elsevier.

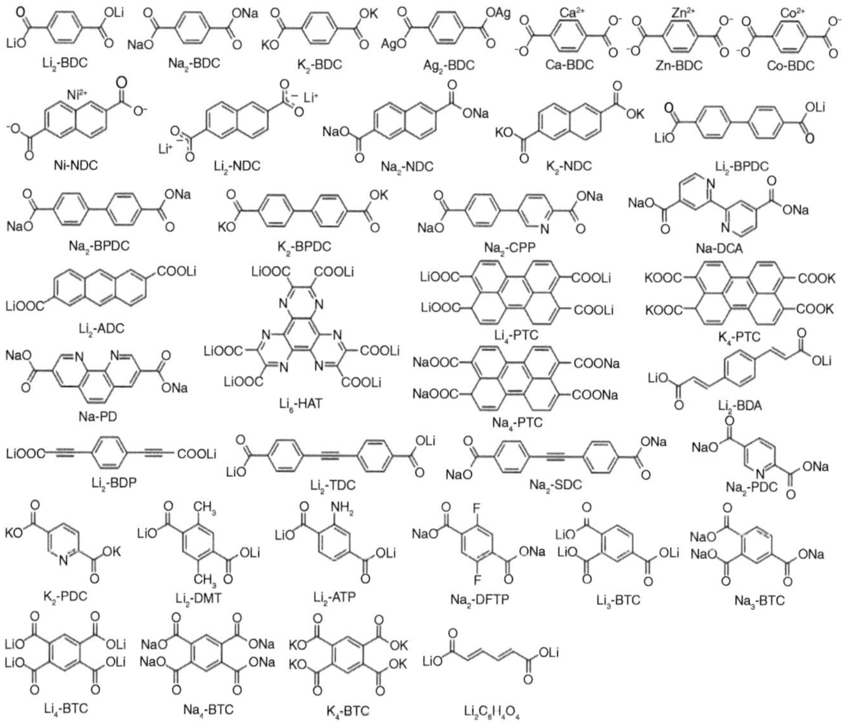

Figure 8.13 The structure of the reported conjugated dicarboxylates in rechargeable batteries.

trans–trans-muconate, $Li_2C_6H_4O_4$) as anode materials in lithium batteries [6]. Compared with aromatic dicarboxylates such as Li_2-BDC, $Li_2C_6H_4O_4$ shows a higher redox potential of 1.4 V versus Li/Li^+ and lower redox reactivity with only one e^- being possible to exchange; the difference is attributed to the conjugated system. The Li/$Li_2C_6H_4O_4$ cell delivers a discharge capacity of 125 mAh g^{-1} after 80 cycles, corresponding to a capacity retention of 73.5% (Figure 8.12a). Later, Walker et al. investigated the effect of trans/cis and length relationships on vinylic compounds [59]. They proved that the cis derivatives exhibit limited reversible reactivity compared with trans derivatives. In addition, the polarization of the cells gradually decreases as the number of vinylic groups increases (Figure 8.12b). Taking into account the practical application, more optimization of aliphatic dicarboxylates needs to be done. The structure and electrochemical performance of the reported conjugated dicarboxylates are summarized in Figure 8.13 and Table 8.1.

8.3 Schiff Bases

Schiff bases are first applied as anode materials in sodium batteries [60], the corresponding redox mechanism is shown in Figure 8.14a. Generally, one azomethine group (C=N) could react with one metal ion and one e^-, and each repeating unit

Table 8.1 Performance comparison of reported conjugated dicarboxylates in rechargeable batteries.

Materials	Metal ions	Output voltage (V)	Initial capacity (mAh g^{-1})	Capacity retention (%)
Li_2-BDC [6]	Li	0.8	300 (0.05 C)	78 (50 cycles)
Na_2-BDC [7]	Na	0.29	255 (0.1 C)	82.7 (60 cycles)
K_2-BDC [9]	K	0.6	205 (1000 mA g^{-1})	94.6 (500 cycles)
K_2-BDC [11]	Li	0.85	157 (0.1 C)	77 (100 cycles)
Ag_2-BDC [12]	Li	0.73	227 (1 C)	50 (500 cycles)
Ca-BDC [14]	Li	0.8	170 (5th)	91.2 (50 cycles)
Zn-BDC [16]	Li	0.79	272.7 (0.5 C)	66 (100 cycles)
Ca-BDC [15]	Na	~0.5	364.8 (25 mA g^{-1})	56.6 (100 cycles)
Ni-NDC [18]	Li	/	~800 (0.1 C)	~35 (500 cycles)
Co-BDC [19]	Li	/	~850 (1000 mA g^{-1})	~79.4 (100 cycles)
Li_2-NDC [21]	Li	0.88	~210 (1 C)	~57 (50 cycles)
Na_2-NDC [26]	Na	0.33	~215 (80 mA g^{-1})	65.1 (100 cycles)
Na_2-NDC [28]	Na	0.38	~260 (25 mA g^{-1})	73.1 (400 cycles)
K_2-NDC [29]	K	0.55	116 (50 mA g^{-1})	85 (200 cycles)
Li_2-BPDC [22]	Li	0.7	~210 (1 C)	87.1 (25 cycles)
Na_2-BPDC [30]	Na	0.5	~200 (0.1 C)	95 (150 cycles)
K_2-BPDC [24]	K	0.3	~205 (50 mA g^{-1})	82.9 (100 cycles)
Na_2-CPP [31]	Na	0.47	144 (5 C)	89.9 (1200 cycles)
Na-DCA [23]	Na	/	~250 (50 mA g^{-1})	49.6 (200 cycles)
Li_2-ADC [32]	Li	0.81	~190 (10 C)	75.8 (50 cycles)
Na-PD [33]	Na	/	252 (2 C)	50 (500 cycles)
Li_6-HAT [34]	Li	/	580.1 (800 mA g^{-1})	50.8 (3400 cycles)
Li_4-PTC [35]	Li	1.1	~200 (25 mA g^{-1})	98 (100 cycles)
Na_4-PTC [37]	Na	0.75	350.5 (25 mA g^{-1})	37.4 (100 cycles)
K_4-PTC [39]	K	0.92	73 (500 mA g^{-1})	68.5 (2500 cycles)
Li_2-BDA [40]	Li	1.2	200 (0.05 C)	85 (20 cycles)
Li_2-BDP [41]	Li	/	~320 (0.1 C)	~56 (100 cycles)
Li_2-TDC [42]	Li	0.65	200 (0.025 C)	90 (50 cycles)
Na_2-SDC [43]	Na	0.39	222 (10th, 50 mA g^{-1})	90 (50 cycles)
Na_2-PDC [45]	Li	/	210 (0.1 C)	95.7 (20 cycles)
Na_2-PDC [46]	Na	0.5, 0.38	270 (0.1 C)	83 (100 cycles)
K_2-PDC [47]	K	~0.6	~240 (0.2 C)	73.3 (100 cycles)
Li_2-DMT [48]	Li	0.65	163.3 (0.05 C)	98 (50 cycles)

(Continued)

Table 8.1 (Continued)

Materials	Metal ions	Output voltage (V)	Initial capacity ($mAh\,g^{-1}$)	Capacity retention (%)
Li_2-ATP [49]	Li	0.9	~200 (0.05 C)	80 (50 cycles)
Na_2-DFTP [51]	Na	0.41	204.8 ($50\,mA\,g^{-1}$)	78.6 (300 cycles)
Li_3-BDC [53]	Li	0.95	154 ($35\,mA\,g^{-1}$)	81.2 (100 cycles)
Na_3-BDC [54]	Na	0.43	300 ($25\,mA\,g^{-1}$)	52.4 (50 cycles)
Li_4-BDC [55]	Li	/	~800 ($600\,mA\,g^{-1}$)	/ (500 cycles)
Na_4-BDC [57]	Na	0.65	~80 ($2\,A\,g^{-1}$)	64.3 (1000 cycles)
K_4-BDC [58]	K	0.63, 0.31	~250 ($500\,mA\,g^{-1}$)	83 (1000 cycles)
$Li_2C_6H_4O_4$ [6]	Li	1.4	170 (0.05 C)	73.5 (80 cycles)

includes the conversion of two metal ions and two e^- during the redox process. Meanwhile, the extended conjugated structure facilitates the intermolecular interaction, thus leading to enhanced charge transport.

In 2014, Castillo-Martinez et al. reported several polymeric Schiff bases in sodium-ion batteries, which were synthesized via the condensation reaction of terephthalaldehyde with aliphatic/aromatic diamines (Figure 8.14b) [60]. The authors first compared the electrochemical performance of the polymers with aromatic or aliphatic conjugation and found that the polymers with aliphatic conjugation show more rapid capacity decay and larger polarization than those with aromatic conjugation. And they also explored the effect of the electron-donating groups ($-CH_3$, $-OCH_3$) on the benzene ring on the redox potential. The polymers substituted with $-CH_3$ group exhibited lower discharge voltage than the one with $-OCH_3$ group although the latter one has stronger electron-donating effect (Figure 8.14c). In addition, the authors found that the Schiff base with the inverse unit –CH=N–Ar–N=CH– (Ar is the benzene ring) does not show similar electrochemical activity due to the nonconjugated unit hindering the reactivity of the active center. The Na/poly[*N,N'-p*(benzylidene)phenylenediamine] cell achieves a reversible capacity of $350\,mAh\,g^{-1}$ with an optimized carbon ratio of 50 wt% KB. Later, they further reported different chain-length oligomeric-Schiff bases functionalized with carboxylate end groups (Figure 8.15a) [61]. The introduction of carboxylates not only contributes additional capacity but also improves electrochemical stability. Unexpectedly, the increased chain length does not significantly affect their electrochemical properties such as redox potential, discharge capacity, and cycling performance, which may be due to factors such as crystallinity or electronic repulsions. They (O2 and O5) all show two redox processes corresponding to the reduction/oxidation of carboxylic groups at low voltage and aromatic active unit at high voltage, respectively, as shown in Figure 8.15b. And they deliver stable performance at 0.1 C and 0.2 C rates (Figure 8.15c). Similarly, they also proved that the inverse unit exhibits lower redox activity.

Wang et al. reported a polyimide Schiff base (NBI-PI) containing an extended π-conjugated naphthalene ring as anode material in LIBs [62]. Benefiting from the

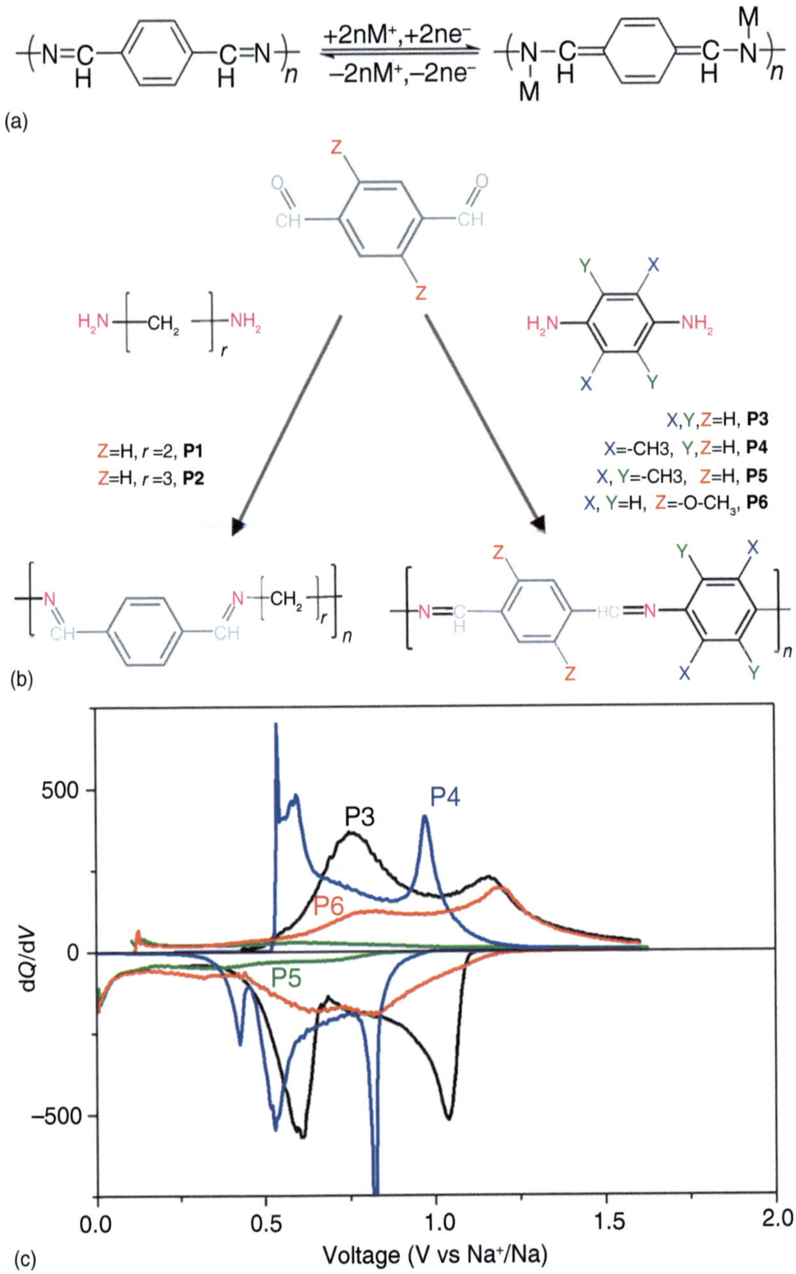

Figure 8.14 (a) The redox mechanism of Schiff bases in rechargeable batteries. (b) The synthesis of polymeric Schiff bases with different monomers. (c) Derivative curves (dQ/dV) of polymeric Schiff bases with different functional groups. Source: Reproduced with permission from Castillo-Martinez et al. [60]/John Wiley & Sons.

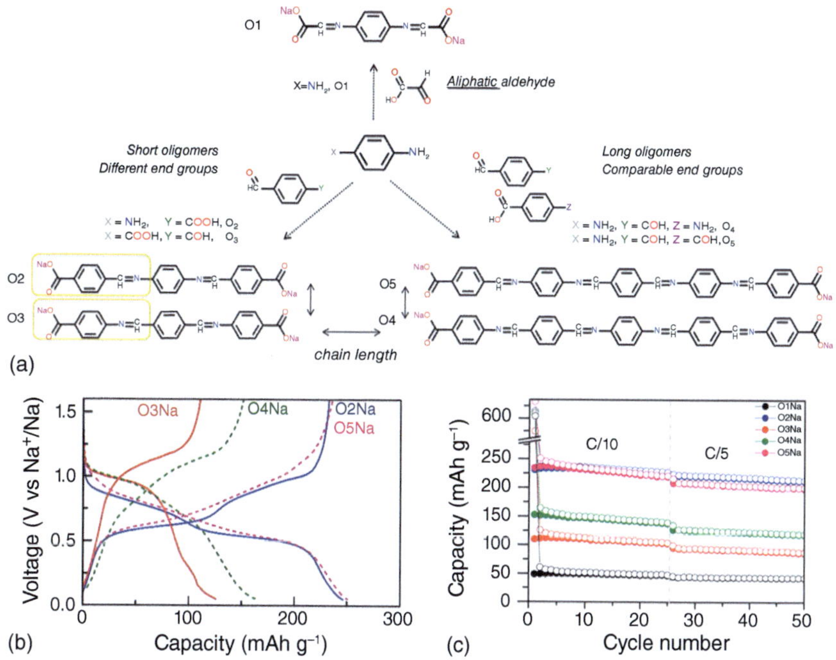

Figure 8.15 (a) The synthesis of oligomeric Schiff-bases vis the condensation reaction of di/aldehydes and aromatic amines. (b) Charge–discharge profiles and (c) cycling performance of oligomers. Source: Reproduced with permission from López-Herraiz et al. [61]/Royal Society of Chemistry.

stable polymer skeleton and high density of active sites, the Li/NBI-PI cell delivers a discharge capacity of 217 mAh g^{-1} at 2 A g^{-1}, with a capacity retention of 85.7% after 1600 cycles.

8.4 Azo Compounds

Azo compounds as a kind of novel active material were first reported by Wang and coworkers [63–67]. The redox mechanism of azo compounds is based on the breakage and reformation from nitrogen–nitrogen double bonds to nitrogen–nitrogen single bonds, which is an n-type conversion. The output potential of azo compounds is mostly at ~1.5 V versus Li/Li^+, ~1.2 V versus Na/Na^+, and ~ 1.25 V versus K/K^+, respectively; therefore, they are usually served as anode materials in rechargeable batteries.

Azobenzene as the model molecular is studied; however, it shows rapid capacity fade and low Coulombic efficiency due to the high solubility of the electrolyte [63]. 4-(Phenylazo) benzoic acid lithium salt (PBALS) with reduced solubility delivers a discharge capacity of 178 mAh g^{-1} after 100 cycles at 0.5 C rate, corresponding to a capacity retention of 80.9%. And PBALS also shows a reversible capacity of 85 mAh g^{-1} at an ultrahigh current density of 20 C rate, which is benefited from the

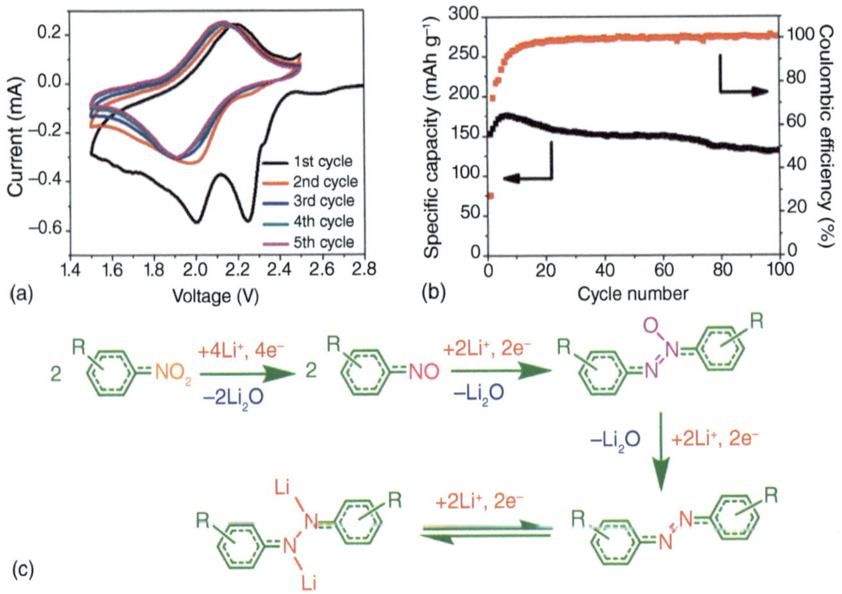

Figure 8.16 (a) CV profiles of NBALS electrode at 0.1 mV s^{-1}. (b) Cycling performance of NBALS at the current density of 0.5 C rate. (c) Reaction mechanism of nitro-based organic materials. Source: Reproduced with permission from Luo et al. [63]/John Wiley & Sons.

extended π-conjugation and strong Li-ion adsorption by nitrogen atoms in the azo group, leading to fast Li-ion diffusion. Meanwhile, they also used nitro compound (4-nitrobenzoic acid lithium salt, NBALS) as the initial active material to explore the electrochemical conversion of nitro compound to azo compound in lithium batteries. The cyclic voltammograms (CV) of NBALS are shown in Figure 8.16a, obviously, the first discharge process is different from the subsequent ones, which proved to be the irreversible reduction from the nitro compounds to azo compounds induced by the lithium ions, and such conversion reaction results in low ICE of only 27.3%. The subsequent reversible reaction happens between Li^+ and the formed azo compounds (Figure 8.16c), a discharge capacity of 131 mAh g^{-1} was obtained after 100 cycles at 0.5 C rate (1 C = 155 mA g^{-1}) with a capacity retention of 85.6 % (Figure 8.16b). Later, they directly used azobenzene-4,4′-dicarboxylic acid lithium salt (ADALS) as active materials in lithium batteries [67]. The Li/ADALS cell delivers a stable electrochemical performance over 5000 cycles at 20 C rate with a capacity retention of 89%.

Furthermore, the excellent electrochemical performances of azobenzene-4,4′-dicarboxylic acid sodium salt (ADASS) [64] and azobenzene-4,4′-dicarboxylic acid potassium salt (ADAPS) [65] are also reported by them in sodium batteries and potassium batteries, respectively. The redox mechanism of ADASS and ADAPS is similar to that of ADALS, and the corresponding CV and long cycling performance are shown in Figure 8.17.

In order to solve the shuttle effect caused by the high solubility of azo compounds, Luo et al. applied solid-state electrolyte (Li_3PS_4, LPS) in PBALS electrode

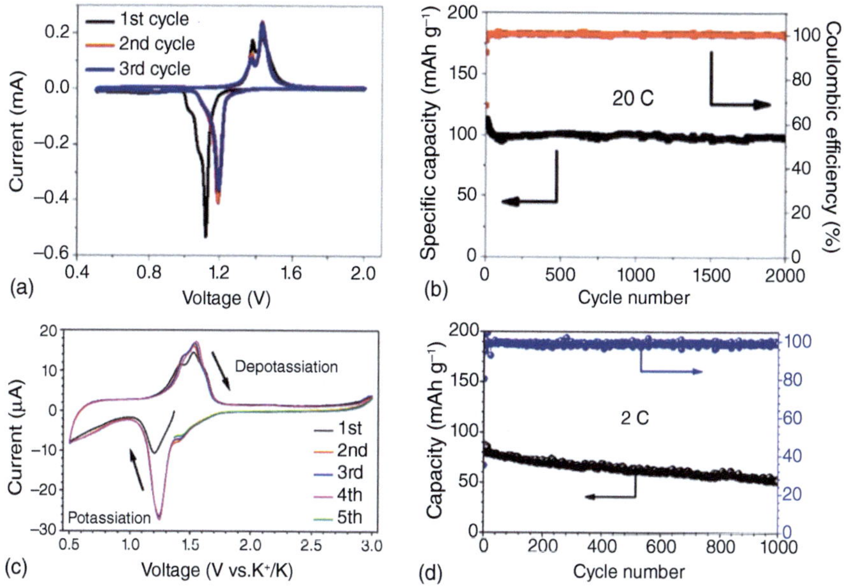

Figure 8.17 The cyclic voltage profiles of (a) ADASS and (c) ADAPS. The corresponding long cycling performance of (b) ADASS and (d) ADAPS. Source: (a and c) Reproduced with permission from Luo et al. [64]/John Wiley & Sons. (b and d) Reproduced with permission from Liang et al. [65]/John Wiley & Sons.

to assemble the all-solid-state lithium batteries (ASSLB) [66]. Interestingly, the LPS solid electrolyte could bond with PBALS through the ionic bonding between lithium ions in LPS and oxygen in PBALS, leading to a stable contact interface for low interface resistance and good electrochemical performance (Figure 8.18a). The PBALS shows one pair of redox plateaus at about 1.5 V in ASSLB (Figure 8.18b and

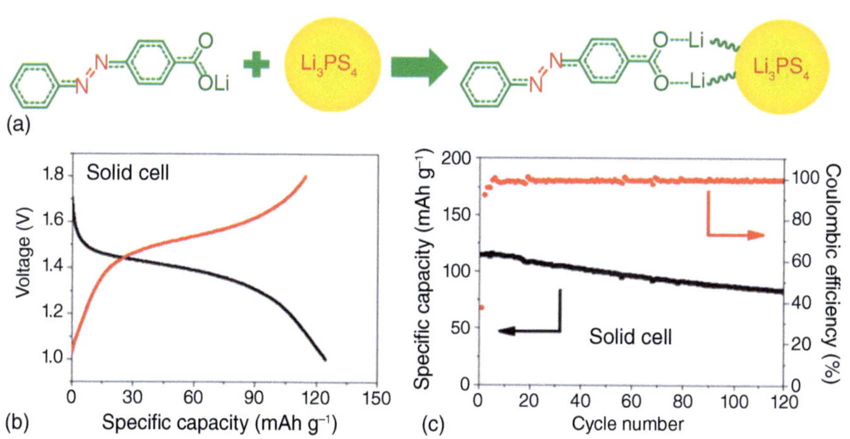

Figure 8.18 (a) Interaction between PBALS and LPS solid-state electrolyte. The discharge–charge curves (b) and cycling performance (c) of PBALS cell. Source: Reproduced with permission from Luo et al. [66]/John Wiley & Sons.

delivers a discharge capacity of 83 mAh g^{-1} after 120 cycles at a current density of 20 mA g^{-1} (Figure 8.18c). In addition, another kind of solution is to prepare soluble small-molecule azo monomers into polymers that are insoluble in the liquid electrolyte, which will be discussed in detail in the Section 8.5.

8.5 Covalent Organic Frameworks

COFs, as one kind of emerging porous material, have been widely explored in energy storage systems due to their tailorable structures, tunable chemistry, well-defined pores, and high specific surface area [68]. In addition, their low solubility in the electrolyte, fast ion transport, and high density of multi-active sites (e.g. C=N groups [69–72], C=O groups [72–75], N=N groups [73, 76], and triazine ring units [77–80] enable COFs to serve as anode materials in batteries. In particular, the proportion of the capacitance contribution in the total capacity contribution at a large rate is high in most of the COF-based anode materials, indicating their excellent kinetic processes.

Liu et al. prepared millimeter-size covalent triazine frameworks (CTFs) (Figure 8.19a) with a lamellar structure and micrometer-size few-layer 2D polymers (2DPs) via further micromechanical cleavage and liquid sonication [77]. The synthesized CTFs and 2DPs all show staggered AB stacking with a pore size of about 6 nm and aligned 1D open channels, which provides smooth and fast diffusion pathways for charge storage and transport. When applied in sodium metal batteries, the bulk CTFs and exfoliated 2DP deliver high capacities of 225 and 262 mAh g^{-1} at 0.1 A g^{-1}, and excellent rate capabilities of 67 and 119 mAh g^{-1} at 5.0 A g^{-1}, respectively. In addition, they also show impressive cycling stability, with a capacity retention of 95% after 1200 cycles at 1.0 A g^{-1}. Later, Zhang et al. introduced fluorine atoms into the CTFs to prepare 2D-layered fluorinated CTF (FCTF) (Figure 8.19b) and correspondingly exfoliated few-layered FCTF (E-FCTF) [78]. The introduction of F atoms significantly decreases the highest occupied molecular orbital (HOMO), lowest unoccupied molecular orbital (LUMO), and the band gap, which benefits improved electrical conductivity and fast ion/electron transfer. The FCTF and E-FCTF are able to be applied as anode materials in Li, Na, and K batteries, and the Li/E-FCTF cell delivers a high reversible capacity of 581 mAh g^{-1} at 2 A g^{-1} after 1000 cycles. Meanwhile, the authors also supposed that the lithium storage mechanism of E-FCTF is a multielectronic redox reaction on the triazine ring and benzene ring units based on their theoretical calculations and experimental probing. Jiang et al. prepared biphenyl-linked CTFs (CTF-2) with one more benzene ring (Figure 8.19c), which delivers an ultrahigh capacity of 1527 mAh g^{-1} at 0.1 A g^{-1} [80].

Moreover, the CTFs containing multi-redox sites are reported. Lei et al. synthesized two types of CTF-based covalent organic nanosheets (CONs) with multielectron redox groups such as triazine ring, benzene ring, piperazine ring, C=N, and –NH– (Figure 8.19d and e, E-CIN-1 and E-SNW-1), and composited them with carbon nanotubes (CNTs) for higher conductivity [69]. The E-CIN-1/CNTs and E-SNW-1/CNTs composites deliver reversible capacities of 1005 and 920 mAh g^{-1} at

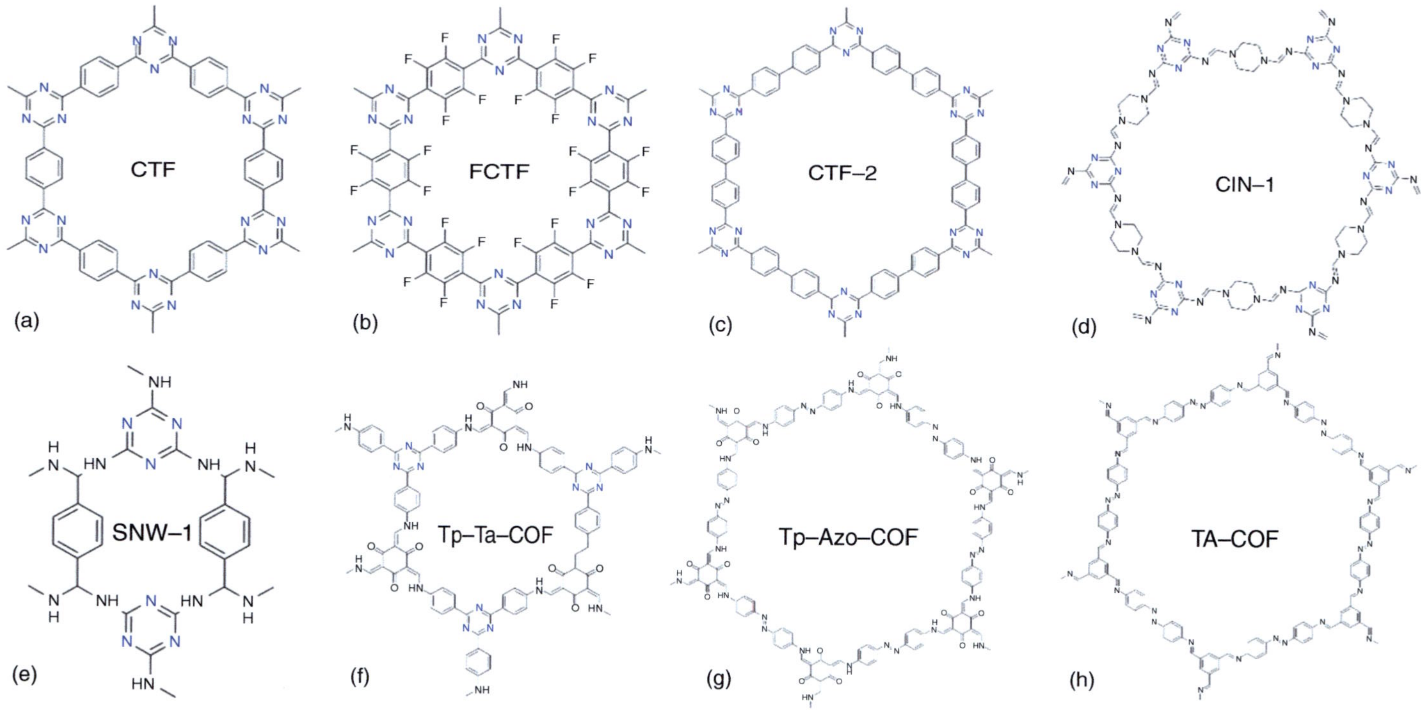

Figure 8.19 The structure of the reported COFs as anode materials in rechargeable batteries. (a) CTF, (b) FCTF, (c) CTF-2, (d) CIN-1, (e) SNW-1, (f) Tp-Ta-COF, (g) Tp-Azo-COF, and (h) TA-COF.

100 mA g^{-1} after 250 cycles, respectively. Zhao et al. reported a C=O-decorated CTF (Figure 8.19f), which is denoted as Tp-Ta-COF [72]. The Li/Tp-Ta-COF cell delivers a capacity of 413 mAh g^{-1} at 200 mA g^{-1} after 800 cycles.

As mentioned in Section 8.4, COFs containing azo groups retain the redox activity of the azo group while reducing their solubility in the electrolyte. Zhao et al. prepared a N=N and C=O chemically modified COF (denoted as Tp-Azo-COF) (Figure 8.19g) via the condensation reaction between 4,4′-azodianiline (Azo) and 1,3,5-triformylphloroglucinol (Tp) in 1,4-dioxane solvent [73]. The electronic conductivity is remarkably improved by the strong π–π interactions among the 2D COF layers. Meanwhile, the carbonyl and azo units play an important role in maintaining the structure stability of Tp-Azo-COF during the redox process, achieving an excellent discharge capacity of 305.97 mAh g^{-1} at 1000 mA g^{-1} after 3000 cycles in lithium batteries. Similarly, Tong et al. prepared a COF containing N=N and C=N active centers (TA-COF) (Figure 8.19h) [76]. Benefiting from the large conjugate structure, rich active sites, and good structural stability, the Li/TA-COF cell delivers a discharge capacity of 227 mAh g^{-1} at 5 A g^{-1} after 2000 cycles.

8.6 Thiophene Compounds

Poly(thiophene) compounds have been employed as anode materials based on the n-doped conjugated structure with reversible redox behavior. However, the poor rate performance, low redox activity, and processing difficulty of the poly(thiophene) compounds inhibit their further development, which can be optimized through structural regulation.

In 2012, Zhu et al. prepared polybithiophene (PBT) via the chemical oxidative method and ball-milling PBT with carbon fibers to ensure PBT is electrochemically workable [81]. The PBT/C composites deliver a redox capacity of ~850 mAh g^{-1} at 40 mA g^{-1} with a potential plateau at 1.25 V versus Li/Li^+. Zhang et al. systematically studied the molecular structure–performance relationship of poly(thiophene) via the comparison of several poly(thiophene) with different thiophene contents and specific surface area and found that these two factors significantly affected the electrochemical performance of poly(thiophene) [82]. The authors prepared two all-thiophene-based conjugated polymers (poly(thiophene) (PT) and poly(3,3′-bithiophene) (P33DT)) via an $FeCl_3$-catalyzed oxidative coupling reaction (Figure 8.20a). PT shows a linear polymer structure, while P33DT shows a crosslinked polymer network due to the difference in the monomer structure and macromolecular aggregation. Meanwhile, P33DT has a larger surface area (696 m^2 g^{-1}) and pore volume (0.53 cm^3 g^{-1}) than that of PT (13 m^2 g^{-1}, 0.07 cm^3 g^{-1}), which enables a larger contact area with electrolytes and shorter ion diffusion distance. The discharge–charge curves of PT and P33DT are shown in Figure 8.20b and c, it is obvious that they have a similar redox process, but P33DT has a higher reversible capacity. The Li/P33DT cell delivers a discharge capacity of 663 mAh g^{-1} after 1000 cycles at 500 mA g^{-1}, corresponding to a capacity retention of 79.9% (Figure 8.20d).

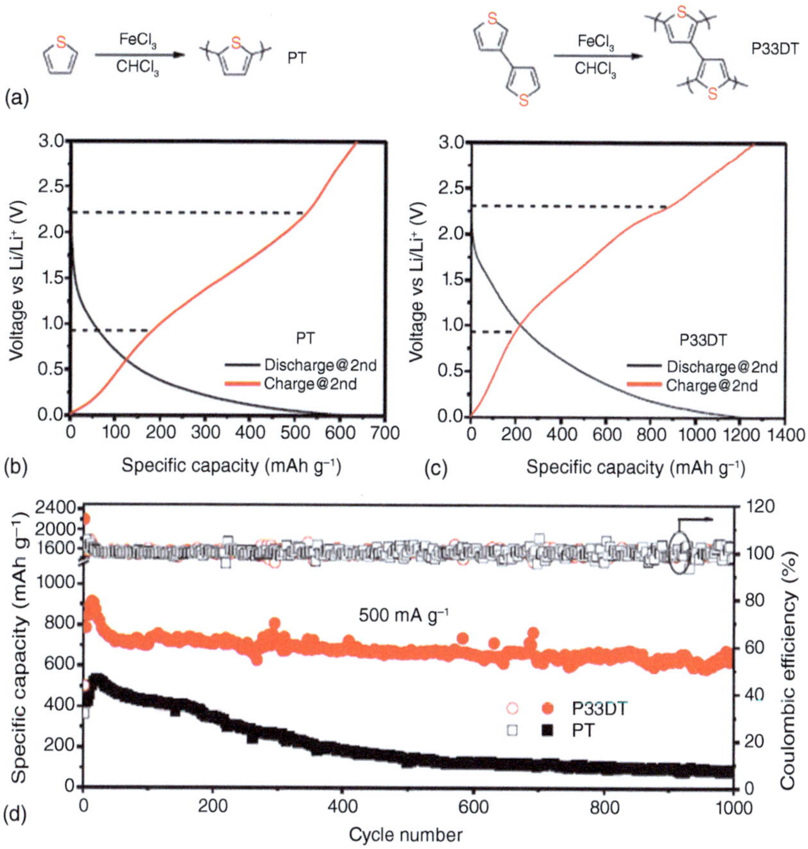

Figure 8.20 (a) Synthetic routes of PT and P33DT. Discharge-charge profiles of (b) PT and (c) P33DT at 45 mA g^{-1}. (d) Cycling performance of P33DT and PT at 500 mA g^{-1}. Source: Reproduced with permission from Zhang et al. [82]/John Wiley & Sons.

Li et al. reported a 1D poly(3-butylthiophene) (P3BT), which is composited with CNTs and serves as anode materials in both lithium and potassium batteries [83, 84]. They reveal that CNTs play an important role in such structure design to improve the electrochemical performance of P3BT. CNTs as a conductive skeleton can enlarge the electroactive surface area, facilitate electron transfer, and shorten the ion diffusion path. In addition, the π–π interaction between CNTs and P3BT not only decreases the energy gap and electron affinity due to the introduction of butyl side chains but also stabilizes the discharged and charged states of P3BT for improving the cycling stability. The P3BT/CNTs composites deliver stable performance over 15 000 cycles and achieve a reversible capacity of 450 mAh g^{-1} at 50000 mA g^{-1}.

8.7 Summary and Outlook

Organic anodes have become one of the intriguing and hot research topic due to the increasing need for renewable energy storage. This chapter provides a comprehensive review of the development of organic anodes in rechargeable metal batteries, which are divided into conjugated carboxylates, Schiff bases, azo compounds, COFs, and thiophene compounds according to the differences in the redox types and structures. Compared with traditional inorganic anodes, organic anodes have shown great advantages, such as multi-redox active sites, high theoretical specific capacity, tunable molecular structure and electrochemical performance, and environmental friendliness. However, they still face several challenges in poor electronic conductivity, low gravimetric and volumetric energy density, and electrode dissolution in the electrolyte.

When considering practical applications, future research on organic anodes should be focused on the following aspects: (i) combining with carbon materials to enhance conductivity and rate capability; (ii) taking full advantage of the structural tunability of organic molecules and introducing electron-donating functional groups to reduce the redox potential and improve the output voltage of the full cell; (iii) introducing multiple redox sites to increase theoretical capacity; (iv) adopting polymerization strategy or solid-state electrolyte to inhibit the dissolution of charged/discharged products; (v) developing new synthetic methods to meet the needs of large-scale preparation; (vi) applying organic anodes in other metal batteries and aqueous battery systems. In conclusion, the research on organic anodes is still in the infancy stage; more comprehensive and in depth explorations are needed to realize their potential.

Acknowledgments

The work was supported by the National Natural Science Foundation of China (Grant Nos. U2004214 and 21975225) and the National Key Research and Development Program of China (2022YFB2402200).

References

1 Zhang, H., Yang, Y., Ren, D. et al. (2021). *Energy Storage Mater.* 36: 147–170.
2 Cheng, X.-B., Zhang, R., Zhao, C.-Z., and Zhang, Q. (2017). *Chem. Rev.* 117: 10403–10473.
3 Ghazi, Z.A., Sun, Z., Sun, C. et al. (2019). *Small* 15: e1900687.
4 Zhang, L., Zhu, C., Yu, S. et al. (2022). *J. Energy Chem.* 66: 260–294.
5 Lakraychi, A.E., Dolhem, F., Vlad, A., and Becuwe, M. (2021). *Adv. Energy Mater.* 11: 2101562.
6 Armand, M., Grugeon, S., Vezin, H. et al. (2009). *Nat. Mater.* 8: 120–125.
7 Zhao, L., Zhao, J., Hu, Y.-S. et al. (2012). *Adv. Energy Mater.* 2: 962–965.
8 Zhang, S.-W., Cao, T., Lv, W. et al. (2020). *Nano Energy* 77: 105203.
9 Lei K., Li F., Mu C., Wang J., Zhao Q., Chen C. and Chen J.Energy Environ. 10, 552–557.
10 Wang, X., Han, K., Wang, C. et al. (2018). *Chem. Commun.* 54: 11029–11032.

11 Deng, Q., Fan, C., Wang, L. et al. (2016). *Electrochim. Acta* 222: 1086–1093.
12 Xue, J., Fan, C., Deng, Q. et al. (2016). *Electrochim. Acta* 219: 418–424.
13 Wang, L., Zhang, H., Mou, C. et al. (2015). *Nano Res.* 8: 523–532.
14 Wang, L., Mou, C., Wu, B. et al. (2016). *Electrochim. Acta* 196: 118–124.
15 He, D., Liu, Y., Fan, Y. et al. (2022). *Chem. Commun.* 58: 4048–4051.
16 Wang, L., Zou, J., Chen, S. et al. (2017). *Electrochim. Acta* 235: 304–310.
17 Lee, H.H., Park, Y., Kim, S.H. et al. (2015). *Adv. Funct. Mater.* 25: 4859–4866.
18 Kim, S.-H., Lee, H.H., Kim, J.-M. et al. (2019). *Energy Storage Mater.* 19: 130–136.
19 Jiang, H., Zhao, S., Ma, X. et al. (2019). *J. Power Sources* 426: 23–32.
20 Fan, C., Zhao, M., Li, C. et al. (2017). *Electrochim. Acta* 253: 333–338.
21 Fedele, L., Sauvage, F.E.E., Bois, J. et al. (2014). *J. Electrochem. Soc.* 161: A46–A52.
22 Fédèle, L., Sauvage, F., Gottis, S. et al. (2017). *Chem. Mater.* 29: 546–554.
23 Holguin, K., Qin, K., Kamphaus, E.P. et al. (2022). *J. Power Sources* 533: 231383.
24 Li, C., Deng, Q., Tan, H. et al. (2017). *ACS Appl. Mater. Interfaces* 9: 27414–27420.
25 Yasuda, T. and Ogihara, N. (2014). *Chem. Commun.* 50: 11565–11567.
26 Medabalmi, V., Kuanr, N., and Ramanujam, K. (2018). *J. Electrochem. Soc.* 165: A175–A180.
27 Cabanero, J.M. Jr., Pimenta, V., Cannon, K.C. et al. (2019). *ChemSusChem* 12: 4522–4528.
28 Desai, A.V., Rainer, D.N., Pramanik, A. et al. (2021). *Small Methods* 5: e2101016.
29 Li, C., Xue, J., Ma, J., and Li, J. (2019). *J. Electrochem. Soc.* 166: A5221–A5225.
30 Choi, A., Kim, Y.K., Kim, T.K. et al. (2014). *J. Mater. Chem. A* 2: 14986–14993.
31 Jia, K., Zhu, L., and Wu, F. (2021). *ChemSusChem* 14: 3124–3130.
32 Fedele, L., Ouari, O., Sauvage, F. et al. (2020). *ChemSusChem* 13: 2321–2327.
33 Jia, K., Liu, H., Huang, G. et al. (2022). *J. Mater. Chem. A* 10: 14917–14922.
34 Xia, S.-B., Liu, T., Huang, W.-J. et al. (2020). *J. Energy Chem.* 51: 303–311.
35 Zhao, R.R., Cao, Y.L., Ai, X.P., and Yang, H.X. (2013). *J. Electroanal. Chem.* 688: 93–97.
36 Fédèle, L., Sauvage, F., and Bécuwe, M. (2014). *J. Mater. Chem. A* 2: 18225–18228.
37 Wang, H.-G., Yuan, S., Si, Z., and Zhang, X.-B. (2015). *Energy Environ. Sci.* 8: 3160–3165.
38 Li, G., Wang, J., Chu, J. et al. (2022). *Energy Storage Mater.* 47: 1–12.
39 Wang, C., Tang, W., Yao, Z. et al. (2019). *Chem. Commun.* 55: 1801–1804.
40 Renault, S., Brandell, D., Gustafsson, T., and Edström, K. (2013). *Chem. Commun.* 49: 1945–1947.
41 Renault, S., Oltean, V.A., Araujo, C.M. et al. (2016). *Chem. Mater.* 28: 1920–1926.
42 Walker, W., Grugeon, S., Vezin, H. et al. (2011). *J. Mater. Chem.* 21: 1615–1620.
43 Wang, C., Xu, Y., Fang, Y. et al. (2015). *J. Am. Chem. Soc.* 137: 3124–3130.
44 Lee, H.H., Park, Y., Shin, K.H. et al. (2014). *ACS Appl. Mater. Interfaces* 6: 19118–19126.
45 Wu, X., Ma, J., Hu, Y.-S. et al. (2014). *J. Energy Chem.* 23: 269–273.
46 Padhy, H., Chen, Y., Lüder, J. et al. (2018). *Adv. Energy Mater.* 8: 1701572.

47 Deng, Q., Pei, J., Fan, C. et al. (2017). *Nano Energy* 33: 350–355.

48 Lakraychi, A.E., Dolhem, F., Djedaïni-Pilard, F. et al. (2017). *J. Power Sources* 359: 198–204.

49 Renault, S., Oltean, V.A., Ebadi, M. et al. (2017). *Solid State Ionics* 307: 1–5.

50 Lakraychi, A.E., Dolhem, F., Djedaïni-Pilard, F., and Becuwe, M. (2018). *Electrochem. Commun.* 93: 71–75.

51 Huang, J., Callender, K.I.E., Qin, K. et al. (2022). *ACS Appl. Mater. Interfaces* 14: 40784–40792.

52 Park, Y., Shin, D.S., Woo, S.H. et al. (2012). *Adv. Mater.* 24: 3562–3567.

53 Maiti, S., Pramanik, A., Dhawa, T., and Mahanty, S. (2017). *Mater. Lett.* 209: 613–617.

54 Gu, T., Gao, S., Wang, J. et al. (2020). *ChemElectroChem* 7: 3517–3521.

55 Guo, R., Wang, Y., Heng, S. et al. (2019). *J. Power Sources* 436: 226848.

56 Cahyadi, H.S., William, W., Verma, D. et al. (2018). *ACS Appl. Mater. Interfaces* 10: 17183–17194.

57 Qin, K., Holguin, K., Mohammadiroudbari, M., and Luo, C. (2021). *Chem. Commun.* 57: 2360–2363.

58 Pan, Q., Zheng, Y., Tong, Z. et al. (2021). *Angew. Chem. Int. Ed.* 60: 11835–11840.

59 Walker, W., Grugeon, S., Vezin, H. et al. (2010). *Electrochem. Commun.* 12: 1348–1351.

60 Castillo-Martinez, E., Carretero-Gonzalez, J., and Armand, M. (2014). *Angew. Chem. Int. Ed.* 53: 5341–5345.

61 López-Herraiz, M., Castillo-Martínez, E., Carretero-González, J. et al. (2015). *Energy Environ. Sci.* 8: 3233–3241.

62 Wang, J., Yao, H., Du, C., and Guan, S. (2021). *J. Power Sources* 482: 228931.

63 Luo, C., Ji, X., Hou, S. et al. (2018). *Adv. Mater.* 30: e1706498.

64 Luo, C., Xu, G.L., Ji, X. et al. (2018). *Angew. Chem. Int. Ed.* 57: 2879–2883.

65 Liang, Y., Luo, C., Wang, F. et al. (2018). *Adv. Energy Mater.* 9: 1802986.

66 Luo, C., Ji, X., Chen, J. et al. (2018). *Angew. Chem. Int. Ed.* 57: 8567–8571.

67 Luo, C., Borodin, O., Ji, X. et al. (2018). *Proc. Natl. Acad. Sci. U.S.A.* 115: 2004–2009.

68 Zhu, D., Xu, G., Barnes, M. et al. (2021). *Adv. Funct. Mater.* 31: 2100505.

69 Lei, Z., Chen, X., Sun, W. et al. (2019). *Adv. Energy Mater.* 9: 1801010.

70 Patra, B.C., Das, S.K., Ghosh, A. et al. (2018). *J. Mater. Chem. A* 6: 16655–16663.

71 Wolfson, E.R., Schkeryantz, L., Moscarello, E.M. et al. (2021). *ACS Appl. Mater. Interfaces* 13: 41628–41636.

72 Zhao, G., Sun, Y., Yang, Y. et al. (2022). *EcoMat* 4: e212221.

73 Zhao, G., Zhang, Y., Gao, Z. et al. (2020). *ACS Energy Lett.* 5: 1022–1031.

74 Zhai, L., Li, G., Yang, X. et al. (2021). *Adv. Funct. Mater.* 32: 2108798.

75 Luo, X.X., Li, W.H., Liang, H.J. et al. (2022). *Angew. Chem., Int. Ed.* 61: e202117661.

76 Tong, Y., Sun, Z., Wang, J. et al. (2022). *SmartMat* 3: 685–694.

77 Liu, J., Lyu, P., Zhang, Y. et al. (2018). *Adv. Mater.* 30: 1705401.

78 Zhang, H., Sun, W., Chen, X., and Wang, Y. (2019). *ACS Nano* 13: 14252–14261.

79 Jiang, F., Wang, Y., Qiu, T. et al. (2021). *ACS Appl. Mater. Interfaces* 13: 48818–48827.

80 Jiang, F., Wang, Y., Qiu, T. et al. (2022). *J. Power Sources* 523: 231041.

81 Zhu, L., Niu, Y., Cao, Y. et al. (2012). *Electrochim. Acta* 78: 27–31.

82 Zhang, C., He, Y., Mu, P. et al. (2018). *Adv. Funct. Mater.* 28: 1705432.

83 Li, T., Wang, L., and Li, J. (2022). *Electrochim. Acta* 408: 139947.

84 Li, T., Lu, C., Li, J. et al. (2022). *Appl. Surf. Sci.* 598: 153862.

9

All-Organic Batteries

9.1 Introduction

As mentioned in the preceding chapters, scholars have explored a variety of organic molecules as organic electrode materials over the years. To date, only inorganic materials have been used as electrode materials for commercial batteries, especially the scarce, expensive, and energy-consuming 3d transition metals [1]. In contrast, organic materials have received more attention for their low-cost and more environmentally friendly advantages because they consist of naturally abundant elements (i.e. C, H, O, N, or S) and are more easily recycled [2]. In addition, the well-established principles of organic chemistry provide a high degree of structural designability for the synthesis of organic electrode materials, making various battery or electrode configurations possible.

In this chapter, the development of all-organic batteries is summarized. A systematic overview of conventional all-organic batteries based on protons or metal ions as charge carriers and the emerging all-organic redox flow battery (ORFB) systems in recent years is presented, respectively. All-organic proton batteries, consisting of a proton storage cathode and anode, are emerging battery systems. Protons are the smallest ions with fast diffusion kinetics that ensure the construction of a battery system that meets the requirements. In addition, protons can be covalently bound to a large number of organic functional groups in organic molecules. Metal-based all-organic batteries using metal ions, such as Li, Na, and K, as charge carriers abandon the use of metal electrodes. As a result, the disadvantages of conventional metal electrodes, such as their environmental unfriendliness and unsustainability, can be improved, and the hazards of high-risk metal electrodes can be eliminated. The all-organic electrode material also eases some concerns about low yield and high price. Therefore, the new concept of an all-organic battery, i.e. using organic molecules or polymers as anodes and cathodes, is ideal for all-organic battery systems. All-ORFBs are a special type of all-organic battery that have received a great deal of attention in recent years. The reactions of organic materials in the electrolyte are typically controlled by the charge states of carbon, nitrogen,

Rechargeable Organic Batteries: Materials, Mechanisms, and Prospects, First Edition.
Yongzhu Fu, Xiang Li, Shuai Tang, and Wei Guo.

oxygen, or sulfur atoms. The operation of ORFBs is dependent on the chemical and electrochemical reactions occurring in the electrolyte while offering the advantages of low cost, effective biodegradability, minimal environmental footprint, and highly tunable chemical and physical properties.

The goal of this chapter is to provide a comprehensive overview of the development of all-organic batteries. First, different types of battery systems and electrode material systems will be briefly introduced, and the origin and development of all-organic batteries will be summarized. Second, three types of organic battery structures are described: proton batteries, metal ion batteries, and redox flow batteries. In each section, the various cell structures and developments reported in the literature are discussed in detail.

9.2 Traditional Batteries

Unlike metal batteries and metal–organic batteries, all-organic batteries utilize organic molecules as both positive and negative electrodes. By replacing the metal electrode with organic molecules, the all-organic battery can resolve the instability issue caused by metal. Organic redox-active compounds are promising candidates for both positive and negative electrodes of all-organic batteries due to their abundant resources, sustainability, eco-friendliness, and structural controllability [3]. The selection of charge carriers in all-organic batteries typically depends on the physiochemical characteristics of the host materials [3]. All-organic batteries employ metal ions as charge carriers but do not generate metal elements on the negative side. Protons, ammonium ions, anions, and other types of ions may also serve as charge carriers, broadening the types of batteries based on different conducted ions. In this section, we divide all-organic batteries into two categories: proton batteries and metal-ion batteries, based on the difference in the charge carrier ions. The developments of each type of all-organic battery are summarized and explained in detail.

9.2.1 Cell Configuration

Traditional all-organic batteries can be grouped into three-cell configurations, depending on the organic redox molecules and charge carriers. When a single cation or anion is used as the charge carrier, the batteries can be categorized as rocking-chair all-organic batteries [4], where the charged species from the electrolyte are involved in the redox process of both the cathode and anode. The most well-known examples of cationic rocking-chair all-organic batteries are proton and lithium-ion batteries. Cationic all-organic batteries employ n-type organic molecules or polymers as the positive and negative electrodes, as shown in Figure 9.1. Typically, protons or alkali metal ions act as charge carriers in a cationic rocking-chair cell. In an anionic rocking-chair all-organic battery, the electrode comprises p-type organic molecules or polymers. The last type is a

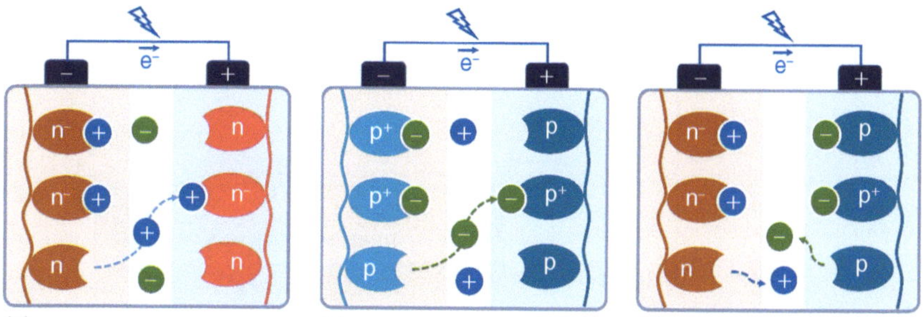

Figure 9.1 Cell configurations of all-organic batteries based on n-type and p-type organic redox couples: (a) Cation-based all-organic battery; (b) Anion-based all-organic battery; (c) Dual-ion-based all-organic battery.

dual-ion cell, in which both cations and anions serve as charge carriers. The anode usually consists of a cationic species undergoing the redox reaction, while the cathode has a p-type material with an anionic species to balance the redox reaction.

Over the past three decades, organic molecules and polymers have garnered significant attention in the energy field owing to the rapid development of organic chemistry. The potential for employing organic components in energy storage systems has been explored to meet the demand for new battery systems [5]. Compared to metal-based batteries, the development of safer, more affordable, and eco-friendly battery systems is imperative, with the requirements of non-toxicity, flexibility, ease of processing, fast charge/discharge, and long life. In this regard, all-organic batteries offer a series of advantages, such as abundant and environmentally friendly precursors and easy tunability. In Sections 9.2.2–9.2.4, we will present a comprehensive review of the different types of all-organic batteries. Firstly, all-organic proton batteries based on various prototypes of molecules as proton ion carriers will be discussed. Then, we will investigate all-organic batteries that utilize metal ions as charge carriers. Finally, the dual-ion all-organic batteries will be discussed.

9.2.2 Proton Batteries

Proton batteries utilize hydrogen ions as charge-balancing agents that are smaller and lighter than Na^+ and Li^+ and bind covalently to anionic organic functional groups. Due to their fast proton diffusion kinetics, all-organic proton batteries with proton storage cathodes and anodes are emerging as high-power energy storage devices with great potential. The first archetype all-organic system to use protons as charge carriers was reported by Surville et al. in 1968 (Figure 9.2a) [6]. Poly-aniline and poly-indole are typically used as cathode and anode materials in all-organic proton batteries with a sulfuric acid electrolyte,

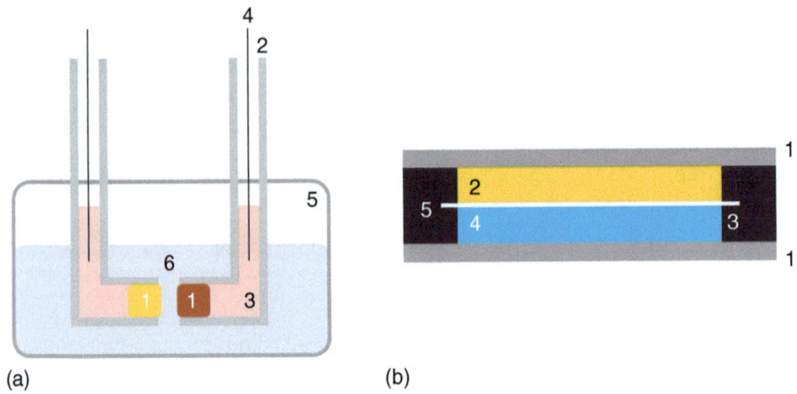

Figure 9.2 (a) Cell unit made of two electrodes of polyaniline. 1, Polyaniline pellets with different redox states; 2, Pyrex cell; 3, Mercury electrode serving as contact; 4, Platinum wire; 5, Outer jar; 6, Sulfuric acid. (b) Cell design of conducting polymers as active material. 1, collector; 2, poly(5-nitroindole); 3, separator; 4, polyaniline; 5, gasket.

exhibiting a fairly flat discharge voltage plateau of 1.1 V [7, 8]. A specific discharge capacity of 79 mAh g^{-1} is achieved at a current density of 1 mA cm^{-2}, which is equivalent to 94% of the theoretical capacity of poly(5-nitroindole). The proton battery has demonstrated an ultra-long lifespan with 80% capacity retention after 32 000 cycles.

Most organic molecules have the unique advantage of structural tunability, making them well-suited for achieving H^+ uptake/removal. Additionally, the voltage of all-organic proton batteries can be effectively regulated by adjusting the LUMO energy levels through the incorporation of desired electron-donating or -withdrawing groups into their structures or skeletons [3, 9]. Normally, all-organic proton batteries involve two different active organics, such as two types of quinones with separable potential [10, 11] or imide with TEMPO [12]. Alternatively, it is possible to design one organic molecule that undergoes two distinct redox couples with a significant LUMO energy gap for H^+ uptake/removal to directly serve as both electrodes of symmetric all-organic proton batteries as the conjugated materials (Figure 9.2b).

9.2.2.1 Two Different Molecules for Anode and Cathode

The most extensively studied proton battery systems are all-organic batteries with quinone compounds as electrode materials. Two different quinones with separable potentials are employed as the positive and negative electrodes and grafted onto the polymer chain as stable and insoluble electrodes, which can enhance the battery's stability. Additionally, ionic liquids, organic acids, and other materials that can provide protons are used as proton donor/receptor materials. The protons shuttle back and forth between the positive and negative electrodes during battery operation, which is also known as rocking-chair proton batteries.

Backbone	Bridging ligand	Electrode materials	References
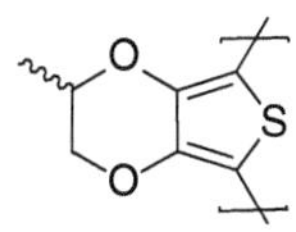	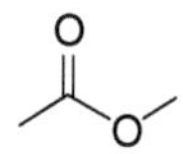	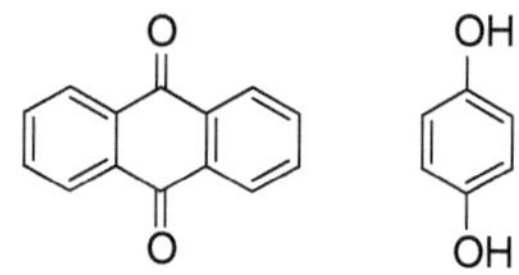	[10]
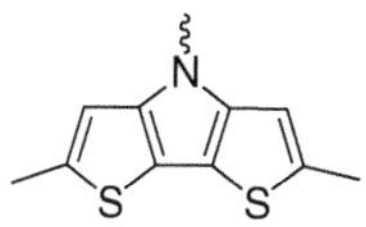	—	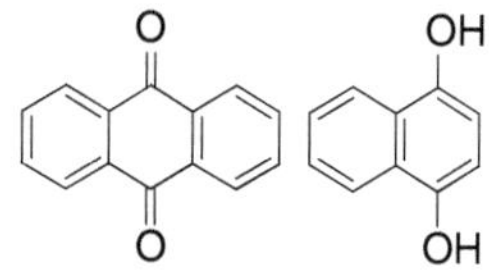	[11]
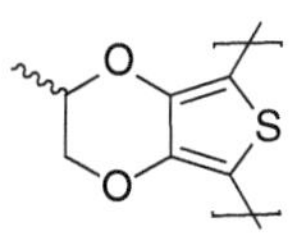	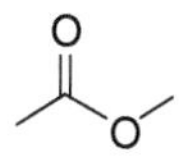	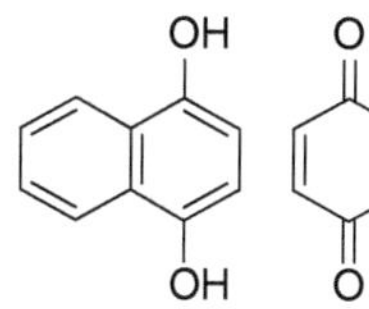	[13]
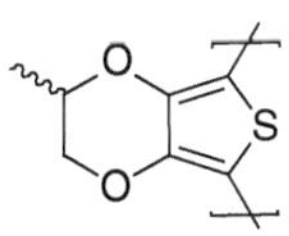	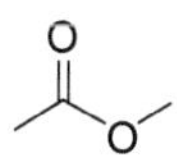	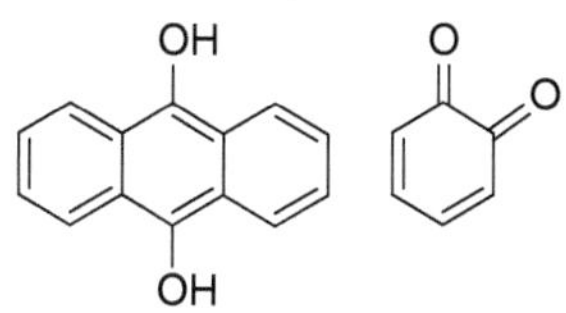	[14]
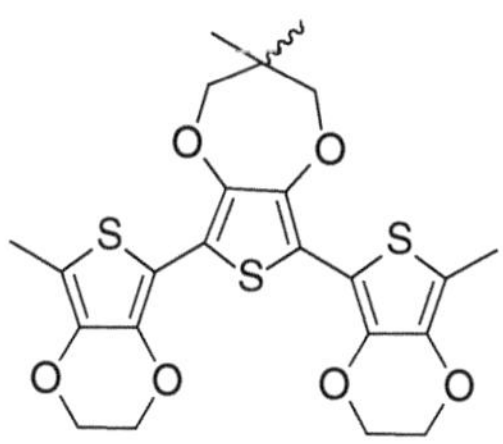	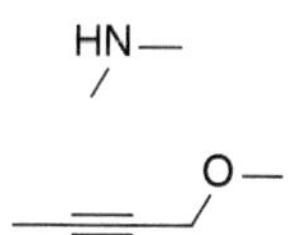	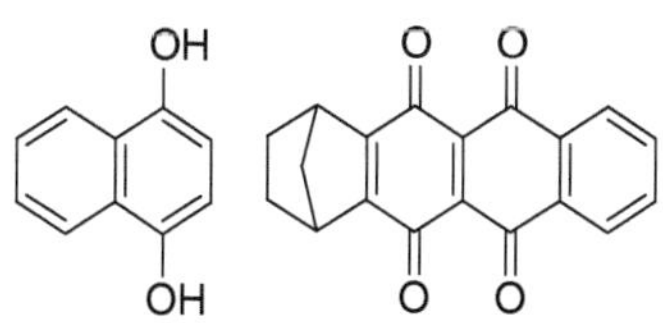	[15]

Sjödin and colleagues reported an all-organic battery utilizing poly(3,4-ethylenedioxythiophene) (PEDOT) functionalized with anthraquinone (PEDOT-AQ) and benzoquinone (PEDOT-BQ) pendant groups as the negative and positive electrodes, respectively [10]. The slurry electrolyte comprises organic acids and bases as an ionic liquid-type electrolyte as a proton donor material, allowing for the $2\,e^-/2\,H^+$ redox reactions of PEDOT-AQ and -BQ, as depicted in Figure 9.3. The proton battery cell corresponds to ~76.5% of the theoretical specific capacity of the materials, exhibiting excellent reversibility for at least 1000 cycles. Nevertheless, the battery has a low average cell potential of 0.5 V. Since the AQ and BQ moieties are located outside of the polymer backbone, there is design space both below and above the AQ and BQ redox reactions that can be utilized to increase cell potential or stability via substitution. For instance, electron-withdrawing substituents could be utilized to increase cell potential above 1 V. In their conception, the electrolyte system should be developed to allow for varying the donor/acceptor strength of the electrolyte through substitution, enabling the formal potential of redox groups such as AQ, BQ, or others to be shifted. Additionally, any polymer backbone and

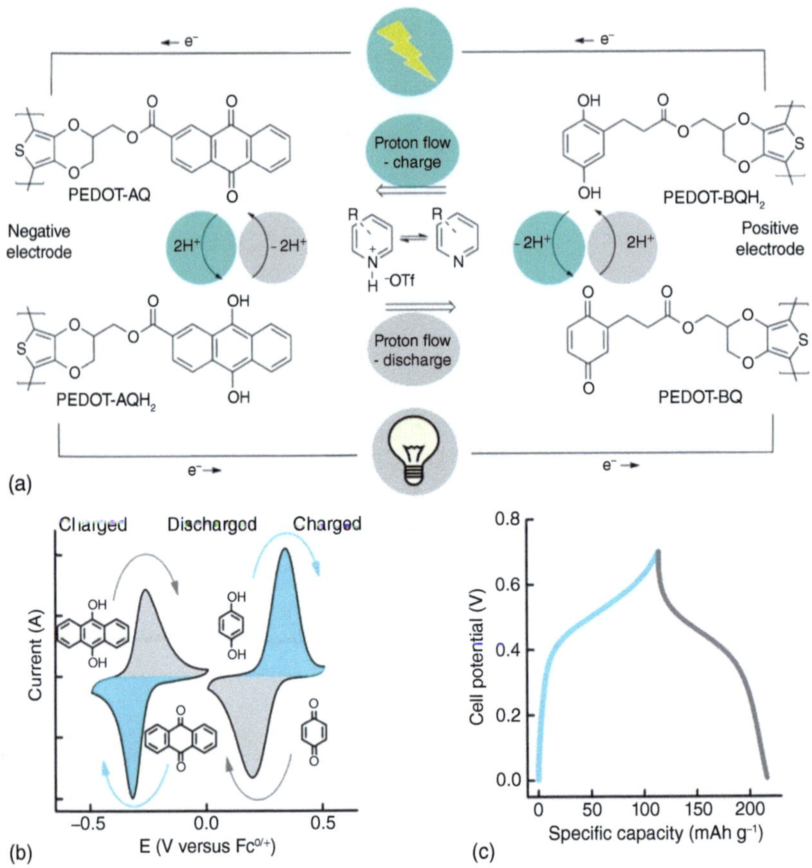

Figure 9.3 (a) The chemical structures of the negative electrode material PEDOT-AQ (G1) and positive electrode PEDOT-BQH$_2$ (G2) and their corresponding redox reaction mechanism, as proposed by Sjödin and coworkers. (b) Three-electrode voltammograms of PEDOT-AQ and PEDOT-BQ. (c) The corresponding charging and discharging curves of the all-organic battery. Source: Reproduced with permission from Emanuelsson et al. [10]/American Chemical Society.

capacity-carrying group combination are feasible. The combination of these factors allows for a vast design space and a high degree of tailoring of organic electrode materials based on proton transport, enabling the identification of the most stable combination of capacity units for targeted applications and maximum battery potential.

They recently proposed a concept for a sub-zero-temperature aqueous proton battery utilizing a conducting redox polymer as both the positive and negative electrodes [16]. Similar to their previous work [10], they employed two active molecules, naphthoquinone and benzoquinone, as the electrodes, coupled with conductive polymer skeletons. The reported thiophene-based structure was processed in solution and then deposited and polymerized in the solid state to form a conductive polymer without any additional additives. A rocking-chair all-organic proton battery built with these materials in an acidic aqueous condition exhibited good cycling stability, with 85% capacity retention after 500 cycles, retaining its favorable features even at −24 °C. Additive-free all-organic aqueous batteries serve as proof-of-concept batteries. The rocking-chair battery utilizes protons as cycling ions, demonstrating a fast-charging process in just a few seconds when using the constant-voltage method, which is advantageous for the stability of the organic electrode compared to constant-current charging. Moreover, the aqueous electrolyte offers a new, safe, affordable, and environmentally friendly energy storage system.

9.2.2.2 Anchoring Type All-Organic Batteries

In contrast to using integral materials with the polymer chain as the backbone for proton batteries, some researchers also employ the adsorption capacity of host materials to prepare electrodes. Xu et al. introduced a new approach toward an all-organic battery utilizing two quinone molecules adsorbed onto porous carbon electrodes to function as the negative and positive electrodes through one-step solution-phase adsorption [17], as depicted in Figure 9.4. Honma et al. demonstrated a proton battery with a high energy density of 30.6 Wh kg^{-1} by functionalizing zeolite-templated carbon (ZTC) with quinone molecules [18]. The all-organic proton battery was constructed from two quinone-ZTC electrodes in sulfuric acid, where energy storage is achieved by a proton rocking-chair mechanism originating from a quinone couple embedded inside the nanopores of ZTC. A simple approach for anchoring organic molecules within the framework of three-dimensional (3D) materials may be an exploratory route for proton batteries. However, the electrical conductivity of the material must still be considered.

9.2.2.3 Bipolar All-Organic Proton Batteries

All-organic proton batteries have gained extensive attention due to their excellent merits. However, the high intercalation energy barrier of H^+ requires a strongly acidic electrolyte (such as H_2SO_4) to provide protons to all-organic proton batteries. Additionally, designing a mild electrolyte-based proton battery is a challenge. For most all-organic batteries that use two different molecules as positive and negative electrodes, a separate synthetic process and composition optimization are needed, resulting in high costs [19]. However, it is indeed possible to design a single molecule that can act as both the anode and cathode and fulfill the requirements of an all-organic proton battery by having two diverse redox potentials with a large gap in LUMO energy to facilitate H^+ uptake and removal. To meet these requirements, a conjugated poly(2,9-dihydroquinoxalino[2,3-*b*]phenazine (PO) molecule was created, which is bipolar and suitable for an all-organic proton battery [3]. The PO molecule exhibits an acceptable voltage between PO and its reduction state, owing to the large LUMO energy gap. Moreover, the conjugated C=N groups possess a pair of electrons that ensure PO molecules have fast H^+ uptake and removal behavior in a mild electrolyte (such as $ZnSO_4$). The two-step redox reactions of the PO molecule are attributed to the electron delocalization effect, which leads to two different chemical environments for the C=N moieties in the PO molecule, as illustrated in Figure 9.5. Moreover, the aromatic structure of the PO molecule exhibits strong

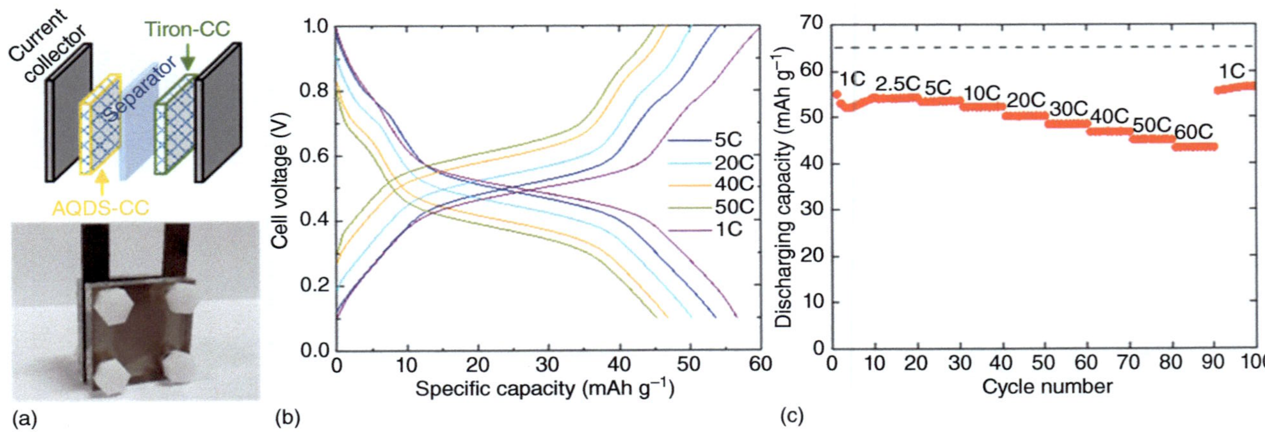

Figure 9.4 (a) Cartoon (top) and photo (bottom) structure of the assembled battery. (b) Galvanostatic charging and discharging curves at different rates. (c) rate performance of the assembled battery. Source: Adapted with permission from Xu et al. [17]/American Chemical Society.

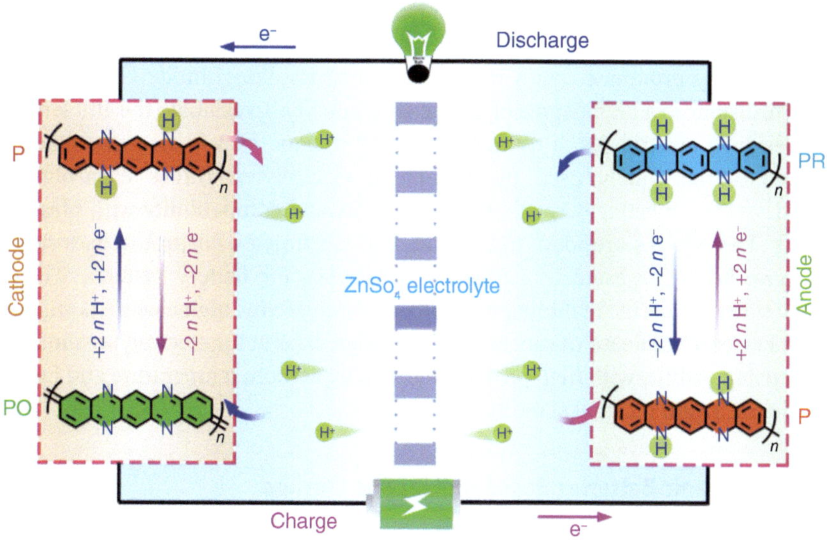

Figure 9.5 The working mechanism and molecule structure of the symmetric all-organic proton battery in a $ZnSO_4$ electrolyte upon discharging and charging (one molecule as both anodes and cathodes) were proposed by Chen and coworkers. Source: Reproduced with permission from Tie et al. [3]/John Wiley & Sons.

stability because of the electron delocalization effect, resulting in a long cycling life when used as a positive or negative electrode in an all-organic proton battery. The use of a single conjugation molecule expands the potential of proton-based all-organic batteries.

9.2.2.4 Other Research on All-Organic Proton Batteries

Organic proton batteries have attracted attention due to the smaller size and faster kinetics of protons and the structural tenability and stability of organic molecules. Recently, Yan et al. reported a one-dimensional rod-like diquinoxalino-phenazine (DPZ) molecule that is suitable for use in organic batteries [20]. When coupled with a polyaniline anode, the DPZ@graphene fiber electrode shows an energy density of 40.88 mWh cm^{-3}. The proton redox process of the DPZ molecule was studied using in-situ Raman technology and theoretical calculations. It was found that organic molecules containing C=N groups should simultaneously coordinate with protons, and the redox reaction between the C=N and C—N bonds is highly reversible upon proton insertion and extraction. Organic electrodes with these features exhibit similar mechanisms of proton intercalation and extraction when used in all-organic proton batteries.

In an all-organic proton battery, the electrolyte used is a crucial parameter that needs to be carefully considered. Although Chen and colleagues have proposed and developed a mild electrolyte system using $ZnSO_4$, the development of electrolytes for use in all-organic proton batteries still requires further investigation. Karlsson et al. reported a novel nonstoichiometric protonic ionic liquid (NSPIL) as an electrolyte for use in all-organic proton batteries [13]. The designed NSPILs contain a mixture of proton acceptors and donors based on either 1,2,4-triazole or 1-methyl-1,2,4-triazole and are partially protonated with bis-(trifluoromethylsulfonyl)imide (TFSI). The electrolyte displayed a high degree of proton conductivity. To evaluate the efficiency of the designed electrolytes, a quinone-based all-organic battery was assembled. Compared to the widely used pyridine-based electrolytes, NSPIL electrolytes exhibited several advantages, such as lower toxicity and compatibility with plastic materials. The NSPILs are identified as efficient electrolytes for proton batteries, which have been suggested as sustainable materials for battery systems. They possess a series of merits, including high proton conductivity, high electrochemical stability, low cost, and ease of synthesis. The design criteria for electrolytes should focus on using NSPILs with high proton conductivity at room temperature and high electrochemical and thermal stability.

9.2.3 All-Organic Batteries Based on Metallic Carriers

Metal-based all-organic batteries typically use metal ions (e.g. Li^+, Na^+, K^+) as charge carriers, as illustrated in Figure 9.6. Conventional metallic rechargeable batteries rely on the redox reaction of metal ions in transition-metal-containing electrodes, which are environmentally unfriendly and unsustainable. Such materials also raise concerns regarding low yield and value. Therefore, the new concept of all-organic batteries, which employ organic molecules or polymers for both the anode and cathode, is desirable to achieve a metal-free electrode. Several all-organic battery systems have been developed in recent years [21–23]. For all-organic batteries that rely on metal ions as charge carriers, the charge storage

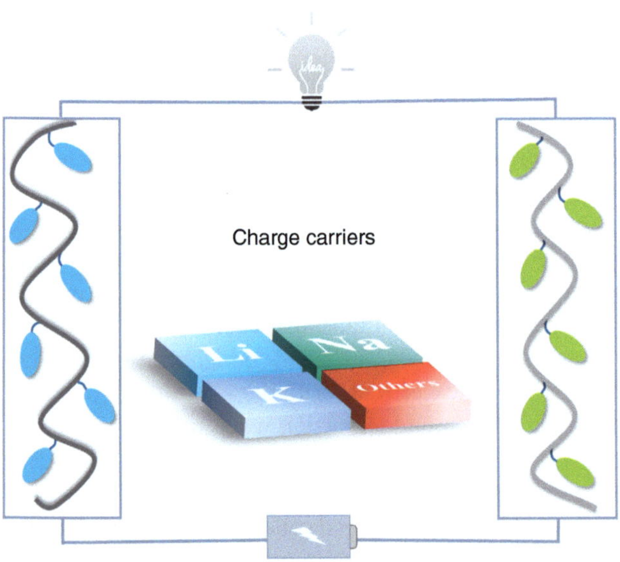

Figure 9.6 Schematic illustration of an all-organic battery with metallic charge carriers (e.g. Li^+, Na^+, K^+, and others) based on organic cathode and anode.

materials also involve an n-type doping state and doping/de-doping mechanisms, such as the reversible reactions of C=O and C-O-M, or a p-type doping state (as in viologen-based materials, which involve a reversible adsorption and desorption process of anions for redox reactions). Alkali metal ions (e.g. Li^+, Na^+, and K^+) have been extensively studied as ion carriers in all-organic batteries due to their small size and easy transport.

In 1981, Heeger et al. reported on an n-type polyacetylene-based organic battery [24], which was also the first reported use of alkali metal ions as charge carriers. They found that polyacetylene could be reversibly electrochemically doped with metal or nonmetal ions to create n- or p-type materials, and the resulting organic metals could be used as anodes and cathodes. However, the doped polyacetylene cathode required lithium metal as the anode. In the same year, Chiang reported on an all-polymeric solid-state battery composed of polyacetylene films that operated via a wholly solid-state reaction at 85 °C (Figure 9.7) [25]. Undoped polyacetylene was used as both sides of the electrode for the all-organic battery, with a high-conductivity polyethylene-oxide-NaI (PEO-NaI) solid electrolyte. The all-polymeric solid-state batteries exhibited an open circuit voltage (OCV) of 2.8–3.5 V and achieved an energy density of approximately 20 Wh kg^{-1} in this cell. In this section, we will discuss the potential for the practical use of all-organic batteries based on metallic carriers.

9.2.3.1 Li-Ion Carrier for All-Organic Batteries

Li-ion batteries are the most researched battery system in recent decades and have gained a lot of attention for electrochemical energy storage technologies. However, many technical challenges must be overcome for Li-ion batteries, such as dendrite growth and unsafe lithium anode battery systems. To eliminate these problems, all-organic batteries based on Li-ion as charge carriers have gained significant interest. Although some small molecules, such as benzoquinone, have a higher theoretical specific capacity, their limited development is circumscribed due to instability caused by the dissolution of the oxidized or reduced states.

Figure 9.7 Schematic structure of an all-organic solid-state battery by Chiang in 1981.

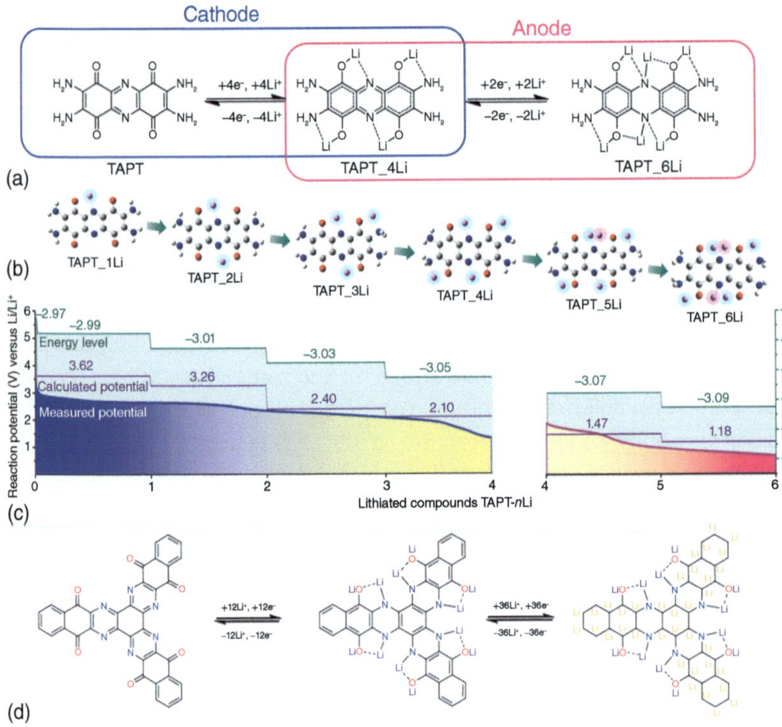

Figure 9.8 (a) The proposed redox mechanism of TAPT; (b) The optimized step-by-step lithiation process and structure of TAPT; (c) the changes in measured potential, calculated potential, and the energy level of TAPT at different states. Source: Reproduced with permission from Li et al. [26]/John Wiley & Sons. (d) Proposed superlithiation reaction mechanism of 3BQ anode for all-organic battery. Source: Adapted with permission from Zhang et al. [27], Royal Society of Chemistry.

In recent years, there has been increasing attention given to the structural design of small molecules. Wang et al. have reported the use of a small molecule (2,3,7,8-tetraaminophenazine-1,4,6,9-tetraone (TAPT), Figure 9.8a–c) in a symmetric all-organic battery [26]. The typical TAPT molecule exhibits two plateaus with high capacity owing to the rich C=O and C=N active centers. The introduction of -NH_2 groups and the abundance of active centers enable robust interactions, including hydrogen bonds and intermolecular forces, resulting in TAPT's specially layered structure and insolubility, as well as high cycling performance. The step-by-step charge transfer process enables the symmetric all-organic lithium battery to exhibit an open-circuit voltage of 2.47 V and an average work voltage of 2 V. The high performance of TAPT can be attributed to the synergetic effect of –NH_2 and C=O/C–N groups, which not only provide rich active centers but also fabricate strong intermolecular interactions. Additionally, the absence of large, non-active groups plays a crucial role. This constructive work highlights the important role of the conjugational effect of rich groups, such as C=O, C—N, and -NH_2 groups, which presents possible design principles of small molecules for symmetric all-organic batteries and stimulates deep comprehension of organic materials for organic electrodes. Like the TPAP molecule, Huang et al. have also assembled an all-organic symmetric battery utilizing the unique structure of tribenzoquinoxaline-5,10-dione (3BQ, Figure 9.8d) [27]. The 3BQ molecule contains plentiful active sites with varied potentials, including C=O and C=N functional groups with high redox potentials and multitudinous unsaturated carbons providing low redox potentials, making it suitable for use in symmetric all-organic batteries. The large conjugated structure of the molecule also contributes to its insolubility and fast electron transfer ability. The all-organic symmetric battery assembled with pristine and pre-lithiated 3BQ molecules has an initial capacity of 483 mA h g^{-1} (approximately 11 Li^+ ions inserted) at 0.2 C between 0.01 and 2.6 V and 172 mA h g^{-1} after 300 cycles. These results demonstrate that 3BQ has immense potential for application as an electrode in all-organic symmetric batteries.

Many other small molecules have been used as electrode materials for all-organic batteries in recent years. For example, conjugated dicarboxylate salts with a close redox potential to metal anodes such as Li and Na are available anode materials for all-organic batteries [21, 28]. However, the performance of an organic anode assembled by such organic metal salts will suffer from the poor electric conductivity and ion diffusivity problems caused by the formation of large crystals. The conjugated system used for organic electrode materials was first proposed by Armand and co-workers in 2009 [28]. Since then, these kinds of organic electrode materials have shown great interest. Conjugated carbonyls have received a lot of attention in organic electrode materials, especially for all-organic batteries, because of their fast kinetics, high theoretical capacity, and structural diversity.

Some studies have shown that when different cations coordinate with terephthalic acid, the crystal morphology can be affected. It has been found that silver and terephthalic acid (Ag_2TP, Figure 9.9a) form smaller crystal structures in comparison to sodium and lithium salts, which also exhibit a very stable discharge profile [21]. Perylene-3,4,9,10-tetracarboxylate (PTC, Figure 9.9b), containing four carboxyl groups in a large π-aromatic core, is used as a negative electrode material for all-organic Li-based batteries [29]. The prelithiated $PTCLi_6$ battery has demonstrated promising cycling stability of up to 200 cycles. Therefore, small organic molecules with conjugated carbonyl groups are considered one of the key candidates for all-organic battery electrode materials.

Organic molecules containing hydroxyl groups, cyano, or heteroatoms, in addition to small molecules containing carboxyl groups, can be used as electrode materials for all-organic batteries. For instance, Fan et al. utilized a novel in-situ electrochemical technique to produce a unique lithium-rich organic cathode for an all-organic battery, namely lithium anthracene-9,10-bis[2-benzene-1,4-bis(olate)] (ABB_4OLi, as shown in Figure 9.9c) [30]. This molecule contains a hydroxyl group that can bind with four lithium ions, resulting in a theoretical specific capacity of 256 mAh g^{-1}. When combined with a Li_2TP anode, the entire battery demonstrates a working voltage above 1.0 V. However, the battery exhibits a low capacity of 70 mAh g^{-1} during 200 cycles at a current density of 50 mA g^{-1}.

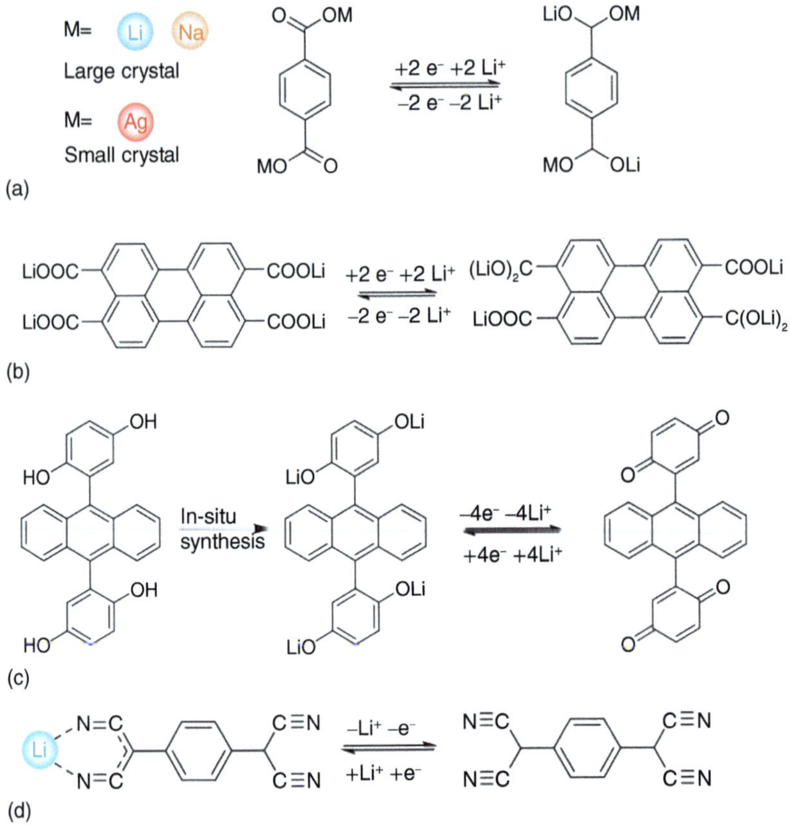

Figure 9.9 Illustration of corresponding redox reaction for (a) terephthalate anode with different cations; (b) electrochemical lithiation and delithiation mechanism expected of $PTCLi_4$; (c) synthesis and redox mechanism of ABB_4OLi; (d) redox mechanism of LiTCNQ.

In-situ synthesis, despite its drawbacks, is one of the most encouraging methods for acquiring lithium-rich electrode materials for all-organic batteries. Pre-lithiated organic molecules utilized for all-organic lithium-ion batteries still face some limitations, such as air stability and high performance. Guo et al. used lithium tetracyanoquinodimethan (LiTCNQ, as shown in Figure 9.9d) as a new and stable Li-ion-containing organic cathode for an all-organic battery [31]. The LiTCNQ molecule is air-stable due to the mismatches of the frontline molecular orbitals with the O_2 molecule, with an energy difference of more than 6 eV. According to research, the Li ions in LiTCNQ are associated with the C=N groups. The cyano in TCNQ is more likely to uptake Li^+; hence, the reaction sites for a cyano-containing molecule could be the N atoms. When combined with a di-lithium terephthalate anode, the full cell exhibits a high capacity of 150 mAh g^{-1} at a current density of 20 mA g^{-1} and capacity retention of 70% after 50 cycles.

The application of small-molecule organics in all-organic batteries has a potential future to be explored, especially in the modification of small-molecule structures by exploiting the tunability of small molecules to meet the desired physicochemical properties. These include low solubility for small-molecule electrode materials in both oxidized and reduced states, high structural rigidity, and thermodynamic and chemical stability during charging and discharging, and considerable electrochemically active functional groups such as carboxyl, carbonyl, and cyano groups. However, small molecule compounds also have their natural disadvantages compared to polymers. To maintain their physicochemical properties, most small-molecule compounds require a rigid conjugation plane, usually a conjugation plane, which greatly reduces the molecule's ratio of electrochemically active functional groups. As a result, small-molecule organic electrode materials have a low theoretical specific capacity (typically less than 200 mAh g^{-1}). In conclusion, small-molecule compounds, whether derived from biological systems or synthesized for application in all-organic batteries, have a thorny road to explore, and there are still many gaps in the field that need to be addressed.

In contrast to small-molecule materials, polymer molecules have garnered significant interest in recent years due to their high molecular weight and intricate structures that can range from one-, two-, and three-dimensional geometries. Polymeric organic electrode materials can be broadly classified into two categories. The first category is comprised of active polymeric native materials that consist of a single molecule that has been polymerized into lengthy chains or other architectures. The backbone structure and active sites of these materials exist within the polymer chains, with the charging and discharging processes depending on the functional groups contained within, such as polyaniline and polyimide (PI) structures, and other polymer molecules with C=N and C=O as active sites. The second category of polymers has a separate structure, in which the active site is separated from the structural skeleton. The polymer family with PTMA (poly(2,2,6,6-tetramethylpiperidinyloxy-4-yl methacrylate)) and polythiophene as the skeleton and quinone material as the active component is an example of this type of polymer.

Zhang et al. detailed a groundbreaking all-plastic electrode-based Li-ion symmetric battery composed of bi-functional ladderized heterocyclic poly(quinone) (PBD, as shown in Figure 9.10a,b) cathode and anode materials [32]. The resultant all-PBD full battery exhibited a competitive energy density (276 Wh kg^{-1}) with a capacity retention of 119 mAh g^{-1} after 250 cycles. This polymer electrode material is connected through double thioether bonds and is insoluble in ether electrolytes, which grants it better stability. Additionally, since quasi-solid or solid polymers are employed as electrolytes, the fabricated flexible battery system offers a stable, safe, and highly malleable energy solution.

Using the design strategy detailed by Zhang et al. [32], Mecerreyes and colleagues created a symmetric all-organic battery composed of a dual-redox-active PI copolymer containing phenothiazine and naphthlene moieties as cathode and anode materials, respectively, which displayed three distinct redox processes at 2.3, 2.5, and 3.7 V (versus Li/Li^+) [33]. The first two redox processes are attributed to the enolization of the naphthalene groups, while the latter is assigned to the oxidation and reduction processes of the phenothiazine moiety. With two redox reactions having a 1.2 V voltage difference and the superior electrochemical

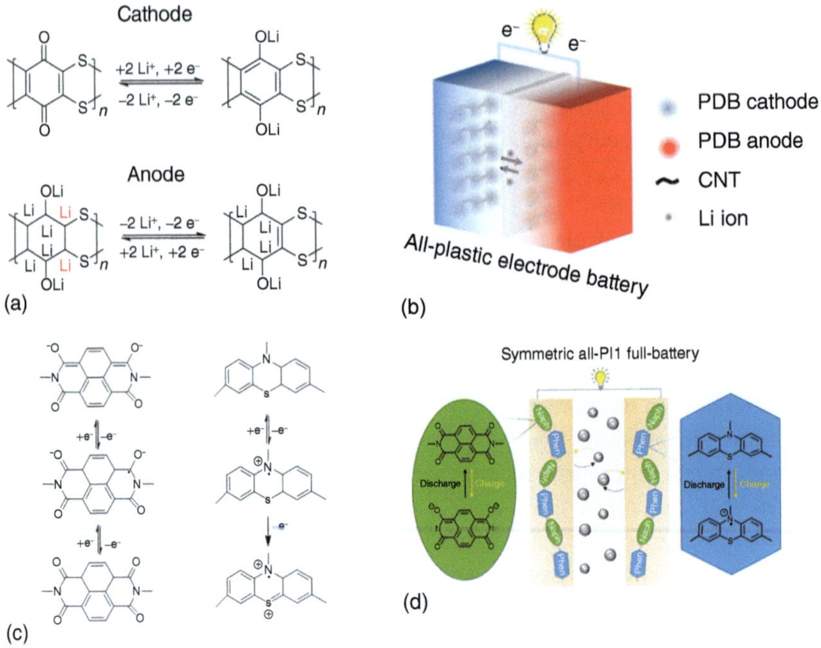

Figure 9.10 (a) Proposed redox mechanism of PDB used as active bi-functional electrodes. (b) Prototype of the all-PDB-electrode battery. Source: Reproduced with permission from Xie et al. [32]/John Wiley & Sons. (c) Redox mechanism of phenothiazine-naphthalene polyimides (PI1). (d) Scheme of the symmetric PI1 battery. Source: Adapted with permission from Casado et al. [33], Wiley-VCH.

performance of homopolymer (PI1, generated using stoichiometric quantities of 3,7-diamino-methylphenothiazine (DAMPT) and naphthalenetetracarboxylic dianhydride (NTDA)), a symmetric battery was created using PI1 homopolymer as both anode and cathode electrodes (Figure 9.10c,d). It's worth noting that an all-organic symmetric battery does not require a prelithiation step since both the anode and cathode materials are initialized in their natural states. During cycling, anions and cations from the electrolyte compensate for the redox process of PI1 to maintain the battery's electrochemical equilibrium. Consequently, this all-organic polymer battery exhibited exceptional cycling stability of 1000 cycles at 800 mA g^{-1}, with high capacity retention of 94% after cycling.

Xia et al. conducted a study on all-organic batteries using an aqueous electrolyte, which showed a high operating voltage of 2.1 V and overcame the safety issues associated with organic-based electrolytes [14]. For their study, they utilized polytriphenylamine (PTPAn, depicted in Figure 9.11a, b), synthesized by chemical oxidation, as the cathode material for all-organic batteries. The active site during charging and discharging is the nitrogen atom in the center of the PTPAn structure. The battery also demonstrated supercapacitor-like characteristics, indicating that the ion diffusion on the electrode does not regulate the kinetic behavior of the electrochemical reaction, a factor that contributes to the battery's high power density.

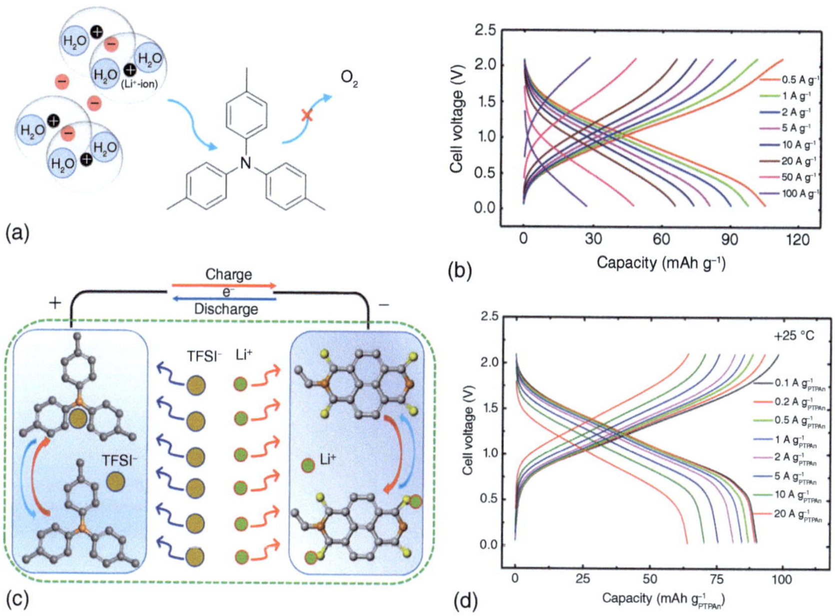

Figure 9.11 (a) Possible mechanism of the possible undesired O_2 evolution reaction elimination (PTPAn). (b) Rate performance of the battery based on the polymers PTPAn and PNTCDA as cathode and anode materials. Source: Reproduced with permission from Dong et al. [14]/John Wiley & Sons. (c) Schematic of the charge/discharge process of the full cell based on organic polymers PTPAn and PNTCDA. (d) Rate the performance of the full cell at room temperature. Source: Reproduced with permission from Dong et al. [34]/Elsevier.

Furthermore, the all-organic battery's water-based electrolyte is advantageous in terms of stability, safety, and cost-effectiveness. It is also environmentally friendly. A year later, Xia et al. designed a low-temperature organic battery using an unfreezing ethyl acetate-based (EA) electrolyte with a sufficient ionic conductivity of $0.2\,mS\,cm^{-1}$ at $-70\,°C$ [34]. The developed all-organic battery contained PTPAn and 1,4,5,8-naphthalenetetracarboxylic dianhydride (NTCDA)-derived PI (PNTCDA) as cathode and anode materials (depicted in Figure 9.11c,d), respectively. Charge storage in solid organic electrode materials primarily occurs on the surface groups or large interstitial spaces, which facilitates the charge storage process. As a result, the all-organic rechargeable battery maintained a significant amount of capacity when operating at such low temperatures. At 0.5 C, the battery achieved discharge capacities of 99 and $69\,mAh\,g^{-1}$ at 25 °C and − 70 °C, respectively. This excellent performance highlights the advantages of organic polymer electrode materials, which exhibit rapid electron transfer kinetics due to the redox reactions being linked to changes in the charge state of electrochemically active sites.

Heteroatom-rich organic materials, such as S and N, are also used as electrode materials for all-organic batteries. Su et al. investigated an all-organic battery using polyisothylène as a redox-active bipolar electrode material [35]. The conjugated polyisothionaphthalene (PITN, depicted in Figure 9.12a) has a small band gap, excellent conductivity, and differential electrochemical behavior, which makes it

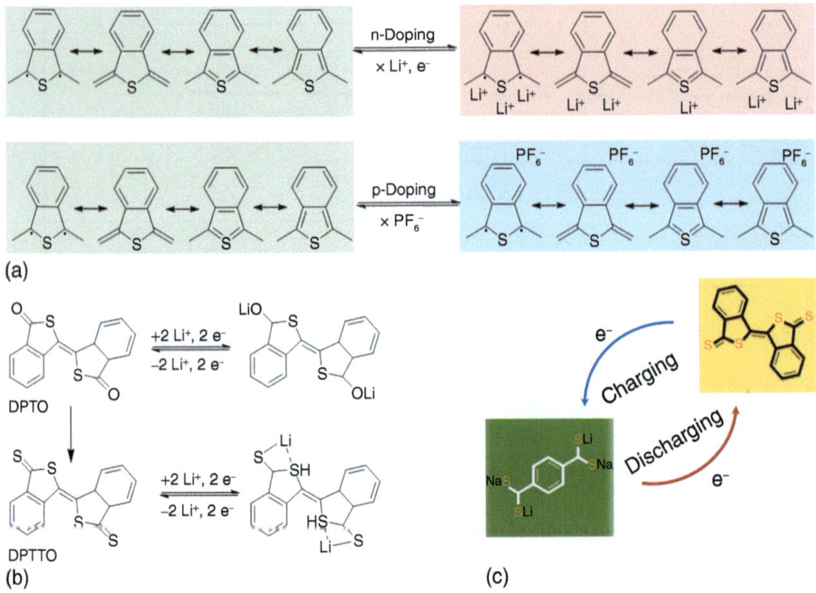

Figure 9.12 (a) The proposed electrochemical reaction mechanism of PITN material in n-doping and p-doping reactions. (b) Optimized structures of lithiated DPTO and DPTTO. (c) Schematics for all-organic coin cells.

a unique bipolar polymer. Its cyclic C—S—C bond and benzene ring can serve as active components for charging and discharging, interacting with cations and anions, leading to p-doping and n-doping processes. During the first electrochemical reaction, lithium polysulfide and lithium sulfide are formed, but compared to other active materials, there is no significant effect on the rate performance and impedance of the cell. When PITN was used as the cathode and anode electrode material, the full cell produced $92\,mAh\,g^{-1}$. Although its low capacity and CE restrictions limit its application, PITN with a low bandgap is highly reversible and can be a promising bipolar organic material for Li-ion batteries. In 2021, He et al. and colleagues investigated sulfur-containing polymers and created thionated benzo[*c*]thiophen-1(3*H*)-one (DPTTO, depicted in Figure 9.12b,c) by replacing oxygen atoms with sulfur [36]. DPTTO has better conductivity and lower solubility in ether electrolytes than un-substituted thionating benzo[*c*]thiophen-1(3*H*)-one (DPTO), and its high performance is supported by its special properties, weakened polarization, high capacitive contribution, and conductivity. The lithiation sites in DPTTO, clamped by the two S atoms, differ significantly from those in normal quinone structures. Such a particular lithiation process in DPTTO has a certain effect on the electrochemical kinetics. Moreover, the sulfur-rich all-organic battery was successfully prepared. The full battery delivers a specific capacity of $75\,mAh\,g^{-1}$ during 100 cycles with an average discharge voltage of 0.65 V. The studies conducted by Su et al., He et al., and their colleagues would inspire fundamental investigations into sulfur-rich all-organic batteries.

Separating polymer electrode materials typically consists of polymer chains for support and stabilization, as well as branched components that possess specific electrochemical activity, resembling polymer molecules with a grapevine structure. In an effort to create an all-organic battery, Jia et al. utilized a TEMPO radical polymer cathode with a terephthalic acid derivative as the anode [21]. Additionally, P(TMA0.95-co-AQ0.05)1610 was synthesized by chemically modifying PTMA with aromatic anthraquinone (AQ). The AQ groups promote binding to CNT at the molecular level via non-covalent interactions, as shown in Figure 9.13a,b. Nitroxide radical polymers such as PTMA boast a high redox potential, i.e. 3.6 V versus Li/Li. Ongoing research has been conducted to develop various PTMA-based cathode composites for organic/metal hybrid or all-organic cells [38]. Similarly, significant efforts have been invested in optimizing the conductivity of the corresponding electrode composites for PTMA [39].

Another suitable strategy for preparing electrode materials is the polymerization of carbon–carbon double bonds. This involves substituting small organic molecules containing active sites with functional groups that contain double bonds and subsequently polymerizing to yield insoluble polymer electrode materials, as depicted in Figure 9.13c,d [37].

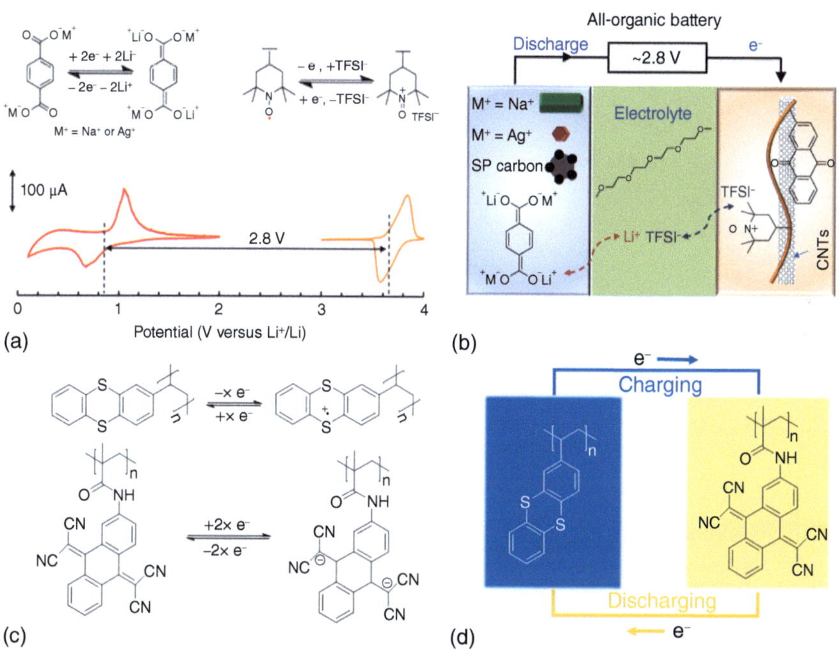

Figure 9.13 (a) The corresponding redox mechanism of terephthalate and nitroxide radical polymers, (b) Illustration of all-organic battery based on P($TMA_{0.95}$-co-$AQ_{0.05}$)$_{1610}$. Source: Reproduced with permission from Jiang et al. [21]/Elsevier. (c) Charging and discharging mechanism and (d) illustration of poly-(2-vinylthianthrene) and poly(2-methacrylamide-TCAQ)-based all-organic battery. Source: Reproduced with permission from Wild et al. [37]/John Wiley & Sons.

9.2.3.2 Na/K Ions Carrier for All-Organic Batteries

The alkali metal family, which includes lithium, sodium, and potassium, has similar properties as carriers in battery applications. Thus, electrode materials used in Li-ion all-organic batteries can generally be applied in Na or K-ion-based all-organic battery systems, including imides and PIs, small-molecule compounds containing carboxyl groups, and polymers with heteroatomic centers such as polytriphenylamine. This section aims to provide insights into the design, application, and progress of organic materials for rechargeable all-organic Na/K-ion batteries.

Chen et al. synthesized an organic compound, 2,5-dihydroxyterephthalic acid (Na_4DHTPA; $Na_4C_8H_2O_6$), with multiple functional groups consisting of phenolic enolate and carboxylate groups at opposite positions of a benzene ring for an all-organic Na-ion battery (Figure 9.14) [40]. Based on the property that enolate and carboxylate groups on the same conjugated benzene ring have different electrochemical potentials, the electron delocalization effect of the carboxylate group leads to a positive shift of the de-sodiation potential, while the electron-donating effect of the phenolic enolate groups leads to a negative shift of the sodiation potential, which results in the widening of the potential gap between the two redox pairs. This allows it to function as a symmetric, all-organic battery. The results demonstrate that two reversible two-Na^+- ion electrochemical reactions occur at 2.3 V for redox couples of Na_2/Na_4 DHTPA as the positive electrode and

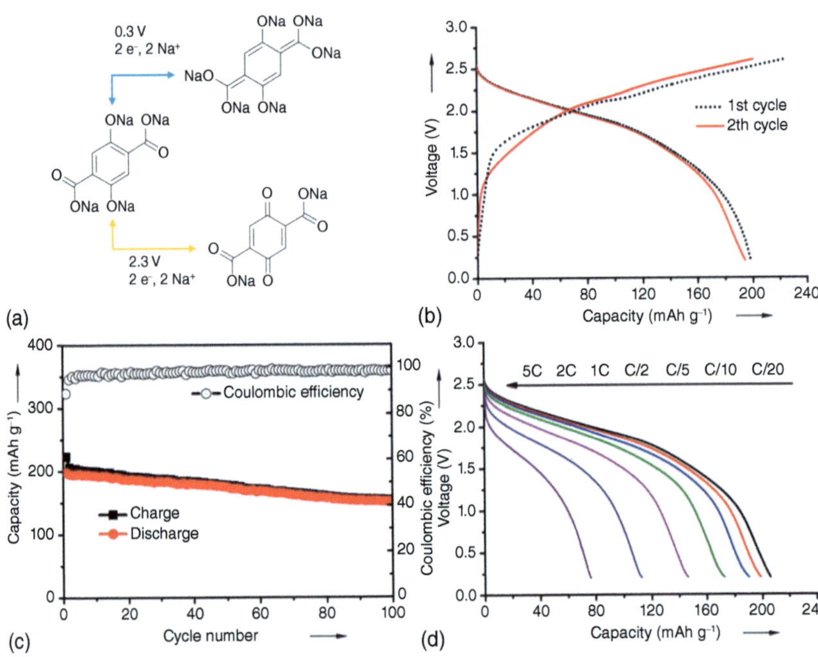

Figure 9.14 (a) Electrochemical redox mechanism of $Na_4C_8H_2O_6$, (b) Galvanostatic charge/discharge curves of full all-organic cells, (c) Cycling performance and Coulombic efficiency under 19 mA g^{-1}, (d) rate performance of the all-organic battery. Source: Reproduced with permission from Wang et al. [40]/John Wiley & Sons.

at 0.3 V for Na_4/N_6-DHTPA as the negative electrode, respectively. Consequently, a rocking-chair sodium-ion all-organic battery was constructed at room temperature using Na_4DHTPA as the initial electrode materials for both the anode and cathode, achieving an average operation voltage of 1.8 V and a practical energy density of approximately 65 Wh kg^{-1}. These findings offer insights into the design of organic compounds with different functional groups as the electrode materials of rechargeable organic-based batteries.

Carboxylate organic molecules are also utilized as electrode materials in all-organic K-ion cells. Potassium naphthalene-1,4,5,8-tetracarboxylate (K_4NTC, 117 mAh g^{-1}) serves as a typical organic anode for K-ion batteries, which contains four strong K—O ionic bonds in a 4-valent NTC^{4-}. Hence, K_4NTC is classified as a polyanionic organic salt [41]. Fan et al. prepared a pure all-organic K-ion battery using perylene-3,4,9,10-tetracarboxydianhydride (PTCDA) as the cathode material and the reduced state (K_6NTC) of K_4NTC as the anode (Figure 9.15a,b). By utilizing ether electrolytes, the assembled battery can achieve a peak discharge capacity of 121 mAh g^{-1} in 0.53 V over 1500 cycles, while the CE (Coulombic efficiency) of the full cell was observed to be almost 100%. It should be noted that the deterioration of organic electrode materials caused by solvent-molecule co-integration is a widespread and unavoidable phenomenon. Therefore, more researchers have focused on developing the desolventization process as a vital and fundamental step to enhance the cycling stability of organic electrodes. This also

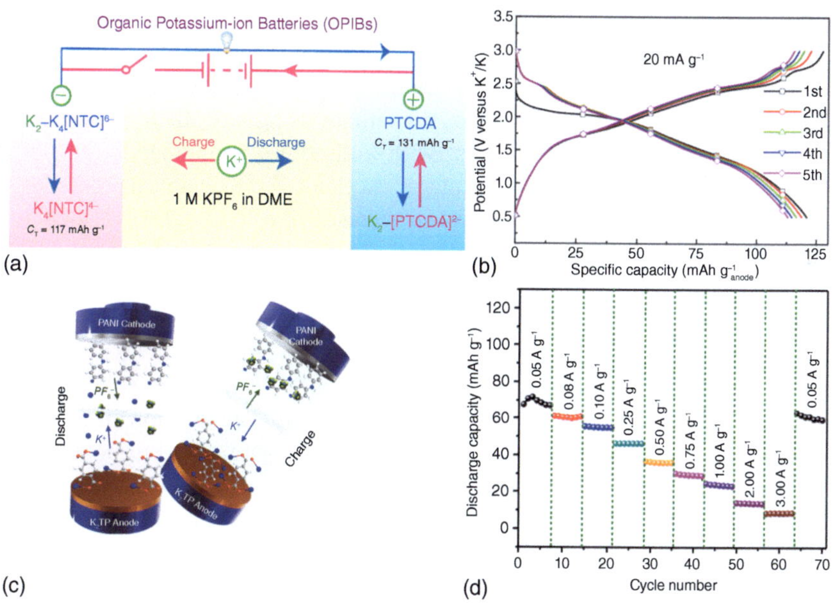

Figure 9.15 (a) Illustration and (b) Charge–discharge curves of all-orgainc K-ion battery based on K_6NTC anode and PTCDA cathode. Source: Reproduced with permission from Liu et al. [41]/American Chemical Society. (c) Illustration and (d) rate performance of all-organic K-ion battery based on K_2TP anode and PANI cathode. Source: Reproduced with permission from Vardhini et al. [42]/American Chemical Society.

provides insight into the design of electrolytes for all-organic cells. Shaijumon et al. reported an all-organic potassium ion hybrid capacitor that employs dipotassium terephthalate (K_2TP) as the anode and is coupled with a polyaniline (PANi) cathode (Figure 9.15c,d) [42].

The fully assembled battery exhibits exceptional electrochemical properties, demonstrating a high energy density of 153 W h kg^{-1} and good cycling performance up to 10 000 cycles. The 3D Bode analysis was employed to understand the charge storage mechanisms of K_2TP and PANi.

So far, various types of organic electrode materials have been reported, including conjugated carbonyl compounds that are widely utilized due to their plentiful electroactive functional groups (C=O) and tunable redox potentials. Moreover, their polymers are frequently employed as organic electrodes in sodium and potassium-ion battery systems. For instance, Liu et al. employed PTCDA as the electroactive material for an all-organic Na-ion battery [43]. However, pure PTCDA presents a series of challenges, such as low electrical conductivity, sluggish reaction kinetics, inferior rate capability, and short cycle life. Consequently, a novel composite nanofibrous organic cathode, namely PTCDA/nitrogen-doped carbon/carbon nanotubes (PTCDA/NC/CNT), was designed to exploit the potential of PTCDA. As a result, the all-organic battery coupled with a conjugated sodium carboxylate/CNT anode delivers a high energy output of 85 Wh kg^{-1}. The high performance of the PTCDA electrode can be attributed to the synergistic effects of the interconnected conductive frameworks and the ultra-strong capillarity derived from the hierarchical micro- and nanotunnels.

A PI-based all-organic sodium-ion battery was developed by Shaijumon and coworkers [44]. Shaijumon and their colleagues [44] developed a sodium-ion battery entirely based on PI. They synthesized the cathode using N,N′-diamino-3,4,9,10-perylenetetracarboxylic PI and paired it with a disodium terephthalate (Na_2TP) (pre-sodiated) anode. The PI cathode displayed excellent electrochemical properties, with an initial capacity of 73 mAh g^{-1} and an average cell voltage of 1.35 V. Notably, Mecerreyes et al. reported an all-organic cell consisting of a PI–polyether anode and a PEDOT-lignin bipolar polymer cathode in another study [45]. The modified PI exhibited excellent stability and electrochemical activity. The full-organic cell retained 85% of its capacity and achieved above 93% CE over 800 cycles at 100 C. However, the full-organic cell's capacity only reached 40 mAh g^{-1}, and further studies are required. Moreover, PI materials can also be applied in K-ion all-organic batteries, as demonstrated by Lu et al., who built an all-organic aqueous potassium dual-ion full battery with a PTPAn cathode and a 3,4,9,10-perylenetetracarboxylic diimide (PTCDI) anode [46]. The all-organic battery combined with a "water-in-salt" electrolyte (21 M potassium bis[fluoroslufonyl]imide) delivered a reversible capacity of 50 mAh g^{-1} over 900 cycles with a high CE of 98.5%.

9.2.4 Metal-Free Carriers for All-Organic Batteries

In the search for all-organic batteries, there has been increasing interest in moving from metal ions to metal-free carriers. In 1981, the first metal-free all-organic battery was demonstrated by conjugating different oxidation states of poly(acetylene) films with an ammonium-based salt in propylene carbonate (PC) or tetrahydrofuran (THF) as the electrolyte [24]. Although the long-term cycling stability of these cells has not been thoroughly studied, poly(acetylene) membranes in various oxidation or reduction states have achieved open-circuit cell voltages ranging from 0.5 to 2.5 V. Furthermore, in 1983, a metal-free all-organic polymer cell with poly(thiophene) films as anodes and cathodes was proposed [47], resulting in cells with an open-circuit voltage of 3.1 V and a specific energy density of 75 $Wh\,kg^{-1}$, corresponding to a specific capacity of 24 $mAh\,g^{-1}$.

In 2006, Wang et al. developed a new flexible textile-based all-polymer battery, which utilized polypyrrole and styryl-substituted dialkoxyterthiophene (poly(OC10DASTT)) as the cathode and anode materials, respectively [48]. Anodes were created by casting poly(OC10DASTT) films onto Ni/Cu-coated nonwoven polyester via the solvent casting method. The cathode consisted of poly(pyrrole) electropolymerized on a stainless-steel mesh (Figure 9.16a). Even though the electrolyte used is still the ethylene carbonate/dimethyl carbonate solution containing 1 M $LiPF_6$, Li-ions do not take part in the redox reactions related to the cathode or anode, thereby making this all-organic polymer battery essentially an anionic rocking-type cell. At a current density of 0.02 $mA\,cm^{-2}$, the battery exhibited a specific discharge capacity of 39.1 $mAh\,g^{-1}$. In 2009, Mihranyan et al. introduced a novel nano-structured electrode material with a high surface area for ultrafast all-polymer paper-based batteries designed for energy storage applications [49]. As both anode and cathode materials, the electrode materials consisted of algal cellulose fibers coated in a 50 nm poly(pyrrole) layer consisting of varying oxidation states, thereby resulting in a symmetric all-organic battery with an open-circuit voltage of 1 V (Figure 9.16b). This paper-based, all-polymer battery had an exceedingly high charging capacity and rate. The composite conductive paper material, with a specific surface area of 80 $m^2\,g^{-1}$, facilitated the rapid mass transfer of charge carriers (chloride ions) in and out of the coating. Even at currents of up to 600 $mA\,cm^{-2}$, the battery made from this material could sustain 70% of its initial discharge capacity and could be charged and discharged for up to 100 cycles. The remarkable performance of the electrodes allowed them to function as secondary

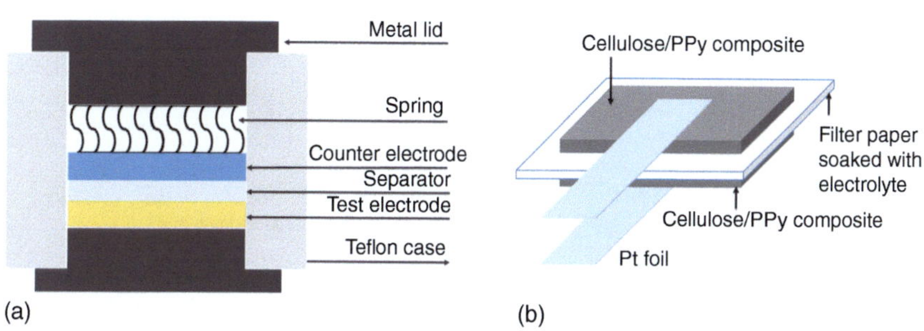

Figure 9.16 (a) Schematic diagram of Teflon test cell based on poly (OC_{10}DASTT) and PPyfilms. (b) Schematic image of all-organic algal cellulose fiber-coated paper-based batteries.

batteries and capacitors simultaneously, with the fastest charging and discharging rates of such battery systems at that time. However, the chemical instability and "capacitive" discharge voltage behavior of conductive polymers partly limit their use in battery applications, given the ultimate degree of oxidation (i.e. 0.3–0.5 charge per unit) that can be achieved.

In 2009, a new class of materials was developed for all-organic batteries by Nishide et al. They created the first metal-free all-organic battery based on organic radical polymers like poly-TEMPO, also known as "organic radical batteries" (ORBs) (Figure 9.17a,b) [12]. Since then, they have extensively researched the radical-based, all-organic battery system. They first examined a styrene-based polymer containing a strong n-type galvinoxyl radical and displayed its redox activity for the first time. An all-organic radical battery that coupled this radical polymer as an anode active material with a p-type radical polymer 2,2,6,6-tetramethylpiperidinyl-*N*-oxy (TEMPO) cathode exhibited a specific capacity of 51 mAh g^{-1}. The resulting double-ion battery, with a 0.5 M $(C_4H_9)_4NClO_4$ acetonitrile solution and a 0.01 M $(C_4H_9)_4NOH$ electrolyte, had a typical output cell voltage of 0.66 V and an intriguing rate capability with an initial specific charge capacity of 32 mAh g^{-1} at 10 C, which also remained at 75% capacity retention after 250 cycles. Nishide et al. then synthesized a p- and n-type bipolar redox-active radical polymer (poly(nitronylnitroxylstyrene),

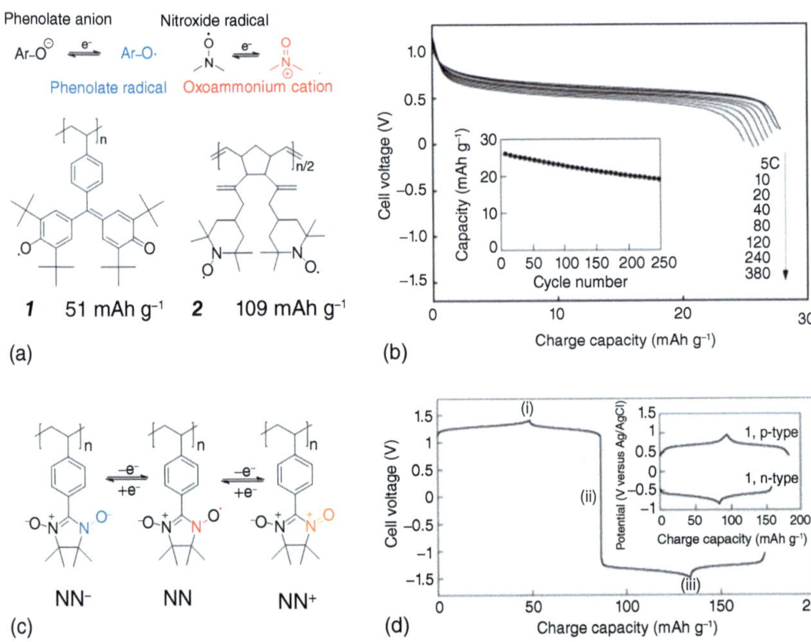

Figure 9.17 (a) Redox reactions mechanism and (b) Rate performance of the all-organic battery based on phenoxyl and nitroxide radicals. Source: Reproduced with permission from Suga et al. [12]/John Wiley & Sons. (c) Redox reactions mechanism of poly(nitronylnitroxylstyrene), (d) Charge–discharge curves of the organic battery composed of poly(nitronylnitroxylstyrene). Source: Reproduced with permission from Suga et al. [50]/John Wiley & Sons.

PNN, Figure 9.17c,d) for both anode and cathode materials in 2010 [50]. The PNN produced two separate redox potentials at −0.61 and 0.72 V versus Ag/AgCl in the presence of tetrabutylammonium hydroxide, which corresponds to the oxidation and reduction of the same nitronylnitroxide at the anode and cathode, respectively. In-situ ESR spectra revealed the mechanisms of p-type oxidation (of NN to NN^+) and n-type reduction (of NN to NN^-), respectively. A symmetric all-organic battery consisting of PNN for both electrodes was constructed, which resulted in a plateau voltage of 1.3 V. Furthermore, the organic base additive eliminated the nucleophilic side reactions of the NN^-, contributing to n-type redox reversibility. Finally, an asymmetric "rocking-chair" type battery was created using PNN and poly(galvinoxylstyrene) as anode and cathode electrodes, respectively, demonstrating the versatility and importance of redox-active radical polymers.

The synthesis of a cathode electrode-active polymer poly(2,2,6,6-tetramethylpiperidinyloxy-4-yl acrylamide) (PTAm) was carried out for an all-organic rechargeable battery with a poly(*N*-4,4-bipyridinium-*N*-decamethylene dibromide) (PV10) film anode [51]. The full cell demonstrated an output voltage of 1.2 V (Figure 9.18a,b). BF_4^- anions served as charge carriers, and the charge and

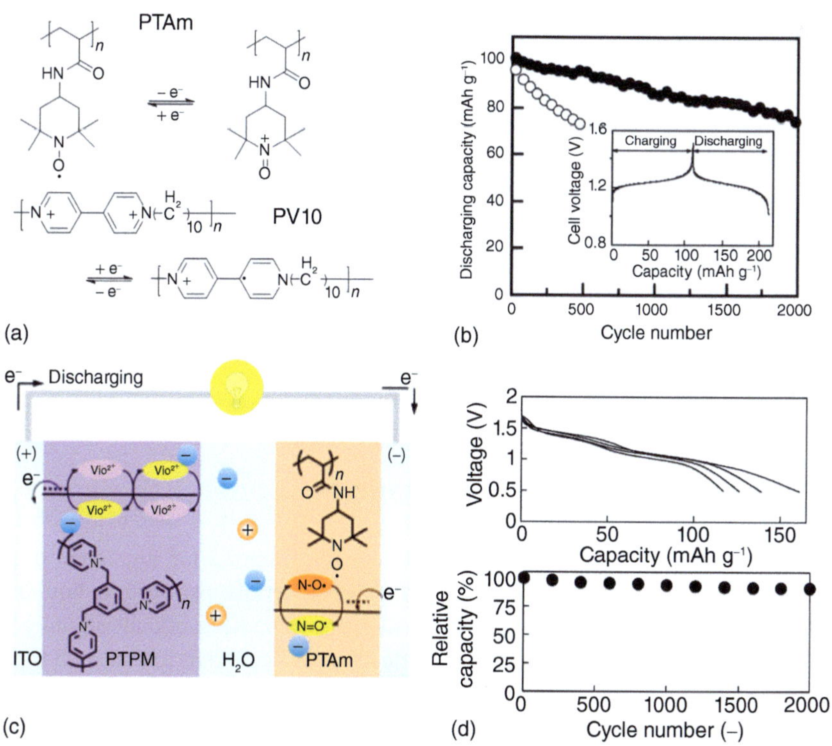

Figure 9.18 (a) Redox reactions mechanism, and (b) Cycling performance of the all-organic battery based on PV10 and PTAm. Source: Reproduced with permission from Chikushi et al. [51]/Springer Nature. (c). Redox reactions mechanism, and (d). Cycling performance of the all-organic battery based on PTPM and PTAm. Source: Reproduced with permission from Sano et al. [52]/American Chemical Society.

discharge processes did not involve any metal ions. The PTAm layer exhibited a discharging capacity of 114 mAh g^{-1} and retained a discharging capacity of 110 mAh g^{-1} (97% of the initial capacity) even after 1000 cycles.

The Nishide Group has presented a new viologen-based polymer that serves as an anode material [52]. They developed poly-(tripyridiniromesitylene) (PTPM, shown in Figure 9.18c,d) with a highly cross-linked polyviologen hydrogel. In an aqueous electrolyte, PTPM exhibits reversible two-electron transfer electrochemical behavior at −0.4 and – 0.8 V versus Ag/AgCl. The PTPM layer is electropolymerized onto a current collector via a 4-cyanopyridinium electro-coupling reaction. By combining a PTPM anode with a PTAm cathode in an aqueous electrolyte, the all-organic battery presents discharging voltages of 1.1 and 1.5 V with cycling stability exceeding 2000 cycles. This approach, utilizing organic-based polymer cathodes and anodes with an aqueous electrolyte, allows for both rapid charging and discharging while maintaining an environmentally friendly design.

In 2018, a radical polymerization reaction was used to produce an anode material for an all-organic battery using vinyldibenzothiophenesulfone (PVZ) [53]. When combined with a cathode made of 2,2,6,6-tetramethylpiperidine-1-oxyl (TEMPO)-substituted poly(methacrylate) and an electrolyte containing tetraethylammonium salt and organic solvent, the resulting battery yielded a high output voltage of 2.6 V. This was due to the electron-withdrawing sulfone group and the stability of the resulting dianion from the two-electron reduction of dibenzothiophenesulfone. The high energy density of the anode active material was found to be 540 mWh g^{-1}. This study demonstrated that an all-organic battery with a high energy density can be achieved and compared to other organic-based materials. However, further investigation is needed to optimize the electrode design and electrolyte of the all-organic cell. Future research will focus on the development of improved cathode and anode materials for all-organic batteries with even higher voltage (>3.0 V), with the belief that this novel anode/cathode-active material design could open up new possibilities for next-generation all-organic batteries with much higher energy densities.

9.3 Flow Batteries Based on Organic Molecules

Redox flow batteries (RFBs) demonstrate significant potential for implementing scaled-up energy storage, as they use electroactive materials dissolved in liquids and stored in external tanks to supply electrical energy. The RFBs operate similarly to fuel cells, requiring a continuous cycle of electroactive materials to maintain the chemical reaction. Unique in their cell structure, RFBs have separate energy and power control, where the energy capacity relies on the electrolyte concentration and power is determined by the collector's size and the number of battery stacks. Notably, RFBs can be recharged promptly by replacing the discharged electrolyte while also recovering the discharged electrolytes, providing excellent flexibility for various energy storage applications. A specific type of all-organic battery, known as all-ORFBs, has gained significant attention in recent years. Unlike metal-based redox species, organic materials' reactions are typically controlled by carbon, nitrogen, oxygen, and sulfur atoms' charge states. Organic molecules are advantageous, boasting low cost, effective biodegradability, a minimal environmental footprint, and highly tunable chemical and physical properties.

Over the past decades, tremendous progress has been made in addressing the limitations of organic molecules in traditional batteries through functional group modifications. However, in optimizing the high solubility of organics in all-ORFBs, there is a need to consider the two primary types: aqueous and nonaqueous ORFBs. Aqueous ORFBs (AORFBs) are more attractive due to their low cost, high safety, non-volatility, ease of processing, high conductivity, and low viscosity, which can reduce pumping costs and increase ionic conductivity. In contrast, nonaqueous ORFBs (NORFBs) have the potential for higher energy density due to their wider potential window, which supplies a wide range of cathode and anode active materials. However, the high cost and flammability of organic electrolytes have limited the development of NORFBs. Despite the growing interest in ORFBs, research in this area is still in its early stages, and many challenges remain to be addressed. This interdisciplinary research area has tremendous potential, and we hope that this section will generate more interest in it.

9.3.1 Cell Configuration

In a typical ORFB, the redox-active electrolytes (anolyte and catholyte) are stored in two external tanks, as illustrated in Figure 9.19. The major components of an ORFB include end plates, electrodes, bipolar plates, membranes, and external tanks. The end plates serve to secure the whole cell, while the electrodes facilitate the electrochemical reaction. The bipolar plates act as the cell's current collector, and the membrane is used to prevent crossover between the anode and cathode. The external tanks store the anode and cathode electrochemically active materials.

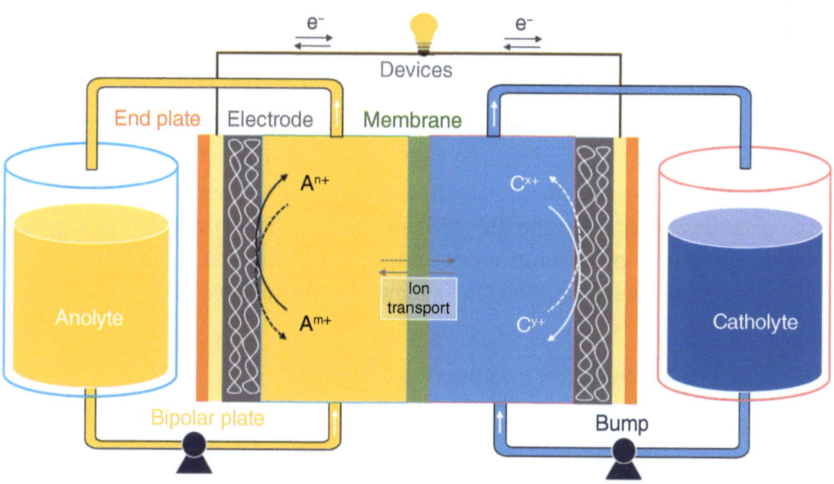

Figure 9.19 Schematic of a typical redox flow battery.

In an ORFB, the most vital component is the electrolyte, which is composed of soluble redox-active organic molecules dissolved in water or organic solvents. The performance of the battery, including voltage, capacity, and cycling stability, can be limited by the energy storage components. Recently reported research has focused on designing high-capacity and stable organic redox materials and understanding the effects of different functional groups on solubility, stability, and redox potential, as well as reversibility. The volumetric energy densities of ORFBs depend on the solubility, specific capacity, and redox potential of redox organic molecules. During operation, the anolyte and catholyte are pumped into electrode chambers. Then, the charge transfer occurs on the electrode, and the charged or discharged organic species are accumulated in the external tanks while the charge-balancing ions penetrate through the membrane to maintain electric neutrality.

9.3.2 Comparison of AORFBs and NORFBs

AORFBs and NORFBs differ primarily in the electrolytes used, which are determined by the intrinsic properties of solvents and salts. AORFBs commonly use water as the solvent and water-soluble acids, bases, and salts as charge carriers, such as sulfuric acid, sodium hydroxide, and sodium chloride for acidic, alkaline, and neutral conditions, respectively. The low viscosity of water allows for high ionic conductivity, enabling AORFBs to be cycled at high current densities, but the narrow electrochemical window limits their use to high voltages. In contrast, NORFBs use organic solvents like acetonitrile and soluble salts like $TEABF_4$ and LiF_6 as electrolytes, providing a wider potential window and the potential for much higher energy densities. However, the high cost and flammability of organic solvents restrict the scale of NORFBs. Additionally, aqueous electrolytes generally have higher conductivity than nonaqueous electrolytes, as shown in Table 9.1 [54–56].

Currently, commercialization of AORFBs has faced technical and economic challenges, as the performance metrics for all-organic redox species require structural modifications. Meanwhile, the cost of raw materials for NORFBs – especially solvents and salts - has hindered their widespread commercialization, despite the potential for higher energy densities. However, the promising potential window of organic solvents has attracted attention from researchers to develop solvents with excellent physicochemical properties for NORFBs. In contrast, AORFBs have shown rapid growth due to the structural diversity and variability of organic redox species, providing guidance for research in realizing ORFBs. The tunability of organic molecules' structures presents an opportunity for further structural tailoring to meet the requirements for long-life ORFBs, making it a key area of study.

9.3.3 Principle of Molecular Engineering for ORFBs

ORFBs use organic molecules, salts, and solvents for both the anode and cathode, separated by a membrane that allows fast transport of charge carriers while preventing the crossover of redox species. Energy density, power density, and efficiency are important metrics to evaluate ORFB performance. Energy density depends on

Table 9.1 Ionic conductivities of different supporting electrolytes in aqueous and nonaqueous RFBs.

	Electrolytes	Ionic conductivity ($mS\,cm^{-1}$)
1	1 M H_2SO_4/H_2O	390
2	2 M H_2SO4/H_2O	630
3	1 M $NaOH/H_2O$	170
4	2 M $NaCl/H_2O$	145
5	1 M $TEABF_4/ACN$	56
6	1 M $LiBF_4/ACN$	16
7	1 M $LiPF_6/DMSO$	10.82
8	1 M $LiPF_6/PC$	7.33
9	1 M $LiPF_6/ACN$	50
10	1 M LiTFSI/ACN	36
11	1 M TEATFSI/DMSO	9

the number of electrons transferred, solubility of the redox species, and potential separation between cathode- and anode-active organic molecules. Power density is influenced by electrolyte conductivity, membrane ionic conductivity, reaction kinetics and diffusion coefficients, cell voltage, and cell structure [57]. CE is a crucial performance metric that estimates the reversibility and stability of organic redox species. Intrinsic properties of the organic molecules themselves, such as solubility, redox potential, molecular size, and chemical reversibility, can be adjusted by modifying the organic core component with functional groups. The relationship between functional groups and solubility-enhancing effects, redox potentials, and redox species sizes must be carefully considered when functionalizing the organic molecules. With sufficient solubility and stability, organic redox species can be used in both aqueous and nonaqueous systems for ORFB applications.

9.3.4 Aqueous all-Organic Redox Flow Batteries

The solubility of organic electroactive materials is essential to the research community of ORFBs, particularly in AORFBs. Unfortunately, water maintains low solubility levels for many organic molecules, including those with large molecular structures. The solubility of organic molecules primarily relies on the chemical and physical properties of both the organic molecule and the solvent. The solubility of any organic substance is theoretically determined by the balance between the intermolecular interactions of the solute and the solvent. To increase the solubility of organic molecules in water, several strategies have been employed over the past few decades. For example, enhancing the solubility, stability, and other properties of organic molecules has included introducing hydrophilic functional groups on organic molecules [58–60], blocking interactions between organic molecules [61], causing organic molecules to form inner salts [62], and changing the polarity of organic molecules have been applied to enhance the solubility, stability, etc., of organic molecules [62]. Ultimately, the implementation of the aforementioned strategies leads to improved performance for AORFBs.

Viologens have been extensively explored as redox-active materials for the anode in AORFBs, utilizing the nitrogen-containing aromatic rings as the redox center [63, 64]. On the other hand, TEMPO (2,2,6,6-tetramethylpiperidine-1-oxyl) [63, 65] and ferrocenyl derivatives [62, 66] are commonly used as cathode materials for AORFBs. However, their low solubility in raw form limits their applications. For instance, the reduced product of methyl viologen is barely soluble, and TEMPO is insoluble in water [63]. To enhance the solubility of these organic molecules used as anode and cathode in polar solvents, hydrophilic functional groups such as hydroxyl [63], sulfonic acid [62, 67], and quaternary ammonium [61, 62, 68] can be attached to their molecular chains to increase the interaction between the organics and water.

Liu et al. have reported on a series of viologen molecules designed rationally as anolytes for pH-neutral AORFBs (Figure 9.20a,b) [62]. The use of synthetic and computational chemistry has opened up new avenues for designing AORFBs with high energy and power density. They obtained viologen derivatives with high electrochemical reversibility using mild synthetic methods. When combined with ferrocene derivatives, the resulting full cell exhibited excellent performance, including a cell voltage of 1.38 V, a power density of 130 mW cm^{-1}, and capacity retention of 99% after 100 cycles. Due to the two-electron storage mechanism

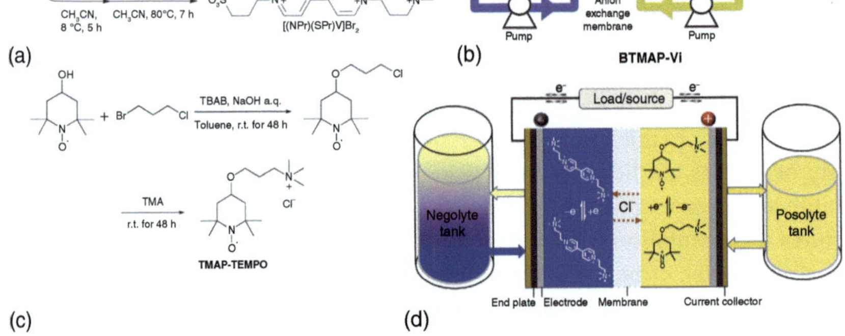

Figure 9.20 (a) Synthesis of two-electron storage viologen molecules, (b) Schematic of the[(Me)(NPr)V]Cl$_3$/FcNCl aqueous organic redox flow battery. Source: Reproduced with permission from DeBruler et al. [62]/Elsevier. (c) Syntheses of TMAP-TEMPO, and (d) Schematic of a flow battery assembled with TMAP-TEMPO and BTMAP-Vi. Source: Reproduced with permission from Liu et al. [60]/Elsevier.

and high solubility, the obtained anolyte has a theoretical capacity of up to 96.5 Ah l^{-1} in water. These strategies broaden the horizons for the application of viologen. Adding hydrophilic functional groups, such as trimethylammonio and 3-sulfonatopropyl enhances the solubility of viologen. Moreover, the intermolecular interaction between hydrophilic functional groups and H_2O can solve the problem of the barely soluble reduced product of viologen derivatives. The asymmetric design of 1-[3-(trimethylammonio)propyl]-10-(3-sulfonatopropyl)-4,40-bipyridinium dibromide ([(NPr)(SPr)V]Br_2), with trimethylammonio and 3-sulfonatopropyl grafted on the N sites of one molecule, has a positive effect on the physicochemical properties of the redox-active viologen molecules. They also constructed AORFBs using quaternary ammonium-grafted ferrocene (FcNCl) and TEMPO (N, N, N, 2, 2, 6, 6 – Tetramethylpiperidinyloxy-4 – trimethyl ammonium chloride, N_{Me}-TEMPO) derivatives as catholytes, respectively, and analyzed the substituent effect of the hydrophilic functional groups on battery performance in detail. Xu et al. improved battery performance even further by developing TEMPO derivatives with longer substitution chains (4-[3-(trimethylammonio)propoxy]-2,2,6,6-tetramethylpiperidine-1-oxyl (TMAP-TEMPO) chloride, Figure 9.20c,d) [60]. The bimolecular interactions and molecular decomposition of HO-TEMPO were suppressed by Coulombic repulsion, and the crossover of active molecules was also weakened by the charge repulsion between the ion-exchange membrane and electrolytes. A concentration-independent temporal capacity retention rate of 99.993% per cycle over 1000 consecutive cycles was achieved in an all-organic AORFB based on TMAP-TEMPO and viologen derivatives. To further improve the performance of AORFBs, stronger Coulomb repulsion due to more charge (terminal phosphoric acid) was used [67]. However, most viologen derivatives exhibit two single-electron reductions separated by about 0.4 V, which is unfavorable for a stable battery system. To address this, Luo and colleagues created a novel two-electron storage anolyte by extending viologen conjugation with a planar thiazolo[5,4-*d*]thiazole (TTz) framework [61]. An AORFB coupled with this anolyte and an ammonium group functionalized TEMPO, which has a 99.97% capacity retention per cycle after being functionalized by the hydrophilic ammonium group. The π-conjugated effect enables a highly reduced reductive potential gap between the two-electron reduction, which delivers a more uniform battery charge/discharge curve.

Monosubstituted catholyte materials, such as TEMPO and ferrocene derivatives, may not always be sufficient for all-organic AORFBs. Therefore, some researchers have improved the performance of these batteries by upgrading the number or structure of substituents. For instance, Hu et al. have reported on the use of a dual-ammonium group-substituted TEMPO molecule, N_2-TEMPO (N^1, N^1, N^1, N^3, N^3, 2, 2, 6, 6-nonamethyl-N^3-(piperidinyloxy) propane-1,3-bis(ammonium) dichloride), as catholyte for AORFBs [68]. This molecule not only increases the solubility of TEMPO but also enables a high redox potential of 1.0 V versus NHE (normal hydrogen electrode), thanks to its dual-ammonium dicationic group. Moreover, Beh et al. have demonstrated an AORFB using a bis((3-trimethylammonio)propyl)-ferrocene dichloride catholyte and a bis(3-trimethylammonio)propyl viologen tetrachloride anolyte [69]. The ferrocene derivative is made 2 M soluble after bis(trimethylammoniopropyl) functionalization, and the dual-ammonium dicationic group suppresses chemical decomposition and crossover of the catholyte. This approach achieved an unprecedented cycling stability of 99.90% per day at a 1.3 M reactant concentration. In conclusion, these strategies could potentially lead to extended lifetimes, making AORFBs commercially viable for grid-scale storage systems.

9.3.5 Nonaqueous all-Organic Redox Flow Batteries

NORFBs have a higher energy density advantage due to the wider voltage window than aqueous systems, but their performance remains limited by the low solubility of redox materials, low cell efficiency, cycling stability, and current density. Research on NORFBs mainly focuses on developing new electrolytes, improving the stability of redox-active species, optimizing the solvent configuration, and so on. Thus far, researchers have mainly applied molecular engineering to develop electrolytes for NORFBs, aiming to obtain scalable energy storage battery systems. For instance, the development of macromolecular oligomers as redox species aims to reduce the high demand for the membrane in nonaqueous systems [70]. The size effect can prevent the crossover of redox-active species, which enables cheap porous membranes that can be used instead of expensive ion exchange membranes. Moreover, the development of single-molecule multi-electron transfer has also received attention in nonaqueous batteries, such as pyridine [71] and phenazine derivatives [72]. Bipolar species have also been explored in all-organic systems [71, 73], i.e. redox species, where a single molecule can act as anolyte and catholyte materials. Candidate redox in organic systems mainly comprises aromatic compounds containing nitrogen or oxygen atoms, e.g. fluorenone derivatives [72], amidine derivatives [74], phenazine derivatives [72, 75], methoxy derivatives, and radicals [73, 76], etc.

To construct NORFBs, 9-fluorenone (FL) is typically used as the anolyte, which is a commercially available redox-active organic material. FL has high solubility and undergoes a one-electron redox reaction with good reversibility in acetonitrile (CAN) [75]. Wei et al. reported an all-ORFB by coupling FL anolyte with 2,5-di-*tert*-butyl-1-methoxy-4-[2′-methoxyethoxy]benzene (DBMMB) catholyte (Figure 9.21a,b) [76]. The cell uses a highly ion-conductive electrolyte containing ACN and tetraethylammonium (TEA^+)-based salts. Due to the high solubility of both anolyte and catholyte and the highly conductive nature of the electrolyte, the cell shows a voltage greater than 2 V with an EE of 70% and a current density greater than 10 mA cm^{-2}. However, the chemical stability of the charged radical species, particularly FL^-, limits the flow cell's cycling stability. The study reveals the capacity-fading mechanism of FL-based NORFBs, which is useful for the development of all-organic flow batteries. Additionally, the optimized electrolyte offers the highest chemical stability of radical species, which can be attributed to the high stabilities of both TEA-TFSI and DME against radical species. The study also provides valuable insight into the degradation mechanism of NORFBs involving radical-based candidates. Phenazine derivatives are also used

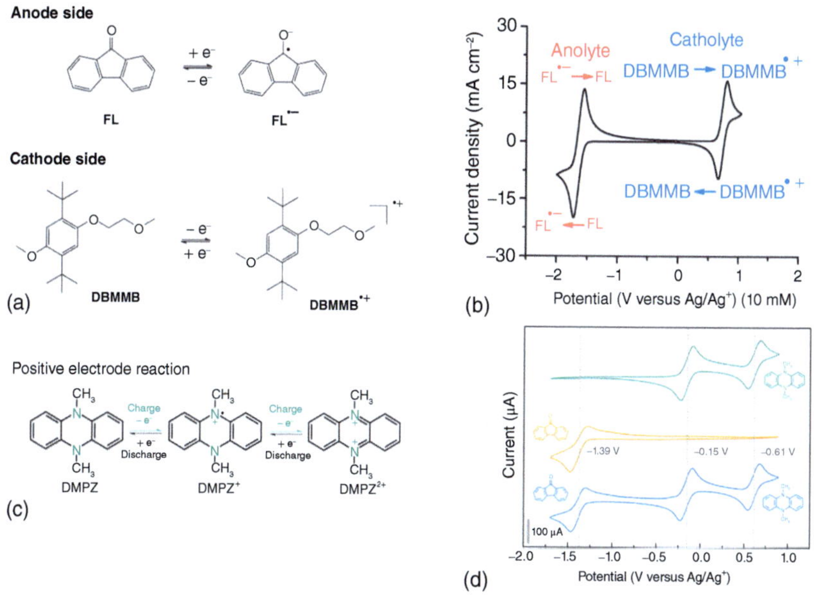

Figure 9.21 (a) Redox reactions and (b) cyclic voltammetry (CV) curve of the nonaqueous FL and DBMMB molecules. Source: Reproduced with permission from Xiaoliang et al. [76]/John Wiley & Sons. (c) Molecular structures and redox mechanisms of DMPZ and FL, (d) CV curves of DMPZ (green line), FL (orange line), and the mixture of FL and DMPZ (blue line). Source: Reproduced with permission from Kwon et al. [72]/Elsevier.

as multi-redox molecule catholyte species coupled with FL for NORFBs [72, 75]. Kwon and coworkers proposed a multi-redox phenazine-based catholyte candidate (5,10-dihydro-5,10-dimethylphenazine[DMPZ]) for NORFBs (Figure 9.21c,d) [72]. Due to the reversible two-step reactions of single-electron redox in the pyrazine ring, DMPZ exploits double-redox activity at 0.15 and 0.61 V versus Ag/Ag^+ with supernormal kinetics and chemical stability. The all-organic DMPA/FL flow cell can reach a high energy density of 85 Wh mol^{-1}. Furthermore, the visible color change of DMPZ at different states makes it possible to visualize the state of charge. The full cell shows two well-defined voltage plateaus at 1.2 and 2.0 V with >90% material utilization. While the solubility of DMPZ (60 mM) is insufficient for the application of energy storage systems, the discovery of multi-electron materials provides a new pathway for the design of feasible redox-active candidates. Subsequently, they present another phenazine derivative with enhanced solubility (0.5 M in ACN) as a catholyte material, 5,10-bis(2-methoxyethyl)-5,10-dihydrophenazine (BMEPZ) [75]. An all-organic cell coupled with FL anolyte exhibits cell voltages of 1.2 and 2.0 V, delivers a capacity retention of 99.94% per cycle over 200 cycles with 0.1 M BMEPZ catholyte, and results in a high energy density (17 Wh l^{-1}).

Alkoxybenzene-based molecules can also be used as anode species in NORFBs, usually coupled with FL or N-methylphthalimide (MePh) to form full cells [76]. Wei et al. reported a NORFB based on 2,5-di-*tert*-butyl-1-methoxy-4-[2′-methoxyethoxy]

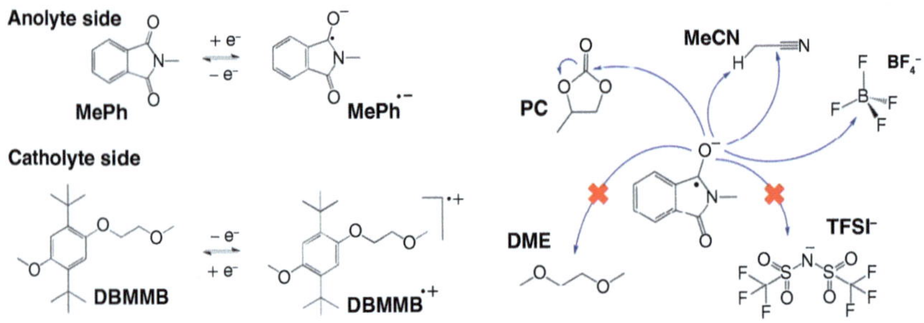

Figure 9.22 (a) The reaction mechanism of the MePh|DBMMB, (b). Possible side reactions of the MePh· − radical anion with PC, ACN, and BF_4^- superoxide radical anion.

benzene (DBMMB) cathode and MePh anode, which achieves a theoretical cell voltage of 2.3 V. The flow battery exhibits excellent cycling stability under current densities as high as 50 mA cm^{-2}, which is significant progress in the development of NORFBs. The charge–discharge mechanisms of the anode and cathode are shown in Figure 9.22. Additionally, the high solubilities of both redox species, 0.7 M for the MePh and high miscibility for the DBMMB (a liquid at room temperature), create an advantage for NORFBs with high volume capacities, broadening the choice of redox species for NORFBs. The suitability of the membranes for the MePh/DBMMB system was studied based on three Daramic® and two Celgard® porous separators. Among these membranes, Daramic-175 exhibits the highest EE of 70% and material utilization of 80%, owing to the tradeoff in thickness and pore size. Also, the electron-withdrawing effect of the imine group reduces the electron density of the MePh, thus providing additional stability, which could be helpful for the long life of NORFBs.

Organic molecules with redox activity can be used as energy storage electrolyte materials in NORFBs, but the crossover of redox species between the anolyte and catholyte, and the resulting chemical degradation, are important factors contributing to the limited electrochemical performance of NORFBs. Bipolar species, which allow cells with symmetrical electrolyte composition, can compensate for such shortcomings. Hendrik et al. reported a rational design of pyridine-based bipolar materials for NORFBs [71], combining synthetic modification and predictive modeling methods to develop a 4-benzoylpyridinium derivative 1^+ as the anolyte, as shown in Figure 9.23a,b. Furthermore, this derivative can also be used as a catholyte species owing to the two-electron transfer ($1^+ \rightleftarrows 1 \rightleftarrows 1^-$) mechanism. The electrochemical performance of the bipolar molecule in different support electrolytes is evaluated by changing the supporting salts, demonstrating that the battery's performance with KPF_6 support electrolyte is excellent compared to that with lithium- and sodium-based electrolytes and is comparable to that with the more expensive $TBAPF_6$-based electrolyte. Indeed, the full cell with KPF_6-based electrolytes cycled for 150 cycles with no obvious degradation. However, the low cell voltage of the symmetrical cell limits the in-depth application of pyridine-based bipolar materials.

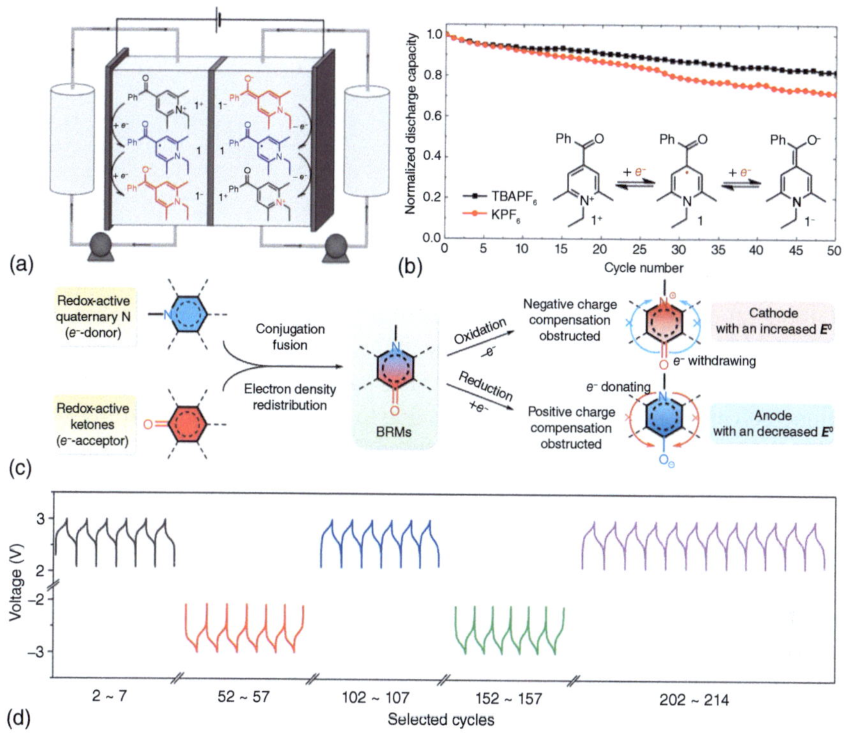

Figure 9.23 Bipolar molecules for NORFBs: (a) Diagram of the symmetric flow cell, and (b) Discharge capacity versus cycle number based on 4-benzoylpyridinium derivative. Source: Reproduced with permission from Hendriks et al. [71]/American Chemical Society. (c) Molecular design strategy of QPT-OMe using quaternary N and a ketone as the redox-active moieties, (d) Representative galvanostatic charge/discharge profiles of QPT-OMe. Source: Reproduced with permission from Liu et al. [77]/American Chemical Society.

Liu et al. introduced a novel approach for designing high-voltage bipolar redox-active small molecules by utilizing fused conjugation to regulate the redox potential of integrated moieties, such as QPT-OMe, as shown in Figure 9.23c,d [77]. The quaternary N and ketone moieties of the redox-active molecule were considered as electron donors and acceptors, respectively, to construct a new molecule with a significant voltage gap and good electrochemical stability. Further studies demonstrated that fused conjugation could facilitate the reconstruction of the electron density in the conjugation structure, redistributing electrons from the ketone to the quaternary N moiety and broadening the potential gap between the two moieties. During an oxidation reaction, the electron-withdrawing effect of the ketone will increase the potential of the quaternary N moiety because of obstruction in negative charge compensation. Similarly, during a reduction reaction, a lower potential is necessary for the ketone due to the contribution from the electron-donating quaternary N moiety, leading to a reduction in the reduction voltage. A symmetrical NORFB based on this molecule showed a high voltage of 2.5 V, with a capacity retention of 63.5% after 900 cycles. This conjugation strategy provides a new horizon for designing small-molecular-weight bipolar redox-active species, which is crucial for restraining the electrolyte crossover and overcoming the low cell voltage issues of NORFBs.

9.4 Summary and Outlook

To summarize, considerable efforts have been made to develop organic molecules and polymers for use in various all-organic systems, such as proton batteries, metal-ion all-organic batteries, dual-ion all-organic batteries, and redox flow batteries. Despite significant research on all-organic batteries, some issues still persist. For instance, some organic materials have slow redox kinetics, such as quinone polymers in aqueous electrolytes, due to their hydrophobic nature. While proton batteries show better compatibility with low-cost aqueous electrolytes, further research is required to improve the design and development of aqueous electrolytes. Metal ion and dual ion-based all-organic batteries, as special "rocking chair batteries," have received attention but require more focus on designing organic molecule structures, as well as improving transport kinetics of ions and affinity behavior with different functional groups of organic molecules compared to proton cells.

The OCV of all-organic batteries is relatively lower than that of inorganic-based ones, and the electrolyte system needs further development to adapt to different organic materials. All-organic batteries still suffer from shortcomings in capacity and cycle life compared to conventional inorganic batteries, highlighting the need to design new organic redox polymers with higher storage capacity, high-rate performance, and longer cycle life. Commercializing the use of all-organic battery electrodes requires significant ongoing work. The wide structural diversity of organics presents an opportunity to develop organic materials with higher voltages, such as quinones substituted with electron-withdrawing groups that demonstrate significantly higher redox potentials compared to their unsubstituted analogs. The development of aqueous electrolytes with a wide electrochemical window provides an avenue for improving the cycling stability of aqueous all-organic batteries while also reducing costs. Additionally, structurally modulating polymer organic molecules can imbue them with electrical conductivity, allowing conductive polythiophene materials to eliminate the need for conductive additives. Theoretically, any combination of conductive polymer backbone and electrochemical active groups is possible, providing a high degree of customization and a large design space to identify stable combinations of energy-carrying units to maximize battery potential for desired applications.

ORFBs have emerged as a promising energy storage system in recent years, attracting the interest of researchers and playing a crucial role in the development of large-scale energy storage devices. However, significant efforts are still required to realize the practical application of all-organic liquid-flow batteries. In particular, research and development of all-organic systems should prioritize the exploration of highly soluble and stable organic molecules that exhibit reversible reactions, as these factors tightly impact ORFB performance. When designing ORFBs, critical performance metrics, including energy density, power density, cycle life, efficiency, and economic and ecological impact, must be carefully considered. Depending on the electrolyte used, ORFBs can be classified into AORFBs and NORFBs. Due to the limitations in membrane and electrolyte stability and ionic conductivity, the power density of NORFBs is significantly lower than that of AORFBs. In contrast, aqueous systems, which employ water as an electrolyte, are highly ionic, conductive, low-cost, and nonflammable. Advances in molecular engineering have facilitated the development of new electroactive molecules, while improvements in characterization techniques for organic compounds have provided valuable insights into the comprehensive study of active components in ORFBs, including reaction reversibility, electron transfer, intermediate stability, and degradation mechanisms.

In summary, all-organic batteries present several advantages compared to traditional metal-electrode batteries, including higher tunability, improved structural design, and wider functional development possibilities, which open up new avenues for the development of all-organic battery systems. Additionally, significant progress in molecular engineering, such as molecular design and synthesis, will drive advances in electroactive organic small molecules and polymers. While research on all-organic batteries is still in its early stages, a promising future lies ahead. Depending on the specific application requirements, all-organic batteries could be an attractive alternative for electrochemical and energy storage devices. We hope that this chapter has provided readers with novel and insightful ideas.

Acknowledgments

The work was supported by the National Natural Science Foundation of China (Grant Nos. U2004214 and 21975225) and the National Key Research and Development Program of China (2022YFB2402200).

References

1 Whittingham, M.S. (2004). *Chem. Rev.* 104: 4271–4302.
2 Poizot, P., Gaubicher, J., Renault, S. et al. (2020). *Chem. Rev.* 120: 6490–6557.
3 Tie, Z., Deng, S., Cao, H. et al. (2022). *Angew. Chem. Int. Ed.* 61: e202115180.
4 Scrosati, B. (1992). *J. Electrochem. Soc.* 139: 2776.
5 Huang, J., Dong, X., Guo, Z., and Wang, Y. (2020). *Angew. Chem. Int. Ed.* 59: 18322–18333.
6 De Surville, R., Jozefowicz, M., Yu, L.T. et al. (1968). *Electrochim. Acta* 13: 1451–1458.
7 Cai, Z., Geng, M., and Tang, Z. (2004). *J. Mater. Sci.* 39: 4001–4003.
8 Zhijiang, C. (2006). *J. Polym. Res.* 13: 207–211.
9 Lu, Y. and Chen, J. (2020). *Nat. Rev. Chem.* 4: 127–142.
10 Emanuelsson, R., Sterby, M., Strømme, M., and Sjödin, M. (2017). *J. Am. Chem. Soc.* 139: 4828–4834.

11 Wang, X., Zhou, J., and Tang, W. (2021). *Energy Storage Mater.* 36: 1–9.
12 Suga, T., Ohshiro, H., Sugita, S. et al. (2009). *Adv. Mater.* 21: 1627–1630.
13 Karlsson, C., Strietzel, C., Huang, H. et al. (2018). *ACS Appl. Energy Mater.* 1: 6451–6462.
14 Dong, X., Yu, H., Ma, Y. et al. (2017). *Chem. Eur. J.* 23: 2560–2565.
15 Wang, H., Emanuelsson, R., Karlsson, C. et al. (2021). *ACS Appl. Mater. Interfaces* 13: 19099–19108.
16 Strietzel, C., Sterby, M., Huang, H. et al. (2020). *Angew. Chem. Int. Ed.* 59: 9631–9638.
17 Xu, Y., Zheng, Y., Wang, C., and Chen, Q. (2019). *ACS Appl. Mater. Interfaces* 11: 23222–23228.
18 Nueangnoraj, K., Tomai, T., Nishihara, H. et al. (2016). *Carbon* 107: 831–836.
19 Dai, G., He, Y., Niu, Z. et al. (2019). *Angew. Chem. Int. Ed.* 58: 9902–9906.
20 Shi, M., He, J., Zhao, Y. et al. (2022). *Mater. Des.* 222: 111043.
21 Jiang, S., Li, W., Xie, Y. et al. (2022). *Chem. Eng. J.* 434: 134651.
22 Song, Z. and Zhou, H. (2013). *Energy Environ. Sci.* 6: 2280–2301.
23 Novák, P., Müller, K., Santhanam, K.S.V., and Haas, O. (1997). *Chem. Rev.* 97: 207–282.
24 MacInnes, D., Druy, M.A., Nigrey, P.J. et al. (1981). *J. Chem. Soc., Chem. Commun.* 317–319.
25 Chiang, C.K. (1981). *Polymer* 22: 1454–1456.
26 Li, Z., Jia, Q., Chen, Y. et al. (2022). *Angew. Chem. Int. Ed.* 61: e202207221.
27 Zhang, Y., Sun, Z., Kong, X. et al. (2021). *J. Mater. Chem. A* 9: 26208–26215.
28 Armand, M., Grugeon, S., Vezin, H. et al. (2009). *Nat. Mater.* 8: 120–125.
29 Iordache, A., Delhorbe, V., Bardet, M. et al. (2016). *ACS Appl. Mater. Interfaces* 8: 22762–22767.
30 Hu, Y., Tang, W., Yu, Q. et al. (2019). *ACS Appl. Mater. Interfaces* 11: 32987–32993.
31 Deng, W., Shi, W., Li, P. et al. (2022). *Energy Storage Mater.* 46: 535–541.
32 Xie, J., Wang, Z., Xu, Z.J., and Zhang, Q. (2018). *Adv. Energy Mater.* 8: 1703509.
33 Casado, N., Mantione, D., Shanmukaraj, D., and Mecerreyes, D. (2020). *ChemSusChem* 13: 2464–2470.
34 Dong, X., Guo, Z., Guo, Z. et al. (2018). *Joule* 2: 902–913.
35 Zhang, B., Yang, X., He, B. et al. (2021). *J. Mater. Chem. A* 9: 14444–14450.
36 Kahsay, B.A., Ramar, A., Wang, F.-M. et al. (2019). *J. Power Sources* 428: 115–123.
37 Wild, A., Strumpf, M., Häupler, B. et al. (2017). *Adv. Energy Mater.* 7: 1601415.
38 Xie, Y., Zhang, K., Yamauchi, Y. et al. (2021). *Mater. Horiz.* 8: 803–829.
39 Hergué, N., Ernould, B., Minoia, A. et al. (2018). *Batteries Supercaps* 1: 102–109.
40 Wang, S., Wang, L., Zhu, Z. et al. (2014). *Angew. Chem. Int. Ed.* 53: 5892–5896.
41 Liu, S., Xiong, M., Tang, W. et al. (2021). *ACS Appl. Mater. Interfaces* 13: 38315–38324.
42 Vardhini, G., Suriyakumar, S., and Shaijumon, M.M. (2022). *ACS Appl. Energy Mater.* 5: 9595–9604.
43 Zhou, G., Mo, L., Zhou, C. et al. (2021). *Chem. Eng. J.* 420: 127597.

44 Banda, H., Damien, D., Nagarajan, K. et al. (2015). *J. Mater. Chem. A* 3: 10453–10458.

45 Hernández, G., Casado, N., Zamarayeva, A.M. et al. (2018). *ACS Appl. Energy Mater.* 1: 7199–7205.

46 Ge, J., Yi, X., Fan, L., and Lu, B. (2021). *J. Energy Chem.* 57: 28–33.

47 Kaneto, K., Yoshino, K., and Inuishi, Y. (1983). *Jpn. J. Appl. Phys.* 22: L567.

48 Wang, C.Y., Ballantyne, A.M., Hall, S.B. et al. (2006). *J. Power Sources* 156: 610–614.

49 Nyström, G., Razaq, A., Strømme, M. et al. (2009). *Nano Lett.* 9: 3635–3639.

50 Suga, T., Sugita, S., Ohshiro, H. et al. (2011). *Adv. Mater.* 23: 751–754.

51 Chikushi, N., Yamada, H., Oyaizu, K., and Nishide, H. (2012). *Sci. China Chem.* 55: 822–829.

52 Sano, N., Tomita, W., Hara, S. et al. (2013). *ACS Appl. Mater. Interfaces* 5: 1355–1361.

53 Oka, K., Kato, R., Oyaizu, K., and Nishide, H. (2018). *Adv. Funct. Mater.* 28: 1805858.

54 Ding, Y., Zhang, C., Zhang, L. et al. (2018). *Chem. Soc. Rev.* 47: 69–103.

55 Leung, P., Shah, A.A., Sanz, L. et al. (2017). *J. Power Sources* 360: 243–283.

56 Su, L., Darling, R.M., Gallagher, K.G. et al. (2016). *J. Electrochem. Soc.* 163: A5253.

57 Noack, J., Roznyatovskaya, N., Herr, T., and Fischer, P. (2015). *Angew. Chem. Int. Ed.* 54: 9776–9809.

58 Li, H., Fan, H., Hu, B. et al. (2021). *Angew. Chem. Int. Ed.* 60: 26971–26977.

59 Hu, S., Li, T., Huang, M. et al. (2021). *Adv. Mater.* 33: 2005839.

60 Liu, Y., Goulet, M.A., Tong, L. et al. (2019). *Chem* 5: 1861–1870.

61 Luo, J., Hu, B., Debruler, C., and Liu, T.L. (2018). *Angew. Chem. Int. Ed.* 57: 231–235.

62 DeBruler, C., Hu, B., Moss, J. et al. (2017). *Chem* 3: 961–978.

63 Liu, T., Wei, X.D., Nie, Z. et al. (2016). *Adv. Energy Mater.* 6: 1501449.

64 Lv, X.L., Sullivan, P., Fu, H.C. et al. (2022). *ACS Energy Lett.* 7: 2428–2434.

65 Pan, M., Gao, L., Liang, J. et al. (2022). *Adv. Energy Mater.* 12: 2103478.

66 Liu, L., Yao, Y., Wang, Z., and Lu, Y.C. (2021). *Nano Energy* 84: 105897.

67 Gao, M., Salla, M., Song, Y., and Wang, Q. (2022). *Angew. Chem. Int. Ed.* 61: e202208223.

68 Hu, B., Hu, M., Luo, J., and Liu, T.L. (2022). *Adv. Energy Mater.* 12: 2102577.

69 Beh, E.S., De Porcellinis, D., Gracia, R.L. et al. (2017). *ACS Energy Lett.* 2: 639–644.

70 Doris, S.E., Ward, A.L., Baskin, A. et al. (2017). *Angew. Chem. Int. Ed.* 56: 1595–1599.

71 Hendriks, K.H., Sevov, C.S., Cook, M.E., and Sanford, M.S. (2017). *ACS Energy Lett.* 2: 2430–2435.

72 Kwon, G., Lee, S., Hwang, J. et al. (2018). *Joule* 2: 1771–1782.

73 Steen, J.S., Nuismer, J.L., Eiva, V. et al. (2022). *J. Am. Chem. Soc.* 144: 5051–5058.

74 Sevov, C.S., Hickey, D.P., Cook, M.E. et al. (2017). *J. Am. Chem. Soc.* 139: 2924–2927.

75 Kwon, G., Lee, K., Lee, M.H. et al. (2019). *Chem* 5: 2642–2656.

76 Xiaoliang, W., Wu, X., Jinhua, H. et al. (2015). *Angew. Chem. Int. Ed.* 54: 8684–8687.

77 Liu, Y., Dai, G., Chen, Y. et al. (2022). *ACS Energy Lett.* 7: 1274–1283.

10

Outlook

With the consecutive consumption of coal and petroleum, people need to hunt for sustainable energy such as solar energy and wind energy. Then the energy needs to be stored by an energy storage device for convenient use when people want to use it. There are many energy storage devices appeared from the view of historical development, such as lead-acid batteries, Ni–Cd batteries, and supercapacitors. The devices are still required aiming to specific markets despite their short mileage or low energy density. Afterward, Li-ion batteries (LIBs) emerge with rapid development and commercialization, which fill our lives with portable equipment and electrical vehicles. With rapid modern social development, we are facing a shortage of resources, which will become a restriction of our sustainable requirement and the development of LIBs due to a low reserve of Li in the earth's crust. Recently, other metal-ion batteries have drawn researcher's attention because of their abundant reserves in the earth's crust, such as sodium-ion batteries, Mg-ion batteries, Zn-ion batteries, and so on.

From a long-term perspective, we require other new types of batteries to break some restrictions such as shortage of resources, energy density, cost, pollution, sealed systems, and organic electrolyte systems. Organic batteries can be a breakthrough due to their rich functional molecules. As introduced in this book, organic materials cover both the positive electrode and negative electrode, with a wide voltage range in both the sealed system and the opened system. The batteries bring various possibilities.

The practical application of organic batteries may include two aspects in the near future. The first one is the traditional pouch cells like the presently popular LIBs. The other one is the redox flow batteries, which are mainly designed for large-scale energy storage. The following is a detailed discussion.

Some positive electrodes deliver a large energy density of over 1300 Wh kg^{-1} based on the mass of the electrode, which is larger than that of the state-of-the-art positive electrodes. However, the materials face difficulties when used in a pouch cell. Given the solubility of the electrolyte of the organic positive electrode material, some special treatment should be considered. A carbon paper, as a typical case, is needed for grabbing the soluble species through the adsorption effect. Meanwhile, the carbon improves the conductivity of the positive electrode side. As a result, the cell sacrifices the energy density when using an amount of carbon. Moreover, the carbon will also reduce the volumetric energy density due to its low density.

Rechargeable Organic Batteries: Materials, Mechanisms, and Prospects, First Edition.
Yongzhu Fu, Xiang Li, Shuai Tang, and Wei Guo.

Apart from carbon, a functional separator is useful for blocking the shuttle effect of the soluble species. However, there is no doubt about increasing manufacturing costs. Therefore, the reported pouch cells are still limited based on the organic electrodes. There is another usage of organic material, as an additive rather than a positive electrode. The organic material as an additive has a synergistic effect, which improves the stability and energy density of the cell. By using this strategy, a pouch cell with high energy density can be achieved. From an application perspective, organic materials as positive electrodes are difficult to use in a pouch cell for now because of their large energy density, especially volumetric energy density. However, rich-type molecules give possibilities; after all, there are organic molecules with high conductivity and insolubility.

The high solubility of the organic species is coincidentally suitable for the redox flow batteries, the energy density of which mainly depends on the solubility. Currently, vanadium-based redox flow batteries are the main kind of redox flow batteries for large-scale energy storage. However, the low variety of the inorganic species supplies the narrow space to tune or design the species chemically. Therefore, the separator has to be carefully prepared to prevent the crossover of the active species on both sides. Meanwhile, the toxicity and high price of the vanadium-based species are also the challenge for redox flow batteries. Organic species have the structural diversity to improve the energy density, besides the advantage of solubility. For example, the introduction of sulfonate or hydroxyl moieties can enhance the solubility of organic species in aqueous electrolytes, ameliorating the energy density. The design of multiple active sites in one single molecule can also improve the capacity, thus improving the energy density. The cyclohexanehexone illustrates the theoretical capacity of 957 $mAh\,g^{-1}$, and the reversible capacity of 912 $mAh\,g^{-1}$ can be achieved. In addition, the theoretical capacity of most of the organosulfides is between 200 and 500 $mAh\,g^{-1}$. Moreover, the capacity can be further enhanced by covalently introducing inorganic sulfur atoms, forming organic polysulfides. The capacity of diphenyl tetrasulfide and dimethyl tetrasulfide is 570 and 1030 $mAh\,g^{-1}$ respectively. The equilibrium potential of the organic species can also be controllably tuned. The electron-withdrawing group can improve the potential, while the electron-donating group can lower the potential, which can be adopted in molecular design to improve the output voltage.

Besides capacity improvement, structural diversity can help to hinder the crossover of organic species. The larger molecular size of organic compounds makes them more difficult to go through the membrane with nanosized channels. The organosulfide cathode is shown to possess a weakened crossover through the Nafion membrane due to its large molecular size. It is like catching fish using the fishnet, in which the larger fishes are more difficult to cross the fishnet. However, the "fishnet" with the ultra-small channel has to be adopted for the inorganic species because of their small sizes, which may lower the power density of the redox flow batteries. The balance between the molecular size and the solubility has to be considered as a larger molecular size may lower the solubility. The absence of expensive metals is also an advantage for the large-scale application compared to the inorganic electrode materials. Therefore, it is believed that organic batteries are a promising candidate for next-generation energy storage, which may be commercialized based on the substantial effort of the scientific and industrial community.

Acknowledgments

The work was supported by the National Natural Science Foundation of China (Grant Nos. U2004214 and 21975225) and the National Key Research and Development Program of China (2022YFB2402200).

List of Abbreviations

$(NPr)_2V$	1,1′-bis[3-(trimethylammonium)propyl]-4,4′-bipyridinium tetrachloride
$(TABQ)_2$-S_8	(tetra(allyloxy)-1,4-benzoquinone)$_2$-S_8
$[(Me)(NPr)V]Cl_3$	1-methyl-10-[3-(trimethylammonio) propyl]-4,4-bipyridinium
$[(NPr)(SPr)V]Br_2$	1-[3-(trimethylammonio)propyl]-10-(3-sulfonatopropyl)-4,40-bipyridinium dibromide
1,4-BDT	1,4-benzenedithiol
1,4-PNQ	2,2′-(1,4-phenylenedithio) bis(1,4-naphthoquinone)
1^+	4-benzoylpyridinium derivative
1D	one dimensional
2,2′-DpyDS	2,2′-dipyridyl disulfide
2,6-DHAQ	2,6-dihydroxyanthraquinone
2D	two dimensional
2F-BDT	2,5-difluorobenzene-1,4-dithiol
3BQ	tribenzoquinoxaline-5,10-dione
3D	three dimensional
3DFNG	three-dimensional nitrogen-doped graphene sponge with Fe_3O_4 nanoparticles
3Q	hexaazatrinaphthylene
4,4′-DpyDS	4,4′-dipyridyl disulfide
A/NORFBs	aqueous and nonaqueous ORFBs
$A^\bullet$	allyl radical
AAO@CNT	anodic alumina@carbon nanotube
AAQ	1-aminoanthraquinone
ABB_4OLi	lithium anthracene-9,10-bis[2-benzene-1,4-bis(olate)]
ADALS	azobenzene-4,4′-dicarboxylic acid lithium salt
ADAPS	azobenzene-4,4′-dicarboxylic acid potassium salt
ADASS	azobenzene-4,4′-dicarboxylic acid sodium salt
Ag_2TP	silver-terephthalic acid
AIBs	aluminum-ion batteries
Al^{3+}	aluminum ions
APh-NQ	2-((4-hydroxyphenyl) amino) naphthalene-1,4-dione

Rechargeable Organic Batteries: Materials, Mechanisms, and Prospects, First Edition.
Yongzhu Fu, Xiang Li, Shuai Tang, and Wei Guo.

AQ	2,6-diaminoanthraquinone
AQ-1,8-3E-OH	1,8-bis(2-(2-(2-hydroxyethoxy)ethoxy)-ethoxy)anthracene-9,10-dione
AQDS	anthraquinone-1,5-disulfonic acid sodium salt
AQSCH	9,10-anthraquinone-2-disulfonic choline salt
$A-S_n-A$	allyl organosulfides
$A-S_n-R$	asymmetric organosulfides
ASSLB	all-solid-state lithium batteries
BBQ	2,2′-bis-*p*-benzoquinone
BBQB	1,4-bis(*p*-benzoquinonyl) benzene
BDPPTS	1,4-bis(diphenylphosphine)tetrasulfide
BDT	1,2-benzenedithiol
BDTD	benzo[1,2-*b*:4,5-*b*′]dithiophene-4,8-dione
BDT-G	4,8-dihydrobenzo[1,2-*b*:4,5-*b*′] dithiophene-4,8-dione/graphene
BDTs	benzenedithiols
BFFD	benzofuro[5,6-*b*]furan-4,8-dione
BMEPZ	5,10-bis(2-methoxyethyl)-5,10-dihydrophenazine
BPDC	4,4′-biphenyldicarboxylate
BQ	benzoquinone
$BQDS(CH)_2$	1,2-dihydroxy-3,5-phenydisulfonic acid dicholine salt
BQP	benzoquinone-formaldehyde polymer
BTA	1,2,4,5-benzenetetramine
BTMAP-Vi	bis((3-trimethylammonio)propyl)-4,4-bipyridinium
BTT	1,3,5-benzenetrithiol
C=N	carbon-nitrogen
C=O	carbon-oxygen
C3Q	calix[3]quinone
C4Q	calix[4]quinone
$C_6H_5S_xC_6H_5$, $4 \leq x \leq 6$	phenyl polysulfides
C_6O_6	cyclohexanehexone
C8Q	calix[8]quinone
Ca	cardanol-based benzoxazine
CE	Coulombic efficiency
CEI	cathode electrolyte interface
Celgard®	Celgard porous separators
CF_3-BQ	2,5-bis(trifluoromethyl)-1,4-benzoquinone
$-CH_2-$	methylene
CHCL	choline chloride
CLA	chloranilic acid
CL-PTGE	crosslinked poly(4-glycidyloxy TEMPO)
CMS	sulfur-containing material
CNT	carbon nanotubes
COF-F	fluorine-containing covalent organic framework

COFs	covalent-organic frameworks
COF-V	COF with densely arranged active vinyl groups
COSPs	covalent organosulfur polymers
CPL	3-carbamoyl-2,2,5,5-tetramethylpyrroline-1-oxyl
cPpy	charged polypyrrole
CS_2	carbon disulfide
CSM	sulfur-containing materials
CTFs	covalent triazine frameworks
CuDEPP	[5,15-bis(ethynyl)-10,20-diphenylporphinato] copper(II)
CuTAPc	Cu tetraaminephthalocyanine
CV	cyclic voltammogram or cyclic voltammetry
DAAQ	1,5-diaminoanthraquinone
DAD	2,3-diaza-anthracene-dione
DADS	diallyl sulfide
DAMPT	3,7-diamino-methylphenothiazine
DANQ	2,3-diamino-1,4-naphthoquinone
DANT	2,7-diaminobenzo[*lmn*][3,8]phenanthroline-1,3,6,8 (*2H,7H*)-tetraone
Daramic®	daramic porous separators
DBMMB	2,5-di-*tert*-butyl-1-methoxy-4-[2′-methoxyethoxy]benzene
DDS, A–S–S–A	diallyl disulfide
DEB	1,3-diethynylbenzene
DFPDS	di-fluorophenyl disulfide
DFT	density functional theory
DHAP	direct (hetero)arylation polymerization
DHAQ	2,6-dihydroxyanthraquinone
DHBQ	2,5-dihydroxy-1,4-benzoquinone
DIB	1,3-diisopropenylbenzene
DIXDS	diisopropyl xanthogen disulfide
DIXPS	diisopropyl xanthogen polysulfide
D_{Li+}	lithium ion diffusion coefficients
DMBQ	2,5-dimethoxy-1,4-benzoquinone
DMcT	2,5-dimercapto-1,3,4-thiadiazole
DMDS	dimethyl disulfide
DME	1,2-dimethoxyethane
DMPZ	5,10-dihydro-5,10-dimethylphenazine
DMTS	dimethyl trisulfide
DMTtS	dimethyl tetrasulfide
DPDS	diphenyl disulfide
DPDSe	diphenyl diselenide
DPDSe-S	phenyl selenosulfide
DPDSe-S_2	phenyl selenodisulfide
DPTO	benzo[*c*]thiophen-1(3*H*)-one
DPTS	diphenyl trisulfide
DPTTO	thionated benzo[*c*]thiophen-1(3*H*)-one

DPyDS	2,2′-dipyridyl disulfide
DpyDSDO	2,2′-dipyridyl disulfide-N′N′-dioxide
DPZ	diquinoxalino-phenazine
DSR	disodium rhodizonate
DTS	diallyl trisulfide
DTT	dibenzo[*b*, *i*]thianthrene-5,7,12,14-tetraone
e^-	electrons
ECOFs	exfoliated COFs
ED-OS	electron-deficient organosulfide
EDP	poly(3,4,9,10-perylenetetracarboxylic dianhydride ethylene diamine)
EES	electrochemical energy storage
EPR	electron paramagnetic resonance
E_{red}	reduction potential
ESP	electrostatic potential
ETT	ethenetetrathiolate
EV-AQ_2	ethylviologen dianthraquinone-2-sulfonate
FcNCl	quaternary ammonium-grafted ferrocene
FCTF-S	fluorinated porous triazine-based framework
FL	9-fluorenone
F-MCDS	macrocyclic 2,5-difluorobenzene-1,4-disulfide
FTIR	Fourier transform infrared spectroscopy
GC-MS	Gas chromatography-mass spectrometry
GO	graphene oxide
GO-PAA	polyacrylic acid-modified porous graphene oxide
GPE	gel polymer electrolyte
$GTMA^+$	glycidyltrimethylammonium cation
H_3IPA	5-hydroxyisophthalic acid
HAT	hexaazatriphenylene
HBQP	hydrobenzoquinone-formaldehyde polymer
HDBCO	2,3,4,6,7,8-hexathia-1,5(1,3)-dibenzenacyclooctaphane
HHTP	2,3,6,7,10,11-hexahydroxytriphenylene
HHTQ	2,3,7,8,12,13-hexahydroxytricycloquinazoline
HKH	hexaketocyclohexane
HOMO	highest occupied molecular orbital
HP	poly(3,4,9,10-perylenetetracarboxylic dianhydride hydrazine)
HTBCO	2,4,5,7,9,10-hexathia-1,3,6,8 (1,4)-tetrabenzenacyclodecaphane
HTP	hexathiapentacene
ICE	initial Coulombic efficiency
ICPSs	intermolecular cyclic polysulfides
IDAQ	1,1′-iminodianthraquinone
IDT	indanedione
IMNQ	1H-naphtho[2,3-*d*]imidazole-4,9-dione

IR	infrared
isoV	1,5-diphenyl-3-isopropyl-6-oxoverdazyl
K^+	potassium ions
K_2TP	dipotassium terephthalate
K_4-BTC	tetrapotassium 1,2,4,5-benzenetetracaboxylate
K_4NTC	potassium naphthalene-1,4,5,8-tetracarboxylate
KB	Ketjin black
lawsone	2-hydroxy-1,4-naphthoquinone
LBDTs	lithium benzenedithiolates
LC	7,8-dimethylalloxazine
LHCE	localized high-concentration electrolyte
Li^+	lithium ion
Li_2-ADC	lithium 2,6-anthracene dicarboxylate
Li_2-ATP	dilithium 2-aminoterephthalate
Li_2-BDA	dilithium benzenediacrylate
Li_2-BDP	dilithium benzenedipropiolate
Li_2-DMT	dilithium 2,5-dimcthylterephthalate
Li2-NDC	dilithium 2,6-naphthalene di-carboxylate
Li_2PDHBQS	poly(2,5-dihydroxy-*p*-benzoquinonyl sulfide)
Li_2S	lithium sulfide
Li_2-TDC	lithium 4,4′-tolane-dicarboxylate
Li_2-ThDC	dilithium thiophene-2,5-dicarboxylate
Li_3-BTC	trilithium 1,2,4-benzenetricarboxylates
Li_4-BTC	tetralithium 1,2,4,5-benzenetetracaboxylate
LIBs	lithium-ion batteries
Li-CP	lithiated carbon paper
Li–S	lithium–sulfur
LiTCNQ	lithium tetracyanoquinodimethan
LiTFSI	lithium bis(trifluoromethanesulfonyl)imide
LMBs	lithium metal batteries
LTO	lithium titanate
LUMO	lowest unoccupied molecular orbital
M^+	alkali metal ions
MALDI-TOF	matrix-assisted laser-desorption ionization time-of-flight
MBA	magnesium bis(diisopropyl)amide
MeCN	acetonitrile
MePh	*N*-methylphthalimide
METAC	[2-(methacryloyloxy) ethyl] trimethylammonium chloride
Mg	magnesium
$Mg(PMDTC)_2$	magnesium pentamethylene dithiocarbamate
Mg–S	magnesium-sulfur
MgS_n, $4 \leq n < 8$	magnesium polysulfide
MHQ	2-methoxy-1,4-hydroquinone
MIBs	magnesium-ion batteries
MOFs	metal-organic frameworks

MQ	2-methoxy-1,4-quinone
MS_x, $x > 2$	transition metal polysulfides
MTT	dimers of disulfide ($n = 2$), trisulfide ($n = 3$), tetrasulfide ($n = 4$) of 5-methyl-1,3,4-thiadiazole-2-thiol
N^+	nitrogen positive ions
N_2-TEMPO	N^1, N^1, N^1, N^3, N^3, 2, 2, 6, 6-nonamethyl-N^3-(piperidinyloxy) propane-1,3-bis(ammonium)dichloride
Na	sodium
Na_2DBQ	disodium salt of 2,5-dihydroxy-1,4-benzoquinone
Na_2-DBTP	disodium 2,5-dibromoterephthalate
Na_2-DCTP	2,5-dichloroterephthalate
Na_2-DFTP	disodium 2,5-difluoroterephthalate
Na_2-PDC	disodium 2,5-pyridinedicarboxylate
Na_2-PzDC	disodium 2,5-pyrazinedicarboxylate
Na_2-SDC	sodium 4,4′-stilbene-dicarboxylate
Na_2S_n, $1 \leq n \leq 2$	low-order polysulfide
Na_2S_n, $4 \leq n \leq 8$	high-order polysulfide
Na_3-BTC	trisodium 1,2,4-benzenetricarboxylates
Na_4-BTC	tetrasodium 1,2,4,5-benzenetetracaboxylate
Na_4DHTPA	2,5-dihydroxyterephthalic acid
Na-CPN	sodium 6-(4-carboxyphenyl) nicotinate
Na-CPP	sodium 5-(4-carboxyphenyl) picolinate
Na-DCA	2,2′ -bipyridine-4,4′ -dicarboxylic acid disodium salt
Na-PD	sodium 1,10-phenanthroline-3,8-dicarboxylate
Na–S	sodium–sulfur
NBALS	4-nitrobenzoic acid lithium salt
NCNTs	nitrogen-doped carbon nanotubes
NDI	naphthalene diimide
NDQ	1,4,5,8-Naphthodiquinone
–NH–	amine
NHE	normal hydrogen electrode
NiNC	norcorrole nickel(II) complex
Ni-NDC	nickel-2,6-naphthalenedicarboxylate
NiNH	$Ni[C_6H_2(NH)_4]_n$
N^{Me}-TEMPO	*N*,*N*,*N*,2,2,6,6-tetramethylpiperidinyloxy-4–trimethyl ammonium chloride
NMP	*N*-methylpyrrolidone
NMR	nuclear magnetic resonance
NNS	benzo[1,2-*d*:4,5-*d*′]bis ([1,2,3]thiadiazole)-4,8-dione
NQ	naphthoquinone
NSN	dithiadiazolo-*p*-benzoquinone
NSPIL	nonstoichiometric protonic ionic liquid
NSS	benzoquino-bis-1,2,3-dithiazole
NTCDA	1,4,5,8-naphthalenetetracarboxylic dianhydride

NTDA	naphthalenetetracarboxylic dianhydride
NTT	2,3,6,7-tetramercapto naphthalene
OCV	open circuit voltage
oIDT	oxidized indanthrone
OMCs	organometallic complexes
ORBs	organic radical batteries
ORFBs	all-organic redox flow batteries
ORPs	organic radical polymers
P14AQ	poly(1,4-anthraquinone)
P15AQ	poly(1,5-anthraquinone)
P15AQS	poly(1,5-anthraquinonyl sulfide)
P18AQS	poly(1,8-anthraquinonyl sulfide)
P33DT	poly(3,3′-bithiophene)
P3BT	poly(3-butylthiophene)
P5Q	pillar[5]quinone
PA	*p*-phenylenediamine
PAAQ	poly-1-amino-anthraquinone
PADAQ	poly(5-amino-1,4-dyhydroxy anthraquinone)
PAN	polyacrylonitrile
PAn	polyaniline
PANi	Polyaniline
PANQ	poly(5-amino-1,4-naphthoquinone)
PAQS	poly(anthraquinonyl sulfide)
PAs	polyanhydrides
PBALS	4-(Phenylazo) benzoic acid lithium salt
PBD	thioether bonded poly(quinone)
PBDTD	poly(benzo[1,2-*b*:4,5-*b*′]dithiophene-4,8-dione-2,6-diyl)
PBDTDS	poly(benzo[1,2-*b*:4,5-*b*′]dithiophene-4,8-dione-2,6-diyl sulfide)
PBI	polybenzimidazole
PBQS	poly(benzoquinonyl sulfide)
PBT	polybithiophene
PC	propylene carbonate
PCTB	poly(2-chloro-3,5,6-trisulfide-1,4-benzoquinone)
P-CTFs	porous phthalazinone-based covalent triazine frameworks
PDA	polydopamine
PDATtSSe	poly(diallyl tetrasulfide)
PDB	poly(2,3-dithiino-1,4-benzoquinone)
PDBM	poly(2,5-dihydroxy-1,4-benzoquinone-3,6-methylene)
PDBS	poly(2,5-dihydroxyl-1,4-benzoquinonyl sulfide)
PDI	perylene diimide
PDIBs	potassium-based dual-ion full batteries
PDI-Bz	perylenediimide-benzidine
PDI-Ur	perylenediimide-urea
PDTDA	poly(dithiodianiline)

PDTV	3-phenyl-1,5-di-*p*-tolylverdazyl
PEDOT	poly(3,4-ethylenedioxythiophene)
PEDOT-AQ (G1)	anthraquinone functionalized PEDOT
PEDOT-BQ (G2)	benzoquinone functionalized PEDOT
PEHS	polyethylene hexasulfide
PEO	poly(ethylene oxide)
PEO-NaI	polyethylene-oxide-nai
PETEA-THEICTA	PETEA-tris[2-(acryloyloxy)ethyl] isocyanurate
PGS	poly(galvinoxyl styrene)
PHATN	poly(hexaazatrinaphthalene)
PhS-SePh	phenyl selenosulfide
PhS-TePh	phenyl tellurosulfide
PhS_xH	phenyl polysulfanes
PI	*N*,*N′*-diamino-3,4,9,10-perylenetetracarboxylic polyimide
PI1	phenothiazine-naphthalene polyimides
PIBN	poly(imide-benzoquinone)
PIBN-G	poly(imide-benzoquinone)-graphene
PIBs	potassium-ion batteries
PID	pyrido[3,4-g]isoquinoline-5,10-dione
PIP	polyisoprcnc
PIs	polyimides
PITN	polyisothionaphthalene
PL-PTGE	poly(2,2,6,6-tetramethylpiperidinyl-*N*-oxyl glycidyl ether)
PMDA	pyromellitic dianhydride
PMDI	pyromellitic diimide
PMTH	dipentamethylenethiuram hexasulfide
PMTT	dipentamethylenethiuram tetrasulfide
PNN	poly(nitronylnitroxylstyrene)
PNP	polypropylene/Nafion/polypropylene
PNQ	2-phenyl-1,4-naphthoquinone
PO	poly (2,9-dihydroquinoxalino[2,3-*b*]phenazine
poly(OC10DASTT)	styryl-substituted dialkoxyterthiophene
poly($Se_{0.1}S_{0.9}$-DIB) 10%	sulfur-selenium hybrid copolymer
poly(S-PETEA)	poly(*S*-pentaerythritol tetraacrylate)
poly(S-r-DIB)	poly(sulfur-random-1,3-diisopropenylbenzene)
polySGN	polysulfur-graphene nanocomposite
PPPS	polyphenyl polysulfide
PPS	phenyl hexasulfide
PPTS	polyphenylene tetrasulfide
PQL	poly(1,6-dihydropyrazino[2,3g]quinoxaline-2,3,8-triyl-7-(2*H*)-ylidene-7,8-dimethylidene)
PQs	polyquinones
PROXYL	2,2,5,5-tetramethyl-2,5-dihydro-1*H*-pyrrol-1-oxyl-3-yl
PS (Chapter 4)	polysulfone
PS (Chapter 5)	poly(styrene)

PSS (Chapter 3)	poly(styrenesulfonate)
PSS (Chapter 5)	4-(propoxy-3-sodium sulfonate)
PT	poly(thiophene)
PTAm	poly(2,2,6,6-tetramethylpiperidinyloxy-4-yl acrylamide)
PTC	perylene-3,4,9,10-tetracarboxylate
PTCDA	3,4,9,10-perylenetetracarboxylic dianhydride or perylene 3,4,9,10-tetracarboxylic dianhydride
PTCDA/NC/CNT	pTCDA/nitrogen-doped carbon/carbon nanotubes
PTCDA/S	3,4,9,10-perylene-tetracarboxylicacid-dianhydride sulfide
PTCDI	3,4,9,10-perylenetetracarboxylic diimide
PTGE	poly(2,2,6,6-tetramethylpiperidinyl-*N*-oxyl glycidyl ether)
PTIO	2-phenyl-4,4,5,5-tetramethylimidazoline-1-oxyl 3-oxide
PTMA	poly(2,2,6,6-tetramethylpiperidinyloxy-4-ylmethacrylate)
PTN	poly(TEMPO-norbornene)
PTO	pyrene-4,5,9,10-tetrone
PTPAn	poly triphenylamine
PTPM	poly-(tripyridiniromesitylene)
PTS	phenyl tetrasulfide
PTVE	poly(2,2,6,6-tetramethylpiperidinyl-*N*-oxyl vinyl ether) or poly(2,2,6,6-tetramethylpiperidinyloxy-4-yl vinyl ether)
PV10	poly(*N*-4,4-bipyridinium-*N*-decamethylene dibromide)
PVDF-HFP	poly(vinylidene fluoridc-*co*-hexafluoropropylene)
PVTMIO	poly(5-vinyl-1,1,3,3-tetramethylisoindolin-2-yloxyl)
PVZ	vinyldibenzothiophenesulfone
Py_2S_x, $3 \leq x \leq 8$	pyridine polysulfide
QFQ	2,5-bis-(*p*-benzoquinonyl) furan
QPQ-1	2,6-bis(*p*-benzoquinonyl) pyridine
QPQ-2	2,5-bis(*p*-benzoquinonyl) pyridine
QPT-OMe	11-methoxy-9*H*-quinolino[3,2,1-*kl*]phenothiazin-9-one
R	organic functional groups
R_1–S_2–R_2, R: organic group	organodisulfides
R_1–S_n–R_2, $n > 2$	organic polysulfides
RABs	Al-ion batteries
Rf_4-BQ	2,5-bis(perfluorobutyl)-1,4-benzoquinone
Rf_6-Cl-BQ	2,5-bis(perfluorohexyl)-3,6-dichloro-1,4-benzoquinone
RFBs	redox flow batteries
rGO	reduced graphene oxide
RIBOTEMPO	riboflavin-TEMPO
ROMP	ring-opening metathesis polymerization
ROP	ring-opening polymerization
RP-CNFs	radical polymers and carbon nanofibers
RSSA	asymmetric organodisulfides
RSSR	organodisulfides
R–S–S–R	symmetric organodisulfides

RSSSR	organotrisulfides
RUB	cis-1,4-polyisoprene
–S–	thioether
S@pPAN	sulfurized-pyrolyzed polyacrylonitrile
S_8	sulfur
S-BOP	sulfur-embedded polybenzoxazine
Se	selenium
Se@C	selenium@carbon
SEI	solid electrolyte interface
semi-IPN	semi-interpenetrating network
SeS_2	selenium disulfide
Se_xSPAN	selenium-doped SPAN
SF-CTF	sulfur-impregnated CTF
SH-DPs	thiol-derived polymers
SIBs	sodium-ion batteries
SLP	sulfur-limonene polysulfide
SOPS	solid polysulfides
SPAN	sulfurized polyacrylonitrile
SPIP	vulcanized polyisoprene
SRVCR	sulfur-rich vulcanized clay rubber
S-SeS_2-DIB@KB600	sulfur-rich polymeric selenium sulfides@Ketjenblack EC600JD
STI	S-triallyl isocyanurate
SWCNTs	single-walled carbon nanotubes
SWNTs	single-wall carbon nanotube
S_xSe_y@C	sulfur-selenium solid solution@carbon
TA	terephthalamide
TABQ (Chapter 3)	tetraamino-*p*-benzoquinone
TABQ (Chapter 7)	tetra(allyloxy)-1,4-benzoquinone
TAIC	triallyl isocyanurate
TAPQ	5,7,12,14-tetraaza-6,13-pentacenequinone
TAPT	2,3,7,8-tetraaminophenazine-1,4,6,9-tetraone
Tb	1,3,5-triformylbenzene
TBBT	4,4′-thiobisbenzenethiol
TCQ	tetrachloro-1,4-benzoquinone
TEA^+	tetraethylammonium
TEG	triethylene glycol
TEMPO	2,2,6,6-tetramethyl-1-piperidinyloxy
TEMPTMA	*N*,*N*,*N*-2,2,6,6-heptamethylpiperidinyl oxy-4-ammonium chloride
TETD	tetraethyl thiuram disulfide
Te_xS_{1-x}@pPAN	tellurium-doped SPAN
TFP	1,3,5-triformylphloroglucinol
$TFSI^-$	bis(trifluoromethanesulfonyl) imide
THF	tetrahydrofuran

THT	2,3,6,7,10,11-triphenylenehexathiol
TiS_2 NSs	titanium disulfide nanosheets
TM	thiuram monosulfide
TMAAcNH-TEMPO	4-[3-(trimethylammonium)acetylamino]-2,2,6,6-tetramethylpiperidine-1-oxyl chloride
TMAP-TEMPO	(4-[3-(trimethylammonio)propoxy]-2,2,6,6-tetramethylpiperidine-1-oxyl
TMTD	tetramethylthiuram disulfide
TMTM	tetramethyl thiuram monosulfide
TOF-SIMS	time-of-flight secondary ion mass spectrometry
Tp	1,3,5-triformylphloroglucinol
TPB	2,3,5,6-tetraphthalimido-1,4-benzoquinone
TPQD	2,3,4,6,8,9,10,12-octathia biscyclopenta[*b*,*c*]-5,11-anthraquinone-1,7-dithione
TRO	truxenone
TT	triptycene tribenzoquinone
TTCA (Chapter 4)	tricyanuric acid
TTCA (Chapter 7)	trithiocyanuric acid
TTF	tetrathiafulvalene
TTF-based	tetrakis(methylthio) derivative cyclohexene-1,4-diylidenes
TTFE	1,1,2,2-tetrafluoroethyl-2,2,2-tri-fluoroethyl ether
TTMD	tetrakis(methylthio) derivative
TTPNQ	the naphthoquinone analog of TTF-TTF-BQ triad
TTPQ	the fused TTF-BQ triad synthesized by a terminal TTF unit in TTPY substituted by BQ
TTPY	2,2′-bi[5-(1,3-dihthiol2-ylidene)-1,3,4,6-tetrathiapentanylidene]
TTT	tetrathiotetracene
TTz	thiazolo[5,4-*d*]thiazole
UB-DPs	unsaturated bond-derived polymers
UP	poly(3,4,9,10-perylenetetracarboxylic dianhydride urea)
UPLC-QTOF-MS	ultrahigh performance liquid chromatography quadrupole time-of-flight mass spectrometry
UV	ultraviolet
UV-vis	ultraviolet-visible
V^{2+}	dicationic form
V-rGO	vinyl group-functionalized reduced graphene oxide
XPS	X-ray photoelectron spectroscopy
XRD	X-ray diffraction
ZIBs	zinc-ion batteries
ZM-CVT	zone-melting chemical-vapor-transport
Zn-mTCPP	Zn(II) 5,10,15,20-tetrakis[(carbazol-9-yl)phenyl]porphyrin
Zn-PTCA	zinc perylenetetracarboxylates
ZTC	zeolite-templated carbon

Index

Rechargeable Organic Batteries: Materials, Mechanisms, and Prospects, First Edition.
Yongzhu Fu, Xiang Li, Shuai Tang, and Wei Guo.

q

u

v

z